Biomedical Materials and Biofabrication *for* Regenerative Medicine

Editors

Naznin Sultana
Texas Undergraduate Medical Academy
Prairie View A&M University
Texas A&M University System
Texas, USA

Sanchita Bandyopadhyay-Ghosh
Department of Mechanical Engineering
Manipal University Jaipur
Rajasthan, India

Chin Fhong Soon
Faculty of Electrical and Electronic Engineering
Universiti Tun Hussein Onn Malaysia
Parit Raja, Batu Pahat, Johor
Malaysia

CRC Press
Taylor & Francis Group
Boca Raton London New York

CRC Press is an imprint of the
Taylor & Francis Group, an **informa** business
A SCIENCE PUBLISHERS BOOK

Cover credit: Cover illustrations reproduced by kind courtesy of the editors.

First edition published 2025
by CRC Press
2385 NW Executive Center Drive, Suite 320, Boca Raton FL 33431

and by CRC Press
4 Park Square, Milton Park, Abingdon, Oxon, OX14 4RN

CRC Press is an imprint of Taylor & Francis Group, LLC

Library of Congress Cataloging-in-Publication Data (applied for)

ISBN: 978-1-032-59423-1 (hbk)
ISBN: 978-1-032-59424-8 (pbk)
ISBN: 978-1-003-45463-2 (ebk)

DOI: 10.1201/9781003454632

Typeset in Times New Roman
by Prime Publishing Services

Preface

Biomedical materials are considered to be the essential components in regenerative medicine applications, and they have shown the roadmap towards reducing the healthcare burden and enhancing the quality of life of many patients. Biofabrication, on the other hand, is an emerging field that consists of technologies that combine cells with biomedical materials. The convergence of biomedical materials science and biofabrication technologies has revolutionized our approach to tissue engineering and organ regeneration. From the development of sophisticated biomaterials that mimic the natural extracellular matrix to cutting-edge biofabrication techniques such as 3D bioprinting, these innovations are paving the way for new biotherapeutic strategies. To this end, the book *Biomedical Materials and Biofabrication for Regenerative Medicine* provides a comprehensive overview of the latest advancements and methodologies in this dynamic and rapidly evolving discipline. The book not only addresses the basic research and presents constraints regarding the usage of diverse biomaterials for tissue engineering and regenerative medicine but also details the existing technologies, advances, and future directions for tissue reconstruction and healing.

Beginning with the basic foundation of biomaterials, tissue engineering, and fabrication, the readers will find expert reviews on various advanced materials, technologies, and processes for emulating human tissue complexity. This book also comprehensively covers the recent advances in 3D bioprinting for tissue engineering, which includes possibilities, challenges, and future aspects of fabrication. It offers readers a deep understanding of regenerative medicine at the intersection of cell biology, molecular biology, engineering, and materials science. It is anticipated that the book will equip readers with the knowledge and tools necessary to advance the field of regenerative medicine and ultimately improve patient outcomes.

Biomedical Materials and Biofabrication for Regenerative Medicine is written by leading experts and a stellar cast of researchers to present a complete story of the development and promise of biomaterials and fabrication techniques in the interdisciplinary domain of regenerative medicine. We extend our gratitude to the contributors of this book who have shared their expertise and insights, making it an invaluable resource for anyone keen on exploring the future of biotherapeutics and biofabrication in patient care. We trust this book will serve as a beacon for a wide variety of multidisciplinary fraternities, including but not limited to researchers, clinical investigators, and advanced undergraduate and postgraduate students related to biomedical engineering, biomaterials, tissue engineering, and bio-additive manufacturing.

Naznin Sultana
Sanchita Bandyopadhyay-Ghosh
Chin Fhong Soon

Contents

Chapter **1**

Overview on the Biomaterials for Regenerative Medicine

Alyaa Idrees Abdulmaged*[1] and Chin Fhong Soon[2]

[1]Human Leukocyte Antigens Typing Laboratory, Al-Hadbaa Hospital for Hematology and Bone Marrow Transplantation, Nineveh Health Department, Iraqi Ministry of Health, 41002 Nineveh, Iraq

[2]Microelectronics and Nanotechnology-Shamsuddin Research Centre, Institute for Integrated Engineering, Universiti Tun Hussein Onn Malaysia, Parit Raja, 86400 Batu Pahat, Johor, Malaysia

1.1 INTRODUCTION

The birth of biomaterials for regenerative medicine is perhaps one of the most revolutionary innovations in the health sector. These materials are poised to solve some of the significant problems facing medical science. These materials enable the replacement of lost tissues and damaged internal organs, something once considered impossible. In regenerative medicine, biomaterials are fabricated to assist in developing different types of biomedical applications (Skardal and Atala 2015, Qu et al. 2019). One of the main goals of regenerative medicine is to reconstruct three-dimensional (3D) tissue structures that have similar functions to living tissues and organs. These structures can be used to replace or restore damaged tissues and organs.

Biomaterials are necessary as regenerative medicine progresses since the incorporation of biomaterials also escalates with the progress of this field. They are not just passive entities (scaffold materials) designed to interact with dynamic biological environments to promote repair and integration into healthy tissue

*For Correspondence: alyaaflayh3@gmail.com

constructs. These materials can provide a glimmer of hope and a chance to find new treatments for chronic illnesses, injuries, and congenital disabilities. So, this may be a shift in how patients will be treated. This promising field can function and enhance the value of the lives of millions of people. It outlines the increased focus on efficiency in addressing healthcare issues and challenges. Furthermore, it is a relatively young field, and new techniques are being introduced, the newest of which is the utilization of decellularized tissues. These tissues are the structural framework that preserves the chemical and architectural cues necessary for tissue engineering and remodeling. This method enhances faster healing and does not exhibit the immune rejection common with synthetic implants.

Recent research shows the increasing development of biomaterials and their numerous categories designed for regenerative medicine in humans. For instance, polysaccharides, proteins, and polyesters can fuse with the framework of the human body. Furthermore, these biomaterials are crucial in tissue and organ repair since they can mimic specific characteristics seen in the extracellular matrix (ECM) of the natural tissues of organs. For instance, hyaluronic acid and alginate materials are well-known for their biocompatibility and potential to re-create the native ECM, providing structural and signaling information crucial for tissue repair and regeneration. Figure 1.1 shows the biological structure of ECM, including collagen fibers, growth factors, and proteoglycans.

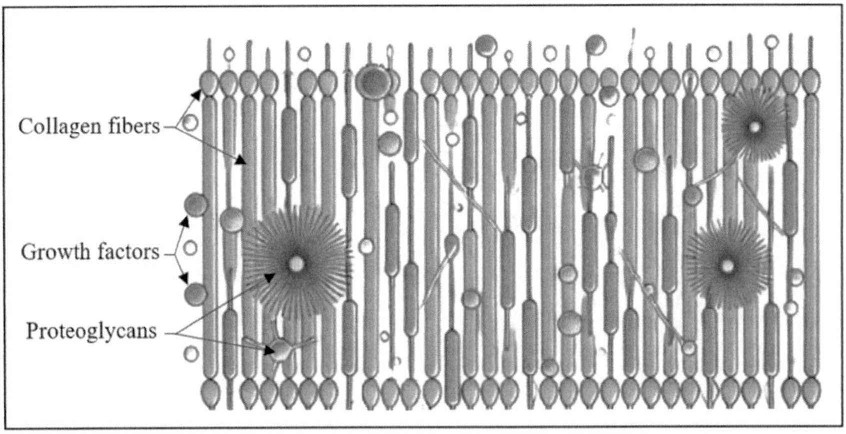

Figure 1.1 Natural ECM.

Since the development of 3D cell culture technology, biomaterials have been increasingly used to generate a 3D microenvironment and become a significant part of the new biosystem model. The combined biomaterials with different cell types in the 3D biosystem model have provided opportunities to produce an artificial microenvironment that can recapitulate the *in vivo* condition of ECM required for native cell function (Lutzweiler et al. 2020). In the field of regenerative medicine, the 3D artificial microenvironment holds excellent promise as many *in vivo* tissues have been mimicked *in vitro*, namely soft tissues, such as those in the brain, and hard tissues, such as those in bones (Lutzweiler et al. 2020, Oleksy et al. 2023).

The unique and controllable biological, chemical, and mechanical properties of biomaterials of natural and synthetic origin have allowed cell growth into more complex tissues (Oleksy et al. 2023). Moreover, applying additive manufacturing approaches such as 3D bioprinting in regenerative medicine enabled the fabrication of customizable and hierarchically complex functional 3D structures for tissue and organ regeneration.

This chapter is divided into four main sections. The following section presents the biomaterial requirements for regenerative medicine. Section 1.3 deliberates the biomaterials for regenerative medicine applications. Section 1.4 discusses the biomaterials commonly used in bioinks fabrication, including natural and synthetic biomaterials.

1.2 BIOMATERIAL REQUIREMENTS FOR REGERANTIVE MEDICINE

The most considerable properties of biomaterials for regenerative medicine are to meet the biological and mechanical requirements that are similar to the native tissues and organs. Biological requirements encompass biocompatibility, bioactivity, and surface tension. The mechanical properties encompass viscosity, shear-thinning feature, crosslink ability, and degradation. The degradation properties encompass degradation rate, degradation byproducts, and environmental sensitivity, as shown in Figure 1.2. These properties play a vital role in the design and development of biomedical products.

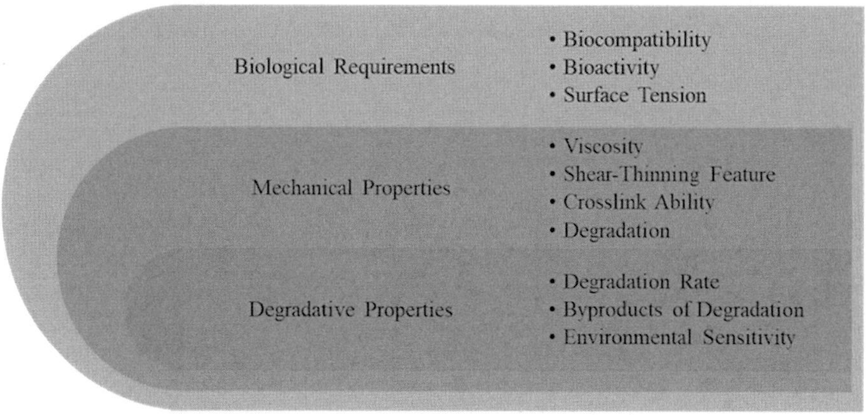

Figure 1.2 Biomaterial requirements for regenerative medicine.

1.2.1 Biocompatibility

Biocompatibility is considered a primary criterion before applying any biomaterial in regenerative medicine. This property is essential in supporting various cellular activities, including migration, adhesion, infiltration, differentiation, and eventually forming functional tissues or organs. Biocompatibility indicates a biomaterial's capability to achieve its required functions without triggering unfavorable effects (You et al. 2017). The biomaterials must be compatible, non-immunogenic, and have nontoxic degradation byproducts to avoid any harmful influences from fabrication to *in vitro* maturation and *in vivo* implantation (You et al. 2017, Zhang et al. 2018). Furthermore, the material must be compatible with the fabrication techniques. For instance, the material requires appropriate rheological properties and crosslinking mechanisms in microextrusion to solidify the printed structures and maintain well-defined shapes post-printing (Di Giuseppe et al. 2018).

1.2.2 Bioactivity

Bioactivity refers to the ability of biomaterial to interact with cells, which leads to a bond between the biomaterial and the tissue (Whitlow and Polini 2016, Joyce et al. 2021). The bioactivity of materials allows cells to adhere, migrate, and stimulates intracellular signals for cell-matrix interactions, differentiation, and formation of new tissues (Babeanu et al. 2022). The bioactivity of materials used in tissue engineering and regenerative medicine applications is either present within the chemical composition of the material used or through surface modifications. For instance, Gelatin is a structural protein widely applied in tissue regeneration scaffolds. It contains GxOGER and RGD sequences that can bind to $\alpha V\beta 3$ and $\alpha 5\beta 1$ integrins present on the surface of cells (Joyce et al. 2021). Polylactic acid and polyglycolic acid are inactive synthetic polymers that have been intensively used in the construction of cartilage scaffolds due to their robust mechanical properties. Although they are not biologically active substances, adding growth factors, such as transforming growth factor-beta 1 (TGF-β1), stimulated chondrogenesis (Ghandforoushan et al. 2022, Farshid et al. 2016).

1.2.3 Surface Tension

The hydrophilicity of a biomaterial surface is an important parameter that affects cellular behavior and protein adsorption. Hydrophilic behavior can vary based on the chemical composition of biomaterials. In tissue engineering and regenerative medicine, hydrophilic behavior is more suitable than hydrophobic behavior for fabricating 3D biostructures (Aiyelabegan and Sadroddiny 2017, Nishimura et al. 2018). Biological adhesion between cells and material surfaces is crucial in controlling cell biological processes such as activation, migration, proliferation, and differentiation to specific cell lineages (Ferrari et al. 2019).

Protein adsorption is the first step in regulating subsequent cellular behavior (Raja et al. 2022). The biomaterial's hydrophilic surface facilitates cell adhesion to the surface via integrin receptors. Thus, it mediates cell-biomaterial interactions, stimulating ECM secretion (Rabe and Seeger 2011). However, studies have shown that an excessive hydrophilic surface may inhibit protein adsorption and drive out protein molecules, and a highly hydrophobic surface may induce strong irreversible protein adsorption (Gun'ko et al. 2017, Ma et al. 2007).

1.2.4 Crosslinking

A biomaterial intended for regenerative medicine should preferably exhibit crosslinking ability from its aqueous solution. Crosslinking provides stabilization to the polymers. The polymers' chains are connected through either ionic or covalent bonds. Crosslinking forms a 3D meshed network with a high affinity for water absorbance, as shown in Figure 1.3. It makes biological structures appear similar to natural tissues. It further offers effective immobilization by releasing active agents and biomolecules (Ermis et al. 2018, Maitra and Shukla 2014). Generally, this network improves the hydrogel's mechanical properties, stability, and heat resistance (Maitra and Shukla 2014).

In tissue engineering and regenerative medicine, hydrogel crosslinking is commonly used to increase the structure's stability for *in vitro* and *in vivo* applications. The crosslinking should have cell-friendly and fast application properties to ensure a conserved shape of the original structure and prevent the collapse of biological constructs. Furthermore, it prevents cell sedimentation during the intricate and more extensive 3D tissue-like structures (Malda et al. 2013). Depending on the presence of specific functional groups in polymers, synthetic and natural polymers can be stabilized through different crosslinking methods. The crosslinking methods can be physical, chemical, or a combination of both (Murphy and Atala 2014, Yang et al. 2017). The schematic process of crosslinking polymer is presented in Figure 1.3, which indicates a brief analysis of how the two networks can crosslink to form a full matrix.

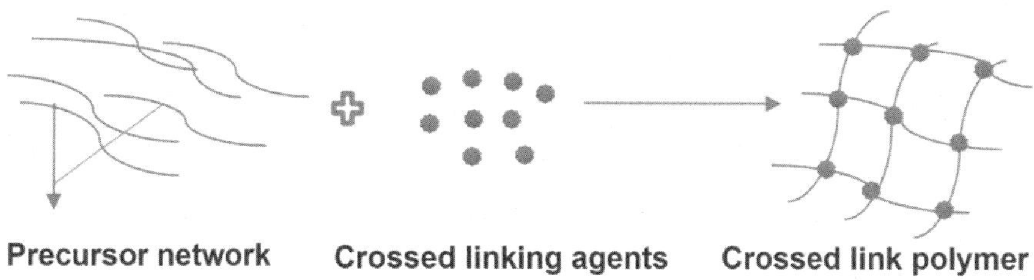

Figure 1.3 Crosslinking in polymers.

Crosslinking methods are classified into two basic techniques: physical and chemical methods. Physical crosslinking entails nonchemical interactions between the hydrophilic polymer chain to undergo phase transition (sol-gel) (Tang et al. 2011). Physical entanglement can prevent hydrogels from being dissolved in the aqueous medium, even though it may not be permanent (Maitra and Shukla 2014). The crosslinking is generated through ionic crosslinking, hydrophobic interaction, hydrogen bonds, and entangled chains. They are among the most frequently used techniques for generating hydrogels for biomedical applications because of their absence of toxicity. It remains rather flexible to be incorporated with living cells and biological molecules (growth factors) (Valot et al. 2019). In addition, physical crosslinking can be regulated by modulating environmental conditions such as temperature, pH, concentration, and ions (Liu et al. 2018).

Chemical crosslinking, therefore, is a process whereby functional groups of different polymer chains are chemically attached to form a three-dimensional polymer network that cannot undergo degradation (Emirs et al. 2018). In this case, the method by which a certain hydrogel can be chemically crosslinked differs from the others. Researchers have proposed that it should contain a crosslinker agent (glutaraldehyde, genuine), radiation (ultraviolet, gamma, electron beam), and enzymatic polymerization (transglutaminase) (Roehm and Madihally 2017). Hydrogels crosslinked through chemical methods are more stable and robust than physically crosslinked hydrogels. However, chemical crosslinking methods may include products that are fatal to cells or reduce cells' survival rate (Valot et al. 2019).

1.2.5 Degradation

The biodegradable behavior of biomaterials is critical for applications in regenerative medicine and tissue engineering. Synthetic and natural hydrogel is broadly used in 3D scaffold formation for tissue engineering applications (Ozcelik 2016, Sarker et al. 2014). Degradation is defined as a subgroup of properties continually distinguished from general biological and mechanical properties. In tissue engineering, the degradation rate is expected to be in balance with the rate of regenerating tissues so that the material degrades properly as the new tissue forms.

The second property is the byproducts of degradation, which must be nontoxic and safely processed by the body. The third property is environmental sensitivity, which means that materials are degraded in response to different body-specific triggers such as enzymes, pH changes, or water presence. This property remains critical as it highlights the importance of striking between material breakdown and biological integration. Thus, scaffolds and devices should be able to support the healing process and reverse consequences by being safely integrated with the entire body system.

In tissue engineering, scaffold degradation is important for new tissue formation. This process allows the cells to secrete their extracellular components and restructure their surrounding environment to guide tissue development. Hydrogels can be degraded hydrolytically or enzymatically (Ozcelik 2016, Nicodemus and Bryant 2008). Hydrolytic degradation in physiological conditions occurs due to esters, anhydrides, and thioesters functional groups in the polymer's backbone or the crosslinked hydrogels (Ozcelik 2016, Nicodemus and Bryant 2008). The hydrolysis of these functional groups leads to the polymer chain's cleavage or crosslinks and hydrogel degradation. Enzymatic degradation occurs during ECM remodeling by encapsulated cells. These cells secrete enzymes such as lysozyme, collagenase, and proteinases to degrade the artificial microenvironment and its components *in vivo* (Fonseca et al. 2014, Nicodemus and Bryant, 2008).

Hyaluronic acid, gelatin, and collagen are examples of naturally derived hydrogels. They can be degraded naturally by enzymes (collagenases, hyaluronidases, and lysozyme) or oxidative processes. Alginate is a natural polymer widely used in tissue engineering applications. Biodegradation in alginate-based hydrogels can occur by reversing the gelation mechanism or partial oxidation of alginate chains (Lee and Mooney 2012). The first process occurs by emancipating the divalent ions, such as sodium ions. It crosslinks the gel into the adjacent medium because of the exchange reactions with monovalent cations (Lee and Mooney 2012, Nicodemus and Bryant 2008). The second process uses sodium periodate, which causes the alginate backbone's degradation by altering the chain's conformation to an open-chain adduct (Lee and Mooney 2012, Nicodemus and Bryant 2008). Hydrogel is based on synthetic polymers, such as PEG, that possess biocompatibility but lack biodegradability. It can be modified with hydrolytic functional groups or by incorporating specific peptide sequences to control the degradation rate for tissue engineering applications (Cruz-Acuna and García 2017).

Generally, the controllable biodegradation of scaffold biomaterials is a highly desirable property that permits the scaffold to degrade at a rate that matches the new tissue formation (Nicodemus and Bryant 2008). The fast degradation rate can cause the cell-laden hydrogel to dissolve, while the slow degradation rate can cause the buildup of the ECM. The latter event influences both cell function and tissue growth. Depending on the targeted tissue applications, the degradation rate of scaffold polymers can be controlled by different mechanisms such as tuning polymer composition, molecular weight, and crosslinking densities (Ozcelik 2016, Nicodemus and Bryant 2008).

1.2.6 Viscosity

Viscosity refers to the internal resistance of materials to flow upon stress application (Zhang et al. 2018). It depends on the materials' concentration, temperature (for temperature-sensitive biomaterial), and molecular weight. Viscosity affects the encapsulated cells' efficacy and shape fidelity. Higher viscosities of a bioink are generally associated with high polymer concentration and molecular weight, resulting in a high printing fidelity. However, high viscosities of bioink solutions may not be favorable for cellular activities such as migration, proliferation, and ECM formation. Therefore, it is important to balance the viscosity of a bioink that supports printability and viscosity suitable for cellular activity.

1.2.7 Rheology

Rheology is a science that deals with the materials' flow and deformation behavior under applied forces (Schwab et al. 2020). A material transition in its rheological behavior to flow like a liquid when exposed to a force and return to its original state (gel) once the force is removed is preferable for scaffold fabrication (Boularaoui et al. 2020, Schwab et al. 2020).

Materials that show shear-thinning and thixotropic behavior are considered preferable for extrusion-based bioprinting. Shear-thinning (time-independent viscosity) behavior is observed when the material rapidly recovers. Its original state is visible after extrusion (i.e. the material will flow like a liquid when a shear force is applied once the force is removed reversely back to its gel state). Thixotropic (time-dependent viscosity) behavior is a time-dependent shear thinning behavior. Material that exhibits thixotropic properties gradually returns to its original state. In this case, the viscosity of the material decreases when exposed to external stress and flows like a liquid, and the recovery of the material to its initial state (gel) takes a limited time. Both shear-thinning and thixotropic properties are essential to maintaining printed constructs' structural integrity (Schwab et al. 2020, Kim et al. 2019).

1.3 BIOMATERIALS FOR REGENERATIVE MEDICINE

Biomaterials have been promoted in research activity in the last few years, especially in searching for applications for these materials in tissue engineering, cancer models, drug delivery, and biosensors. Current research focuses mainly on the bioinks for tissue engineering; this is the most advanced field. In this section, the pros and cons of a range of natural and synthetic materials that can be employed to fabricate bioinks via 3D bioprinting are presented.

Bioink is a material that supports cells and is an integral part of the 3D bioprinter, which produces functional living tissue and organ mimics for tissue engineering and regenerative medicine purposes (Ahadian and Khademhosseini 2018). Bioinks is a 3D biomaterial precursor temporary structure that interacts with the cells, promotes their differentiation, and stimulates tissue formation. It is used to form specific characteristics of functional grafted tissues that are similar to the primary tissues (Zhang et al. 2018).

The type of bioink used in most of the 3D bioprinting technologies include scaffold categories such as hydrogels, decellularized ECM (DECM), and synthetic polymers-based bioinks (Mobaraki et al. 2020 Hospodiuk et al. 2018, Jia et al. 2017). Based on the above examples, a variety of scaffold-based bioinks is depicted in Figure 1.4. The scaffold-based bioinks are incorporated into the exogenous material, and then the bioinks are bioprinted into a 3D construct. Various natural and synthetic materials have been employed as bioinks to create tissue constructs with biomimetic organizational architecture. Furthermore, selecting biomaterial to be used as bioink is crucial, and it varies with the kind of tissue to be produced and the application of the developed structure. The key elements of ideal bioinks for 3D bioprinting include printability, biocompatibility, biodegradability, high mechanical integrity, high crosslink ability, nonimmunogenic, and nontoxicity (Mobaraki et al. 2020, Hospodiuk et al. 2017).

In vivo, cellular behaviors and functions are influenced by stimuli surrounding the ECM. Likewise, the structures and physicochemical properties of bioinks provide important markers for controlling the functions of the encapsulated cells, thus directing their morphology, proliferation, and differentiation (Benwood et al. 2021, Li et al. 2018).

Hydrogels are unique in their properties, such as bioactivity, biocompatibility, porosity, and mechanical properties, making them a suitable choice as bioinks for fabricating 3D scaffolds. Therefore, cell-laden hydrogel is one of the heavily involved materials for bioink formulation of tissue and organs prepared from synthetic and/or natural polymers (Spicer 2020, Malda et al. 2013). Materials like alginate, gelatin, chitosan, poly-ε-caprolactone, and polyvinyl alcohols are the most popular materials in 3D-printed cell-laden hydrogels (Rijal 2017).

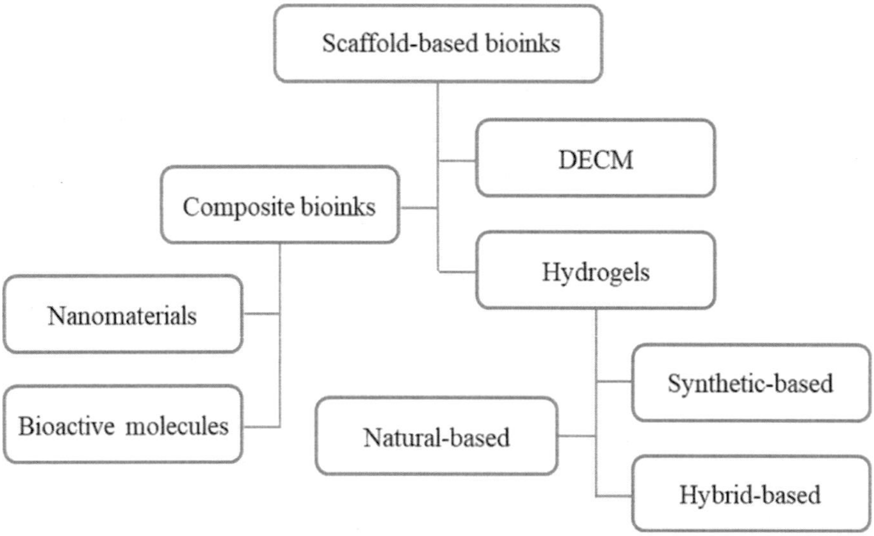

Figure 1.4 Scaffold-based bio-inks.

The chemical composition of the composite bioink is founded on proteins and carbohydrates, giving it a complex structure that replicates natural ECM (Fang et al. 2023). Some of the bio-composite inks that have been used in the 3D printing of bioinks include alginate/gelatin, cellulose/agarose/gelatin, alginate/gelatin/cholesteryl ester liquid crystals, and others (Othman et al. 2020, Alyaa et al. 2022). In another approach, a shear-thinning composite bioink has been prepared by incorporating nano biomaterials and different biomaterial solutions, including cellulose nanofibril (CNF), for 3D bioprinting. It aims to maintain robust structural stability in the post-printing phase (Han et al. 2020). Additionally, bioactive materials such as calcium phosphate, silica nanoparticles, and bioactive glass may be used to enhance the development of cells in the composite bioink for the formulation of tissues (Heid and Boccaccini 2020).

Another type of natural biomaterial widely used in regenerative medicine as a cell-laden scaffold is decellularized extracellular matrix (dECM). A key characteristic of the dECM is that it incorporates the hierarchical complexity of native tissues as it is derived from organs/tissues of the body, such as cardiac, lung, and cartilage, through physical, chemical, or enzymatic decellularization methods (Brown et al. 2022, Yao et al. 2019). The chemical composition of ECM consists of structural protein, carbohydrate, and bioactive molecules, making it a promising biomaterial for bioink applications (Rijal 2017, Hoshiba 2019). For this reason, one emerging approach involves converting the natural ECM into materials that can be used alone or in combination with other materials to produce functional structures (McInnes et al. 2022, Brown et al. 2022). Hence, dECM contains these compounds that can promote *in vivo* tissue remodeling and healing.

In 3D bioprinting, dECM-based bioinks have been at the front lines in fabricating various tissue substitutes and *in vitro* cancer models (Brown et al. 2022). Cellular studies have shown that liver extracellular matrix-based bioink (dECM Bioink) can create a liver-specific micro-environment essential for the bioprinting of liver tissues. A dECM and gelatin-based bioink was formulated and assessed by examining various printability parameters for the 3D printing process. This demonstration of the liver cells and the endothelial cells to be highly alive and mature indicates the excellent quality of the dECM bioink. For instance, a bioprinted construct can simulate the liver lobule and show excellent liver functions. The bioprinted liver architectonics displayed the characteristics of the native liver both structurally and functionally (Kim et al. 2023). Scaffold-based bioinks can be engineered by fabricating biomaterials according to specific cell types and then processed by 3D bioprinting techniques. The developments of viable solutions occurred during the past decade, including those for liver, kidney, skin, and heart tissues (Mandriki et al. 2016, Ozbulat et al. 2017).

1.4 BIOMATERIALS FOR BIOINKS

Biomaterials used in tissue engineering and regenerative medicine are categorized as either natural or synthetic polymers. The fourth part of this chapter provides a general overview of typical polymers; Section 1.4.1 is dedicated to natural and synthetic polymers used in the 3D bioprinting of bioinks.

Natural polymers constitute one of the most important applications of polymeric materials in regenerative medicine, and they can be obtained from natural resources (Spicer 2020, Malda et al. 2013). When these polymers are hydrated, they can achieve a highly spongy structure, ensuring a workspace for cell attachment and migration. Natural biomaterials can be classified into two subclasses: proteins and polysaccharides. In this part, we will deal with hydrogel ingredients that are naturally derived hydrogels (polysaccharides and protein) used for bioink fabrications.

1.4.1 Alginate

Alginate is a polyanion (an anionic polymer) (Axpe and Oyen 2016). The negative charge of alginate arises from the presence of carboxylic groups in the polymer backbone. In these instances, the carboxylic groups are ionically bonded to monovalent cations such as sodium (Na^+), forming alginate salt (Freeman and Kelly 2017). Alginate is a linear carbohydrate, which is referred to as a natural polysaccharide extracted from brown seaweed and various types of bacteria. The structure of carbohydrate polymers is based on a mixture of α-L-guluronic acid and β-D-mannuronic acid linked together by a chemical bond (Bidarra et al. 2014). The M and G blocks are sequentially positioned alternatingly built on a repeating group of GGGs, a repeating group of MM, or an alternating pattern of MGMG. The amount and distribution of each unit depend on the sources from which alginate was extracted (Lee and Mooney 2012). The proportion and the order of arranged M and G monomers will decide the most functional properties of alginate and its hydrogels. For example, gels prepared from an alginate gel with blocks at the M site (M-G-M-G-M) are more easily flowing than a gel with a high content of G blocks (G-G-G-G), which are stiffer chains (Cao 2011). Alginate has favorable properties, including shear-thinning properties, high compatibility, nonimmunogenicity, easily modulated with other polymers, and entrap water and other molecules (Gopinathan and Noh 2018, Liu et al. 2017). It has been widely used in 3D bioprinting of tissue and organs (Lee and Mooney 2012).

In 3D bioprinting, alginate-based hydrogels have substantial advantages as bioinks for cell printing. They have been widely applied in the 3D printing of tissues such as bone, blood vessels, cartilage, muscle, nerve, and heart tissues (Park et al. 2017, Axpe and Oyen 2016, Lee and Mooney 2012). However, using pure alginate bioinks has several limitations to engineer effective 3D-printed constructs. Studies reported that the printing of pure alginate hydrogel alone yields poor mechanical properties, making the printed constructs unable to retain the overall shape of the fabricated structure during and after 3D printing (Liu et al. 2018, Di Giuseppe et al. 2018, Chung et al. 2013). Therefore, alginate hydrogels are usually modified or blended with other polymers to enhance their mechanical strength.

Kim et al.'s (2019) work shows that through rheological testing of Alginate-$CaSO_4$ and Alginate/Carrageenan/$CaSO_4$, hydrogels that crosslinked alginate solution (Alg-$CaSO_4$ and Alg-Carr-$CaSO_4$) can exhibit gel-like behavior. Alginate solution behaves as a viscous fluid at 0% $CaSO_4$ concentration. Alg-Carr-$CaSO_4$ hydrogels demonstrated better resolution of 3D-printed scaffolds, and different percent weight-in-volume ratios of Carrageenan exhibited high viscosity compared to pre-crosslinked alginate hydrogels. Similarly, Wu et al. (2017) employed 2% Alginate-4% CNC to build a biomimetic liver-like honeycomb matrix with fibroblast and HepG2 hepatoma cells. Consequently, their study showed that the bioink with 2% Alginate-4% CNC had the desired shear thinning characteristics, high printing resolution, and shape fidelity. Thus, when the concentration of CNC is increased, the printed structure's geometries are also enhanced. Preliminary viability tests of fibroblast and hepatoma cells have indicated that hybrid bioinks are biocompatible and may be used in 3D bioprinting.

One disadvantage of using alginate-based bioinks in 3D bioprinting is the absence of a molecular receptive site for cell adhesion that is essential in directing and controlling cellular integration (Wu et al. 2020, Panwar and Tan 2016). This issue could be rectified by using Alginate–RGD cell attachment promoter or incorporating bioactive molecules (Park et al. 2017, Lee and Mooney 2012). In addition, incorporating other polymers that possess inherent RGD sequences, like collagen, in the formulation while preparing the alginate bioinks has also been found to enhance the nature of the material to support cell functionality (Chung et al. 2013).

1.4.2 Chitosan

Chitosan is a natural derivative of chitin. The chemical structure of polymer chains consists of D-glucosamine and N-acetyl-D-glucosamine with β (1-4) glycosidic linkages (Croisier and Jérôme 2013). The polysaccharide chains of chitosan share biological features with glycosaminoglycan, similar to the ECM of native tissue (Roehm and Madihally 2017, Liu et al. 2017). Chitosan is widely utilized in tissue engineering and 3D bioprinting due to its favorable features, such as high porosity, nontoxicity, and antimicrobial activity (Yang et al. 2017, Panwar and Tan 2016). Chitosan forms a highly viscous hydrogel, which makes it optimum for 3D bioprinting.

Nevertheless, chitosan has low mechanical strength and slow gelation issues, which makes the 3D-printed structure soft and unstable. The blending of chitosan with other polymers like gelatin, alginate, hyaluronic acid, and synthetic polymers has been applied to improve its printability and, hence, its mechanical properties. Ng et al. (2016) printed a combination of chitosan and gelatin hydrogels for skin tissue engineering. The resulting complex polyelectrolyte achieved good printability, high structural stability of the 3D printed hydrogels, and good biocompatibility with fibroblast skin cells.

1.4.3 Hyaluronic Acid

Hyaluronic acid (HA) is a natural, unbranched polysaccharide constructed of repeating units of D-glucuronic acid and N-acetyl-D-glucosamine residues linked by β (1-3) bond (Ashammakhi et al. 2019, Yang et al. 2017). HA acid naturally exists in the ECM of native tissues, like neural, epithelial, and connective tissues (Liu et al. 2017, Hamman 2010). HA plays a substantial role in maintaining structural integrity and hydration of ECM, cell-matrix interaction, and tissue homeostasis (Liu et al. 2017, Panwar and Tan 2016). Therefore, HA is considered an ideal biomaterial for bioink in the 3D bioprinting of tissue and organs. Antich et al. (2020) have developed HA and alginate-based bioink for articular cartilage (AC) regeneration with excellent mechanical properties. Encapsulating cells in the HA/alginate bioink demonstrated high viability, which increased during culture period. It results in enhancing the capacity of the bioink to induce the formation of chondrogenesis in the printed construct. They thus may overcome several current challenges related to cartilage tissue engineering.

1.4.4 Carrageenan

Carrageenan is a natural, water-soluble carbohydrate extracted from the class Rhodophyceae red algae (Necas and Bartosikova 2013). It is unbranched, sulfated polyglactin. It is chemically composed of alternate D-galactose disaccharide units connected by α-1,3 and β-1,4 glycosides linkage (Kim et al. 2019, Necas and Bartosikova 2013). Carrageenan is not nutritional and is used for gelling and food product stabilizers. There are three main sorts of Carrageenan identified based on the number and position of sulfate groups, namely, kappa (κ), iota (ι), and lambda (λ) (Kim et al. 2019). The sulfate groups in the backbone of the polymer chains positively affect cellular behavior in areas such as attachment, infiltration, and differentiation (Lim et al. 2020). In a study by Lim et al. (2020), a class of biocompatible and biodegradable carrageenan bioink has been developed for 3D bioprinting using Kappa-carrageenan crosslinked with methacrylate (MA-κ-CA). Assessment of rheological behavior, swelling ratio, degradation profile, and mechanical properties proved that the MA-κ-CA satisfied the essential requirements of the bioink. Encapsulated NIH_3T3 fibroblast cells showed good *in vitro* cytocompatibility and formed cellular spheroid in the MA-κ-CA. The results indicated that the MA-κ-CA bioink forms a preferable material for applications in tissue engineering.

1.4.5 Cellulose

As opposed to conventional materials applied in the biomedical field, the high interest in utilizing nano-cellulose is due to its favorable characteristics, such as biocompatibility, high mechanical strength, and tunable chemical composition and structure (Huang et al. 2019, Wang et al. 2018). Cellulose nanocrystals (CNC) and cellulose nanofibril (CNF) were employed to form hydrogel structures to provide scaffolding; they were applied in tissue engineering (Sultan et al. 2017). Thus, as nano-cellulose particles can shear thin and possess unique rheology, nano-cellulose is a valuable material for developing bioinks for 3D bioprinting. The bioinks are a mixture of nano-cellulose only or a blend of other biopolymers such as polyethylene glycol, gelatin methacrylate, or alginate (Han et al. 2020). Some of the advantages of CNF include a high degree of purity, biocompatibility, crystalline structure, wettability, and tunable surface chemistry. Such attributes are useful in environmental applications, especially when scaffold fabrication is incorporated and biocompatibility issues are essential.

1.4.6 Gelatin

Gelatin is a natural hydrophilic protein produced by the denaturation of collagen. Gelatin can be extracted from many sources, such as pork, calfskin, and coldwater fish skin (Aljohani et al. 2018, Das et al. 2017). Based on the hydrolysis process of collagen, there are two types of Gelatin polymer, namely Gelatin A and Gelatin B. Gelatin type A is produced through acid-based hydrolysis, and type B Gelatin is obtained through base-based hydrolysis of collagen (Panwar and Tan 2016). Gelatin is characterized by natural cell adhesion moieties that can support cell growth and function (RGD peptide sequences) (Yang et al. 2017, Panwar and Tan 2016, Skardal and Atala 2015). Besides that, gelatin is a biodegradable, noncytotoxic polymer that can form physically crosslinked hydrogels at low temperatures (Panwar and Tan 2016, Thiele et al. 2014). All these features make it an appropriate material for bioinks (Aljohani et al. 2018, Skardal and Atala 2015).

Gelatin forms thermo-reversible hydrogels that undergo a sol-gel transition below 25°C, making it suitable for deposition layer by layer (Schuurman et al. 2013).

However, physically crosslinked pure gelatin hydrogel forms weak stability structures at the physiological temperature (37°C), which hinders its application in 3D bioprinting (Yang et al. 2017, Skardal and Atala 2015). Due to this limitation, gelatin hydrogel often requires additional modification for implementation in the living system. One strategy to improve the low viscosity of gelatin solution is through the chemical modification of gelatin with methacryloyl groups using photo-polymerization, which results in covalently forming crosslinked hydrogels of Gelatin Methacrylamide (GelMA) under mild conditions (Valot et al. 2019, Skardal and Atala 2015). GelMA hydrogel is a biocompatible biomaterial with a tunable degradation rate, cell-affinitive domains, tailorable mechanical properties, and safety for native tissue (Yang et al. 2017, Aljohani et al. 2018). The combination of GelMA hydrogels with other polymers has been effectively used to fabricate bioinks for 3D bioprinting (Adib et al. 2020, Jia et al. 2016). For instance, GelMA, hyaluronic acid, and chondrocytes were used to engineer cartilage constructs using 3D bioprinting. HA increased the viscosity, facilitated the printing process, allowed the controlled, homogeneous deposition of bioink filaments, and improved the printed constructs' shape fidelity. The encapsulated chondrocytes in the gelMA/HA mixtures exhibited high viability, and cartilaginous tissue was formed after 28 days of cultivation (Schuurman et al. 2013).

1.4.7 Collagen

Collagen is one of the most attractive natural polymers broadly used in bioink development for tissue engineering applications. Collagen is the most abundant structural protein in the ECM of mammalian tissues. Collagen is a triple-helical protein with the amino acid sequence RGD in its structure, which integrin receptors on the cell membrane can recognize and provide binding support to the cells (Yang et al. 2017). The living organism has about 28 kinds of collagen proteins, such as Collagen types I, II, III, and V. However, Collagen type I is mainly used in tissue engineering because of its high availability and ability to undergo physical gelling under physiological conditions (37°C), exhibiting preferable features in tissue bioprinting areas (Panwar and Tan 2016).

Collagen types I and II are widely used as promising 3D scaffolds in skin and cartilage tissue engineering. However, pure collagen hydrogels' low mechanical properties and slow gelation hinder their application in 3D bioprinting (Liu and Ma, 2003, Panwar and Tan 2016). Several strategies have been deployed to enhance collagen hydrogels' mechanical properties, such as blending with other polymers like fibrin, chitosan, HA, and agarose.

1.4.8 Fibrin

Fibrin is a natural structural protein that interplays fibrinogen and protease thrombin during blood coagulation. *In vivo* has an essential role in coagulation, tumor growth, and wound healing (Aljohani et al. 2018, Liu et al. 2018). Fibrin is easily polymerized by incubating calcium ions, fibrinogen, and thrombin (Panwar and Tan 2016). Fibrin hydrogel is considered an appealing biopolymer for tissue engineering applications because of its biocompatibility and the presence of cell adhesion sites that can enhance the biological features (functionality) of the 3D scaffold (Liu et al. 2017). Besides, the stiffness of fibrin hydrogel can be easily controlled by changing the concentration of its components. It becomes a favorable material in bioink preparation, especially for inkjet bioprinting techniques (Liu et al. 2017, Panwar and Tan 2016). In bioinks fabrication, fibrinogen gels are low in viscosity. They are enhanced by adding other biomaterials because of their excellent biological activity and suitable mechanical properties after polymerization (Włodarczyk-Biegun and Del Campo 2017).

1.5 SYNTHETIC BIOMATERIALS

Synthetic hydrogels are frequently used in bioink development due to their better performance, controlled structure, and flexibility than naturally derived materials (Bidarra et al. 2014). The following section summarizes some frequently used synthetic polymers for tissue engineering applications.

1.5.1 Poly(ε-caprolactone)

Poly(ε-caprolactone) (PCL) is a degradable, semicrystalline synthetic polymer with hydrophobic properties and an appropriate melting point of 60°C (Taghavi et al. 2019). Indeed, the mechanical properties and structural strength of PCL have made it a popular choice for tissue engineering covering material, especially for bone tissue engineering as a supportive polymer (Koupaei and Karkhaneh 2016). A novel class of hybrid biomaterials, including hydroxyapatite, dicalcium phosphate dihydrate, and tricalcium phosphate, has formed from modifying PCL, which shows promising results in bone tissue engineering (Koupaei and Karkhaneh 2016). The problem with PCL is its low bioactivity, which prevents cells from adhering to the surface. PCL will provide stronger adherence of the cells combined with the RGD peptide sequences or blended with collagen and gelatin (Mota et al. 2014, Zhang et al. 2009). Researchers, namely Duymaz et al. (2019), have used PCL-based hydrogel and Gelatin and Halomonas levan (HLh) as novel bioinks systems, which allowed for HOb cell differentiation. The composite bioink comprising PCL/GT/HLh achieved suitable degradation rates, high mechanical properties, and functionality for bone tissue engineering.

1.5.2 Polyethylene Glycol

Polyethylene Glycol (PEG) is a non-immunogenic, highly hydrophilic, and biocompatible synthetic polymer with a branched and linear structure (Hamid and Lim 2016, Hospodiuk et al. 2016). Considering that its biocompatibility is adjustable and the mechanical properties are tunable, PEG and its derivatives have received considerable attention recently for cell encapsulation, drug delivery, and scaffold materials in tissue engineering applications (Burke et al. 2019). PEG lacks sufficient mechanical strength for microextrusion bioprinting. It is observable that its mechanical stiffness needs to be higher. Therefore, PEG-based derivatives are commonly combined with other polymers, especially crosslinkers, to increase the mechanical properties of the bioink and improve the printing resolution.

1.5.3 Pluronic Acid

Pluronic acid or poloxamers is a class of synthetic biomaterials that consist of tri-block copolymers. The copolymer comprises water-loathing PPO in the core and two water-craving PEO chains from outside. Pluronic acid forms a reversible thermosensitive gel which, during the sol-gel transition under the influence of temperature, can work only in a concentration of more than 25% (Panwar and Tan 2016).

The drawback of poloxamers for 3D bioprinting includes their low mechanical strength, reducing the quality of printed structures. However, photo-polymerization of pluronic acid can overcome this drawback (Zarrintaj et al. 2020). Another weakness of pluronic is the low interaction with cells, which eventually leads to low cell viability (Panwar and Tan 2016, Zarrintej et al. 2020).

1.5.4 Polyvinyl Alcohol

Polyvinyl Alcohol (PVA) is a linear synthetic polymer obtained from polyvinyl acetate through partial or full hydroxylation (Meng et al. 2020). PVA has been widely used for biomedical applications (Baker et al. 2012, Meng et al. 2020), due to its biodegradability, low tendency for protein adhesion, biocompatibility, and nontoxicity (Baker et al. 2012). The significant advantage of using PVA in bioink development is its high hydrophilicity (supports the diffusion of nutrient and gas exchange), which is considered an attractive feature for tissue engineering, especially for cartilage replacement (Baker et al. 2012). Meng et al. (2020) employed extrusion-based technology in printing PVA, graphene oxide, and hydroxyapatite hydrogel for artificial cartilage replacement. The 3D-printed PVA/GO/HA hydrogel showed high stiffness, excellent lubrication, and biological activity as a cartilage replacement.

1.6 CONCLUDING REMARKS

This chapter offers an overview of the biomaterials used in regenerative medicine and tissue engineering, including their biological and mechanical requirements as bioinks for 3D bioprinting. It also discusses the bioink types and their challenges in 3D bioprinting technologies. It explores the development of bioinks with the right balance of biological and mechanical requirements for 3D bioprinting of tissue constructs. Furthermore, it provides a simplified and comprehensive explanation of various biomaterial types, from natural (polysaccharides and proteins) to synthetic origins. This chapter also summarizes the most common bioinks for preparing scaffolds, such as cell-laden hydrogels, composite bioinks, and dECM-based bioinks.

REFERENCES

Ahadian, S. and A. Khademhosseini. 2018. A perspective on 3D bioprinting in tissue regeneration. Bio-design and Manufacturing. 1(3): 157–160.

Aiyelabegan, H.T. and E. Sadroddiny. 2017. Fundamentals of protein and cell interactions in biomaterials. Biomedicine and Pharmacotherapy. 88: 956–970.

Aljohani, W., M.W. Ullah, X. Zhang and G. Yang. 2018. Bioprinting and its applications in tissue engineering and regenerative medicine. International Journal of Biological Macromolecules. 107: 261–275.

Ashammakhi, N., S. Ahadian, C. Xu, H. Montazerian, H. Ko, R. Nasiri, et al. 2019. Bioinks and bioprinting technologies to make heterogeneous and biomimetic tissue constructs. Materials Today Bio. 1: 100008.

Axpe, E. and M.L. Oyen. 2016. Applications of alginate-based bioinks in 3D bioprinting. International Journal of Molecular Sciences. 17(12): 1976.

Babeanu, N., N. Radu, C.E. Enascuta, E. Alexandrescu, M. Ganciarov, M.S.O. Mohammed, et al. 2022. Obtaining and Characterizing Composite Biomaterials of Animal Resources with Potential Applications in Regenerative Medicine. Polymers. 14(17): 3544.

Baker, M.I., S.P. Walsh, Z. Schwartz and B.D. Boyan. 2012. A review of polyvinyl alcohol and its uses in cartilage and orthopedic applications. Journal of Biomedical Materials Research Part B: Applied Biomaterials. 100(5): 1451–1457.

Benwood, C., J. Chrenek, R.L. Kirsch, N.Z. Masri, H. Richards, K. Teetzen, et al. 2021. Natural biomaterials and their use as bioinks for printing tissues. Bioengineering. 8(2): 27.

Bidarra, S.J., C.C. Barrias and P.L. Granja. 2014. Injectable alginate hydrogels for cell delivery in tissue engineering. Acta Biomaterialia. 10(4): 1646–1662.

Boularaoui, S., G. Al Hussein, K.A. Khan, N. Christoforou and C. Stefanini. 2020. An overview of extrusion-based bioprinting with a focus on induced shear stress and its effect on cell viability. Bioprinting. 20: 00093.

Brown, M., J. Li, C. Moraes, M. Tabrizian and N.Y. Li-Jessen. 2022. Decellularized extracellular matrix: New promising and challenging biomaterials for regenerative medicine. Biomaterials. 121786.

Burke, G., V. Barron, T. Geever, L. Geever, D.M. Devine and C.L. Higginbotham. 2019. Evaluation of the materials properties, stability and cell response of a range of PEGDMA hydrogels for tissue engineering applications. Journal of the Mechanical Behavior of Biomedical Materials. 99: 1–10.

Cao, N. 2011. Fabrication of alginate hydrogel scaffolds and cell viability in calcium-crosslinked alginate hydrogel (Doctoral dissertation, University of Saskatchewan).

Chung, J.H., S. Naficy, Z. Yue, R. Kapsa, A. Quigley, S.E. Moulton, et al. 2013. Bioink properties and printability for extrusion printing living cells. Biomaterials Science. 1(7): 763–773.

Croisier, F. and C. Jérôme. 2013. Chitosan-based biomaterials for tissue engineering. European Polymer Journal. 49(4): 780–792.

Cruz-Acuña, R. and A.J. García. 2017. Synthetic hydrogels mimicking basement membrane matrices to promote cell-matrix interactions. Matrix Biology. 57: 324–333.

Das, M.P., P.R. Suguna, K. Prasad, J.V. Vijaylakshmi, M. and Renuka. 2017. Extraction and characterization of Gelatin: a functional biopolymer. International Journal of Pharmacy and Pharmaceutical Sciences. 9(9): 239.

Di Giuseppe, M., N. Law, B. Webb, R.A. Macrae, L.J. Liew, T.B. Sercombe, et al. 2018. Mechanical behaviour of alginate-gelatin hydrogels for 3D bioprinting. Journal of the Mechanical Behavior of Biomedical Materials. 79: 150–157.

Duymaz, B.T., F.B. Erdiler, T. Alan, M.O. Aydogdu, A.T. Inan, N. Ekren, et al. 2019. 3D bio-printing of levan/polycaprolactone/gelatin blends for bone tissue engineering: Characterization of the cellular behavior. European Polymer Journal. 119: 426–437.

Ermis, M., S. Calamak, G.C. Kocal, S. Guven, N.G. Durmus, I. Rizvi, et al. 2018. Hydrogels as a new platform to recapitulate the tumor microenvironment. pp. 463–494. *In*: João Conde (ed.). Handbook of Nanomaterials for Cancer Theranostics. Elsevier.

Fang, W., M. Yang, L. Wang, W. Li, M. Liu, Y. Jin, et al. 2023. Hydrogels for 3D bioprinting in tissue engineering and regenerative medicine: current progress and challenges. International Journal of Bioprinting, 9(5).

Farshid S., M. Youseffi, S.A. Khaghani, C.F. Soon and F. Javid. 2016. Effect of transforming growth factor-β3 on mono and multilayer chondrocytes. Cytokine. 83:118–126.

Ferrari, M., F. Cirisano and M.C. Morán. 2019. Mammalian cell behavior on hydrophobic substrates: Influence of surface properties. Colloids and Interfaces. 3(2): 48.

Fonseca, K.B., P.L. Granja and C.C. Barrias. 2014. Engineering proteolytically-degradable artificial extracellular matrices. Progress in Polymer Science. 39(12): 2010–2029.

Freeman, F.E. and D.J. Kelly. 2017. Tuning alginate bioink stiffness and composition for controlled growth factor delivery and to spatially direct MSC fate within bioprinted tissues. Scientific Reports. 7(1): 17042.

Ghandforoushan, P., J. Hanaee, Z. Aghazadeh, M. Samiei, A.M. Navali, A. Khatibi and S. Davaran. 2022. Enhancing the function of PLGA-collagen scaffold by incorporating TGF-β1-loaded PLGA-PEG-PLGA nanoparticles for cartilage tissue engineering using human dental pulp stem cells. Drug Delivery and Translational Research. 12(12): 2960–2978.

Gopinathan, J. and I. Noh. 2018. Recent trends in bioinks for 3D printing. Biomaterials Research. 22: 1–15.

Gun'ko, V.M., I.N. Savina and S.V. Mikhalovsky. 2017. Properties of water bound in hydrogels. Gels. 3(4): 37.

Hamid, Z.A. and K.W. Lim. 2016. Evaluation of UV-crosslinked poly (ethylene glycol) diacrylate/poly (dimethylsiloxane) dimethacrylate hydrogel: properties for tissue engineering application. Procedia Chemistry. 19: 410–418.

Hamman, J.H. 2010. Chitosan based polyelectrolyte complexes as potential carrier materials in drug delivery systems. Marine Drugs. 8(4): 1305–1322.

Han, C., X. Wang, Z. Ni, Y. Ni, W. Huan, Y. Lv., et al., 2020. Effects of nanocellulose on alginate/gelatin bioinks for extrusion-based 3D printing. BioResources. 15(4): 7357–7373.

Heid, S. and A.R. Boccaccini. 2020. Advancing bioinks for 3D bioprinting using reactive fillers: a review. Acta Biomaterialia. 113: 1–22.

Hoshiba, T. 2019. Decellularized extracellular matrix for cancer research. Materials. 12(8): 1311.

Hospodiuk, M., M. Dey, D. Sosnoski and I.T. Ozbolat. 2017. The bioink: a comprehensive review on bioprintable materials. Biotechnology Advances. 35(2): 217–239.

Huang, L., X. Du, S. Fan, G. Yang, H. Shao, D. Li, et al., 2019. Bacterial cellulose nanofibers promote stress and fidelity of 3D-printed silk based hydrogel scaffold with hierarchical pores. Carbohydrate Polymers. 221: 146–156.

Joyce, K., G.T. Fabra, Y. Bozkurt and A. Pandit. 2021. Bioactive potential of natural biomaterials: Identification, retention and assessment of biological properties. Signal Transduction and Targeted Therapy. 6(1): 122.

Kim, M.H., Y.W. Lee, W.K. Jung, J. Oh and S.Y. Nam. 2019. Enhanced rheological behaviors of alginate hydrogels with Carrageenan for extrusion-based bioprinting. Journal of the Mechanical Behavior of Biomedical Materials. 98: 187–194.

Kim, M.K., W. Jeong and H.W. Kang. 2023. Liver dECM–Gelatin Composite Bioink for Precise 3D Printing of Highly Functional Liver Tissues. Journal of Functional Biomaterials. 14(8): 417.

Koupaei, N. and A. Karkhaneh. 2016. Porous crosslinked polycaprolactone hydroxyapatite networks for bone tissue engineering. Tissue Engineering and Regenerative Medicine. 13: 251–260.

Lee, K.Y. and D.J. Mooney. 2012. Alginate: properties and biomedical applications. Progress in Polymer Science. 37(1): 106–126.

Lim, W., G.J. Kim, H.W. Kim, J. Lee, X. Zhang, M.G. Kang, et al., 2020. Kappa-carrageenan-based dual crosslinkable bioink for extrusion type bioprinting. Polymers, 12(10): 2377.

Liu, M., X. Zeng, C. Ma, H. Yi, Z. Ali, X. Mou, et al., 2017. Injectable hydrogels for cartilage and bone tissue engineering. Bone Research. 5(1): 1–20.

Liu, Q., Q. Li, S. Xu, Q. Zheng and X. Cao. 2018. Preparation and properties of 3D printed Alginate–Chitosan polyion complex hydrogels for tissue engineering. Polymers. 10(6): 664.

Lutzweiler, G., A. Ndreu Halili and N. Engin Vrana. 2020. The overview of porous, bioactive scaffolds as instructive biomaterials for tissue regeneration and their clinical translation. Pharmaceutics. 12(7): 602.

Ma, Z., Z. Mao and C. Gao. 2007. Surface modification and property analysis of biomedical polymers used for tissue engineering. Colloids and Surfaces B: Biointerfaces. 60(2): 137–157.

Malda, J., J. Visser, F.P. Melchels, T. Jüngst, W.E. Hennink, W.J. Dhert, et al., 2013. 25th anniversary article: engineering hydrogels for biofabrication. Advanced Materials. 25(36): 5011–5028.

Małgorzata, O., K. Dynarowicz and D. Aebisher. 2023. Advances in biodegradable polymers and biomaterials for medical applications: a review. Molecules. 28(17): 6213.

McInnes, A.D., M.A. Moser and X. Chen. 2022. Preparation and use of decellularized extracellular matrix for tissue engineering. Journal of Functional Biomaterials. 13(4): 240.

Meng, Y., J. Cao, Y. Chen, Y. Yu and L. Ye. 2020. 3D printing of a poly (vinyl alcohol)-based nano-composite hydrogel as an artificial cartilage replacement and the improvement mechanism of printing accuracy. Journal of Materials Chemistry B. 8(4): 677–690.

Mobaraki, M., M. Ghaffari, A. Yazdanpanah, Y. Luo and D.K. Mills. 2020. Bioinks and bioprinting: a focused review. Bioprinting. 18: e00080.

Murphy, S.V. and A. Atala. 2014. 3D bioprinting of tissues and organs. Nature Biotechnology. 32(8): 773–785.

Necas, J. and L. Bartosikova. 2013. Carrageenan: a review. Veterinarni Medicina. 58(4): 187–205.

Ng, W.L., W.Y. Yeong and M.W. Naing. 2016. Polyelectrolyte gelatin-chitosan hydrogel optimized for 3D bioprinting in skin tissue engineering. International Journal of Bioprint. 2(1): 53–62.

Nicodemus, G.D. and S.J. Bryant. 2008. Cell encapsulation in biodegradable hydrogels for tissue engineering applications. Tissue Engineering Part B: Reviews. 14(2): 149–165.

Nishimura, T., Y. Ogino, Y. Ayukawa and K. Koyano. 2018. Influence of the wettability of different titanium surface topographies on initial cellular behavior. Dental Materials Journal. 37(4): 650–658.

Ozcelik, B. 2016. Degradable hydrogel systems for biomedical applications. pp. 173–188. *In*: L. Poole-Warren, P. Martens and R. Green (eds). Biosynthetic Polymers for Medical Applications. Woodhead Publishing.

Panwar, A. and L.P. Tan. 2016. Current status of bioinks for micro-extrusion-based 3D bioprinting. Molecules. 21(6): 685.

Qu, H., H. Fu, Z. Han and Y. Sun. 2019. Biomaterials for bone tissue engineering scaffolds: a review. RSC Advances. 9(45): 26252–26262.

Rabe, M., D. Verdes and S. Seeger. 2011. Understanding protein adsorption phenomena at solid surfaces. Advances in Colloid and Interface Science. 162(1–2): 87–106.

Raja, I.S., S.H. Lee, M.S. Kang, S.H. Hyon, A.R. Selvaraj, K. Prabakar, et al., 2022. The predominant factor influencing cellular behavior on electrospun nanofibrous scaffolds: wettability or surface morphology. Materials and Design. 216: 110580.

Rijal, G. 2017. The decellularized extracellular matrix in regenerative medicine. Regenerative Medicine. 12(5): 475–477.

Roehm, K.D. and S.V. Madihally. 2017. Bioprinted chitosan-gelatin thermosensitive hydrogels using an inexpensive 3D printer. Biofabrication. 10(1): 015002.

Sarker, B., R. Singh, R. Silva, J.A. Roether, J. Kaschta, R. Detsch, et al. 2014. Evaluation of fibroblasts adhesion and proliferation on alginate-gelatin crosslinked hydrogel. PloS One. 9(9): e107952.

Schwab, A., R. Levato, M. D'Este, S. Piluso, D. Eglin and J. Malda. 2020. Printability and shape fidelity of bioinks in 3D bioprinting. Chemical Reviews. 120(19): 11028–11055.

Sekar, M.P., H. Budharaju, S. Sethuraman and D. Sundaramurthi. 2023. Carboxymethyl cellulose-agarose-gelatin: a thermoresponsive triad bioink composition to fabricate volumetric soft tissue constructs. SLAS Technology. 28(3): 183–198.

Skardal, A. and A. Atala. 2015. Biomaterials for integration with 3-D bioprinting. Annals of Biomedical Engineering. 43(3): 730–746.

Spicer, C.D. 2020. Hydrogel scaffolds for tissue engineering: The importance of polymer choice. Polymer Chemistry. 11(2): 184–219.

Sultan, S., G. Siqueira, T. Zimmermann and A.P. Mathew. 2017. 3D printing of nano-cellulosic biomaterials for medical applications. Current Opinion in Biomedical Engineering. 2: 29–34.

Taghavi, M.A., S.M. Rabiee, M. Jahanshahi and F. Nasiri. 2019. Electrospun poly-ε-caprolactone (PCL)/dicalcium phosphate dihydrate (DCPD) composite scaffold for tissue engineering application. Molecular Biotechnology. 61: 345–354.

Tang, Y., C.L. Heaysman, S. Willis and A.L. Lewis. 2011. Physical hydrogels with self-assembled nanostructures as drug delivery systems. Expert Opinion on Drug Delivery. 8(9): 1141–1159.

Thiele, J., Y. Ma, S.M. Bruekers, S. Ma and W.T. Huck. 2014. 25th Anniversary article: designer hydrogels for cell cultures: a materials selection guide. Advanced Materials. 26(1): 125–148.

Valot, L., J. Martinez, A. Mehdi and G. Subra. 2019. Chemical insights into bioinks for 3D printing. Chemical Society Reviews. 48(15): 4049–4086.

Wang, Q., J. Sun, Q. Yao, C. Ji, J. Liu and Q. Zhu. 2018. 3D printing with cellulose materials. Cellulose. 25: 4275–4301.

Whitlow, J., A. Paul and A. Polini. 2016. Bioactive materials: definitions and application in tissue engineering and regeneration therapy. Biocompatible Glasses: From Bone Regeneration to Cancer Treatment. 1–17.

Xing, Q., Z. Qian, W. Jia, A. Ghosh, M. Tahtinen and F. Zhao. 2017. Natural extracellular matrix for cellular and tissue biomanufacturing. ACS Biomaterials Science and Engineering. 3(8): 1462–1476.

Yang, J., Y.S. Zhang, K. Yue and A. Khademhosseini. 2017. Cell-laden hydrogels for osteochondral and cartilage tissue engineering. Acta Biomaterialia. 57: 1–25.

Yao, Q., Y.W. Zheng, Q.H. Lan, L. Kou, H.L. Xu and Y.Z. Zhao. 2019. Recent development and biomedical applications of decellularized extracellular matrix biomaterials. Materials Science and Engineering: C. 104: 109942.

You, F., Eames, B.F. and X. Chen. 2017. Application of extrusion-based hydrogel bioprinting for cartilage tissue engineering. International Journal of Molecular Sciences. 18(7): 1597.

Zarrintaj, P., J.D. Ramsey, A. Samadi, Z. Atoufi, M.K. Yazdi, M.R. Ganjali, et al., 2020. Poloxamer: A versatile tri-block copolymer for biomedical applications. Acta Biomaterialia. 110: 37–67.

Zhang, H., C.Y. Lin and S.J. Hollister. 2009. The interaction between bone marrow stromal cells and RGD-modified three-dimensional porous polycaprolactone scaffolds. Biomaterials. 30(25): 4063–4069.

Zhang, B., Y. Luo, L. Ma, L. Gao, Y. Li, Q. Xue, et al., 2018. 3D bioprinting: an emerging technology full of opportunities and challenges. Bio-Design and Manufacturing. 1: 2–13.

Chapter 2

Polymers and Composites in 3D Bioprinting for Regenerative Medicine

Mrunmayi Gadre[1], Divya Mallya[1], Meghana Kasturi[2] and Kirthanashri S. Vasanthan*[1]

[1]Manipal Centre for Biotherapeutics Research, Manipal Academy of Higher Education, Manipal-576104, Karnataka, India

[2]Department of Mechanical Engineering, University of Michigan, Dearborn, USA

2.1 INTRODUCTION

An emerging field which involves combination of technologies known as additive manufacturing allows designing and developing three-dimensional models from desired computer aided designs (Stansbury and Idacavage 2016, Wohlers Associates 2019). In case of 3D bioprinting, biomaterial is the base for the development of models as the end product generally results in *in vitro* models representing complex tissues which should be able to provide physical and chemical signals. These biomaterials are chosen based on the properties they can exhibit which includes cellular activities like cellular adhesion, metabolic activities, cell proliferation, differentiation and migration (Gasiorowski et al. 2013, Liu and Wang 2020, Fu et al. 2022, Arefin et al. 2021, Liu et al. 2022, Xiang et al. 2022, Yang et al. 2023). Polymers in Greek means "many parts" and they are macromolecules that are built by repeating small and identical subunits called monomer linked to

*For Correspondence: Kirthanashri.sv@manipal.edu

each other by covalent bonds (Guarino et al. 2010). A term known as degree of polymerization of polymer is the number of subunits repeated in a polymer chain. Polymers are majorly classified as natural, hybrid and synthetic polymers based on their origin. Natural polymers are those which occur naturally in nature, are extracted and are water-soluble, pliable and cell-friendly. Synthetic polymers are human-made polymers which are produced by processing with different reaction for polymerization. Synthetic polymers have varied properties (thermos-resistant and toughness) based on their processing and their natural properties. Hybrid polymers are the combination of two or more polymers synthesized in labs to obtain desired properties in a single polymer. Their properties depend on the polymers that are combined, but overall, more effective than natural polymers as they can be customized based on the requirement. In 3D bioprinting, these polymers are being applied to make bioinks in order to bioprint 3D models for various studies and applications. Polymers have been utilized in 3D bioprinting as many studies have proved that polymers meet key requirements for *in vitro* and *in vivo* applications (cell proliferation, migration and differentiation) (Wang et al. 2005, Hou et al. 2015, Yaser et al. 2016, Wang et al. 2017, Li et al. 2019).

2.2 HISTORY

3D bioprinting, which is a part of additive manufacturing, has been evolving over the past four decades and their application has been widening ever since. Many different techniques have been developed since 1983, first being stereolithography, after which came the fused deposition modeling, also known as 3D printing, which was patented by 1989 (Crump and Wales 1991). Majority of these techniques are based on the principle where the materials (generally polymers) are formulated in a semi-liquid form and extruded via filament by moving in X, Y and Z plane which results in a layer-by-layer deposition leading to formation of 3D models (Vanaei et al. 2020a, b, 2021). The production of 3D models is established based on the desired end product and directly designed accordingly by computer-aided design (Stansbury and Idacavage 2016, Wohlers and Campbell 2017). These models have a wide range of applications including generation of complex geometrical, low cost and easy operational parts in medical field and industries like automobile and aerospace (Boparai et al. 2016, Zeltinger et al. 2001, Grida and Evans 2003, Woodfield et al. 2004, Tsai et al. 2017, Singh et al. 2017, Pranzo et al. 2018, Vanaei et al. 2020c, d). One of the recent and most researched field is the development of novel biodegradable bioinks consisting of polymers for clinical applications (Arslan-Yildiz et al. 2016, Munaz et al. 2016, Tappa and Jammalamadaka 2018). One of the first fabrication of 3D model involving cells was developed by laser based bioprinting in the 1990s (Odde and Renn 1999). Major classification of polymers includes natural, synthetic and hybrid polymers, to highlight the emergence of synthetic polymers that began in the mid-nineteenth century (Flory 1953). It was only by 1920 when Staudinger had proposed the idea of polymers as macromolecules that slowly led to the acceptance in the next decade by scientific community (Staudinger 1920). By 1929, Carothers confirmed the hypothesis proposed by Staudinger on macromolecules by synthesising polymers and carrying out various experiments to analyse high molecular weight molecules (Carothers 1929). Morawetz in 1995 published a book emphasizing on the history of polymers including the origin and growth of polymers (Morawetz 2002).

2.3 THE POLYMERS

The most vital component of 3D bioprinting and the bottleneck for designing and applying is choosing the polymer to formulate the bioink. Polymer matrix along with the loaded cells and bioactive signals mainly constitute the bioinks (Boland et al. 2006). Alteration in the physical and chemical properties of the biological components like the extra cellular matrix, and the behavior of cells can be regulated and enhanced to accelerate the rate of construct integration and functionality (Chung et al. 2013). The polymers are classified into three major types, which are discussed in detail; they are used in the bioink because of one of the most essential characters which is the tunability that provides suitable microenvironment and favors cellular interactions (migration, differentiation and proliferation). Various studies exhibiting the application of different types of polymers have been summarized in detail in the Tables 2.1, 2.2 and 2.3. In many research, natural polymers are mostly preferred as it mimics native tissue easily (Mano et al. 2007). With the growing advancement in the additive manufacturing practices, the application of synthetic polymer has been recognized and their combination with adhesive materials provides researchers with improved mechanical properties (Gunatillake and Adhikari 2003). Various factors of polymers are involved which are evaluated for a study. Figure 2.1 briefly depicts the major classification of polymers used in bioink formulation.

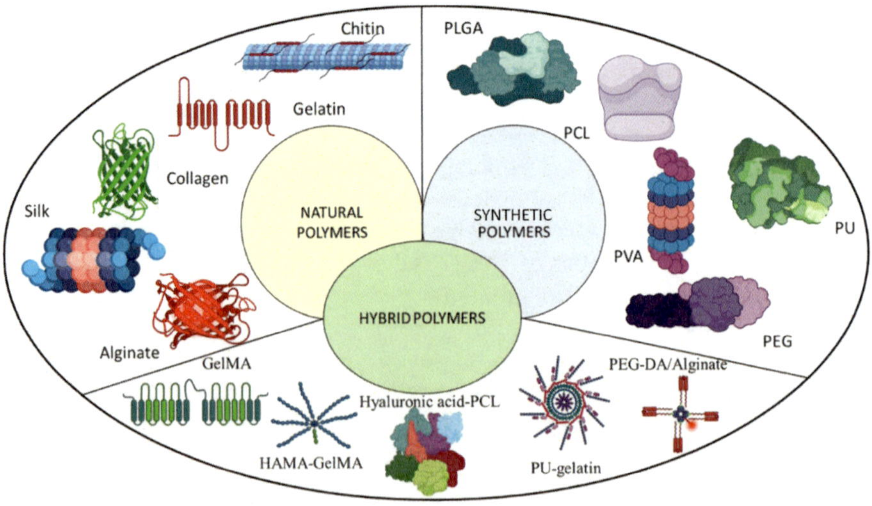

Figure 2.1 Polymers and major classification.

Table 2.1 Natural polymers and their applications in 3D bioprinting

Polymer	Fabrication Technique	Cells Used	Application	Key Features	References
Alginate	3D bioplotting	PRSCs[1]	Nerve TE[2]	Better printability of low concentration alginate Improved cell functionality	Naghieh et al. (2019)
Collagen	Extrusion based	C2C12[3], hESC-CMs[4], cardiac fibroblasts, HUVECs[5]	Heart components	Enhanced mechanical strength Promotes microvascularization	Lee et al. (2019)
Gelatin	Extrusion based	Hepatocytes	Liver TE	Long-term cell survival Promotes nutrient exchange and cell growth	Wang et al. (2006)
Hyaluronic acid	Extrusion based	HDFs[6]	Wound healing	Fast gelation High swelling capacity	Si et al. (2019)
Decellularized extracellular matrix	Extrusion based	NIH/3T3[7]	General TE applications	Fast gelation Cells showed lineage-specific morphology early in culture	Toprakhisar et al. (2018)
Cellulose	Extrusion based	—	Drug delivery	Good mechanical strength Shape-responsive and flexible	Mohan et al. (2021)
Fibrin	Microfluidics based	hiPSC[8]-derived NPCs[9]	Neural TE	Promoted neural differentiation High cell viability	Sharma et al. (2020)
Chitosan	Extrusion based	MC3T3-E1[10]	Bone TE	Promotes osteoblastic differentiation Increased mineralization and ECM[11] production	Demirtaş et al. (2017)
Agarose	Extrusion based	—	Blood vessel	Hollow channels were created Printing can be done at room temperature	Forget et al. (2019)
Collagen/alginate	Extrusion based	Chondrocytes	Cartilage TE	Promotes cartilage-specific gene expression Increased GAG[12] production	Yang et al. (2018)

PRSCs[1] – primary rat schwann cells, TE[2] – tissue engineering, C2C12[3] – hESC-CMs[4] – human embryonic stem cell-derived cardiomyocytes, HUVECs[5] – human umbilical vein endothelial cells, HDFs[6] – human dermal fibroblasts, NIH/3T3[7] – embryonic mouse fibroblast hiPSC[8] – human induced pluripotent stem cells, NPCs[9] – neural progenitor cells, MC3T3-E1[10] – preosteoblast cells, ECM[11] – extracellular matrix, GAG[12] – glycosaminoglycans.

Table 2.2 Hybrid polymers and their applications in 3D bioprinting.

Polymer	Fabrication Technique	Cells Used	Application	Key Features	References
GelMA[1]	3D[2] pneumatic dispensing	BMSCs[3]	General TE[4] applications, drug delivery	Rapid crosslinking and high resolution Better cell spreading and viability	Yin et al. (2018)
HAMA-GelMA[5]	Stereolithographic	Chondrocytes	Cartilage TE	Customizable constructs Can be used for disease modeling	Lam et al. (2019)
Hyaluronic acid-PCL[6]	Extrusion based	hDPSCs[7]	Pulp regeneration	Promotes odontogenic differentiation Promotes cell adhesion	Nejad et al. (2021)
PEG-DA[8]/Alginate	Extrusion based	Porcine aortic valve interstitial cells	Aortic valve	Cells remained viable for long duration Desirable mechanical properties	Hockaday et al. (2012)
PU[9]-gelatin	Extrusion based	HiPSC[10]-derived MSCs	General TE application	Can be printed at near room temperature Desirable elastic properties	Hsieh and Hsu (2019)

GelMA[1] – Gelatin methacryloyl, 3D[2] – Three dimensional, BMSCs[3] – Bone marrow mesenchymal stromal cells, TE[4] – tissue engineering, HAMA–GelMA[5] – hyaluronic acid–methacrylated gelatin, PCL[6] – polycaprolactone, hDPSCs[7] – human dental pulp stem cells, PEG–DA[8] – poly(ethylene glycol) diacrylate, PU[9] – polyurethane, hiPSC–derived MSCs[10] – human induced pluripotent stem cell–derived mesenchymal stem cells.

Table 2.3 Synthetic polymers and their applications in 3D bioprinting.

Polymer	Fabrication Technique	Cells Used	Application	Key Features	References
PLA[1]	Fused deposition modeling	MC3T3[2]	Osseus tissue graft	Compressive strength similar to native tissues Good thermal stability	Vasconcelos et al. (2023)
PVA[3]	Extrusion based	RSC96[4]	Neural TE[5]	Antioxidant properties Cell carrier properties	Rizwana et al. (2023)
PLGA[6]	Fused deposition modeling	—	Development of mAb[7]-loaded dosage forms	Sustained release of mAbs Good stability and affinity	Carlier et al. (2021)
PCL[8]	Extrusion based	UCs[9] and SMCs[10]	Desirable mechanical properties Promoted expression of cell-specific biomarkers	Desirable mechanical properties Promoted expression of cell-specific biomarkers	Zhang et al. (2017)
PEG[11]	Extrusion based	MSCs[12]	Customizable properties of bioink	Customizable properties of bioink Cytocompatibility	Xin et al. (2019)
PU[13]	Extrusion based	Fibroblasts	Potential drug screening applications	Thermoresponsive Potential drug screening applications	Ho and Hsu (2018)
PHB[14]	Fused deposition modeling	MSCs	Bone TE	Promotes osteogenic differentiation High porosity	Volkov et al. (2020)

PLA[1] – polylactic acid, MC3T3[2] – preosteoblast cells, PVA[3] – polyvinyl alcohol, RSC96[4] – rat Schwann cell 96, TE[5] – tissue engineering, PLGA[6] – poly–lactic–co–glycolic acid, mAb[7] – monoclonal antibodies, PCL[8] – polycaprolactone, UCs[9] – urothelial cells, SMCs[10] – smooth muscle cells, PEG[11] – poly(ethylene glycol), MSCs[12] – mesenchymal stem cells, PU[13] – polyurethane, PHB[14] – Polyhydroxybutyrate

2.3.1 Natural Polymers

Natural polymers, derived from various sources like microbes, plants, and animals, can be categorized as protein- and polysaccharide-based polymers. They are biocompatible, biodegradable, and exhibit similarities to the extracellular matrix (ECM) of animal tissues with minimized immunogenicity risks. These polymers can undergo enzymatic degradation in the body, resulting in nontoxic byproducts that are metabolized naturally. Their easy availability makes them attractive and versatile for use in tissue engineering applications.

2.3.1.1 *Collagen*

Collagen is a protein which is extracted from animal tissues like skin, bone, tendons and ligaments, etc. by various processing methods and is used in tissue engineering applications because of its tensile strength and ability to provide structural support (Zhang et al. 2018, Copes et al. 2019). Since collagen is a major component of the ECM, collagen-based scaffolds can be fabricated to mimic the biological and structural properties of native ECM (Matthews et al. 2002). Also, they can be used as cell-free as well as cell-based scaffolds (Glowacki and Mizuno 2008). Compared with other proteins, collagen has relatively lower immunogenic properties, and the immunogenicity of collagen-based scaffolds can be further reduced by proteolytic removal of the non-helical regions or by suitable crosslinking methods (Chevallay and Herbage 2000). Pure collagen has disadvantages such as low mechanical strength and rapid degradation. This drawback is overcome by crosslinking or combining collagen with other polymers to produce hybrid polymers (Martínez et al. 2015).

2.3.1.2 *Gelatin*

Gelatin is derived from porcine or bovine collagen after its hydrolysis with acid and alkaline methods (Malafaya et al. 2007). Like collagen, gelatin gets degraded naturally in the human body by metalloproteinases (Lukin et al. 2022). Also, the antigenic property of gelatin is lower than collagen due to the distinct differences in their chemistry (Rose et al. 2014). An attractive property of gelatin is the presence of RGD (Arg-Gly-Asp) sequence which plays a key role in cell recognition and adhesion (Wang et al. 2012). Along with biocompatibility and biodegradability, its properties are highly modifiable and versatile due to the many accessible functional groups (Elzoghby 2013). Some of the drawbacks of gelatin include low thermostability, inferior mechanical properties and shorter degradation time, which can be improved by modifying its properties by crosslinking or by incorporating it with other materials (Bello et al. 2020).

2.3.1.3 *Chitin and Chitosan*

Chitin is derived from animal, plant and fungal sources. The positive charge of the amino groups of chitin-based materials exerts a bactericidal effect—interactions of the positive charge of the polymer with the negative charge of the bacterial cell wall lead to breakage and release of the bacterial cell contents (Chung and Chen 2008).

The structural similarity of chitin to glycosaminoglycans of the body is a unique feature of the polymer (Islam et al. 2016). Deacetylation of chitin yields chitosan, which has better solubility in common solvents compared to chitin (Schmitz et al. 2019). The drawbacks of chitosan are brittleness, low mechanical strength, and susceptibility to degradation by lysozyme (Ahsan et al. 2018). Combining chitosan with other polymers during scaffold fabrication overcomes some of these drawbacks.

2.3.1.4 *Silk*

Silk is composed of two proteins—fibroin and sericin. Silk fibroin is highly biocompatible and is used for tissue engineering applications after complete removal of the sericin coating by the process of degumming (Mahmoodi et al. 2010). Silk-based scaffolds allow lower cell infiltration and vascularization and have a lower rate of biodegradation, which limits their applications to softer tissues (Leal Egaña and Scheibel 2010). However, their biocompatibility and higher mechanical strength make them more suitable for tissue engineering of hard and avascular tissues (Kasoju and Bora 2012). A standout feature of silk is its excellent mechanical properties. Silk has various desirable properties such as high tensile strength, toughness and breaking strain, due to the hierarchical structure of the silk fibers (Koh et al. 2015). Silk also has good elasticity due to the strength provided by the crystalline β-sheet structures and the flexibility and stretchability provided by the amorphous random coil regions (Kundu et al. 2013). The cell adhesion properties of silk scaffolds depend on the source of silk. It is known that the silk produced by wild silkworms (*Antheraea pernyi*) has the RGD sequence for cell adhesion, which is not present in the silk produced by domesticated silkworms (*Bombyx mori*) (Minoura et al. 1995).

2.3.1.5 *Hyaluronic Acid*

The proteoglycan hyaluronic acid is used for tissue engineering because of its biocompatibility and biodegradability and promotion of cell adhesion and migration. Since it is naturally found in the body, its intrinsic functions of lubrication, promotion of cell adhesion and growth, and wound healing properties are inherent in the developed scaffolds (Collins and Birkinshaw 2013). The large number of carboxyl and hydroxyl groups give hydrophilic properties to hyaluronic acid, making it gel-like in aqueous conditions (Chircov et al. 2018). This property further improves the wound healing and cell infiltration properties of hyaluronic acid-based scaffolds. However, it gets degraded rapidly in the body, and the water-absorbing quality can cause shrinkage or collapse of the scaffold. Crosslinking of hyaluronic acid or combining it with other polymers can overcome these challenges and widen their scope of applications (Collins and Birkinshaw 2007, Fakhari and Berkland 2013).

2.3.2 Synthetic Polymers

Synthetic polymers are developed through the polymerization of monomers. Their production processes are highly reproducible, and their inherent properties such as microstructure uniformity and rate of degradation will be retained, which makes them well-suited for large scale production. Some of the properties of synthetic polymers can be easily modified, such as tensile strength and elastic modulus, and the addition of functional groups can bioactivate them. Thus, they are highly adaptable for specific tissue engineering applications.

2.3.2.1 *Poly(Lactic-Co-Glycolic Acid (PLGA)*

PLGA is prepared by ring-opening copolymerization of lactide (LA) and glycolide (GA), and based on the ratio of the monomers used, different forms of PLGA can be synthesized (Pan and Ding 2012). It is approved by the Food and Drug Administration for clinical applications. The adjustable degradation rate, good toughness and processability make PLGA well-suited for tissue engineering of hard tissues. These properties can be easily tuned by varying the ratio of LA and GA during polymer synthesis (Makadia and Siegel 2011). However, PLGA has low cell affinity, due to which they are used in combination with other polymers and after relevant surface modifications (Gentile et al. 2014).

2.3.2.2 *Poly(E-Caprolactone) (PCL)*

PCL is developed by ring-opening polymerization of ε-caprolactone monomers (Hoskins and Grayson 2009). PCL-based scaffolds can be fabricated by various techniques due to its high processability. The processability is due to its favorable rheological properties such as glass transition temperature of −60°C, low melting point of 60°C, and decomposition temperature of 350°C (Patrício et al. 2014). Additionally, it has high toughness due to its semi crystalline structure (Mishra et al. 2008). Due to these properties, PCL-based scaffolds are suitable for regeneration of hard and soft tissues. However, PCL has lower biocompatibility, and it is highly hydrophobic, and these drawbacks are overcome by using PCL in combination with other scaffolds for tissue engineering applications.

2.3.2.3 *Polyurethane (PU)*

PU is a synthetic polymer synthesized by the reaction of isocyanates with polyols in the presence of a chain extender. Based on the type and molecular weight ratios of monomers used, the properties of PU can be significantly changed (Sonnenschein et al. 2005). Its advantages include elasticity and mechanical properties, such as elasticity, similar to the native tissues (Tsai et al. 2015). However, the poor cell adhesion and slow degradation rate with the release of toxic degradation products necessitate the modification of PU scaffolds before extensive use in tissue engineering applications (Naureen et al. 2021).

2.3.2.4 *Polyethylene Glycol (PEG)*

PEG is used in tissue engineering applications after copolymerization. Not only is it biocompatible and nonimmunogenic, but also has the unique property of resistance to protein adsorption, which is used to create inert polymer surfaces for further modification by the addition of appropriate proteins or peptides. The bioinert property of PEG is due to the hydrated polymer coils formed due its hydrophilic and uncharged structure (Tessmar and Göpferich 2007). Due to the lack of cell adhesion and slow degradation rate of PEG, it is modified before use (Zhu 2010).

2.3.2.5 *Poly(Vinyl) Alcohol (PVA)*

The hydrolysis of polyvinyl acetate yields PVA, and the degree of hydrolysis determines the properties of the resultant PVA. PVA is used in various forms in biomedical applications because it is biodegradable and biocompatible with tunable mechanical properties. It also has high strength, resistance to deformation and water retention capacity (Nathan et al. 2023). Thus, PVA-based scaffolds have good capacity to absorb strain while retaining their flexibility, broadening their potential applications (Teixeira et al. 2019). The major limitation of this polymer is the lack of cell-signaling and cell-interacting motifs (Rodríguez-Rodríguez et al. 2019).

2.3.3 Hybrid Polymers

Hybrid polymers are a class of materials formed by combining natural polymers with synthetic polymers. In these polymers, the desirable characteristics of natural polymers, such as biocompatibility, are combined with the desirable characteristics of synthetic polymers, such as enhanced mechanical strength and controlled degradation rates. This blending of polymers results in versatile scaffolds that can be tailored to suit the specific requirements of the target tissue or organ. Some commonly used hybrid polymers are combinations of collagen, chitosan, or alginate with synthetic polymers like PLGA, PEG, or PU.

Utilizing hybrid polymers in tissue engineering offers many advantages over employing only natural or synthetic polymers. Firstly, hybrid polymers can be tailored to meet the basic and most important requirements of biocompatibility and biodegradability, which may not be fulfilled if the constituent polymers are used alone. These constructs exhibit enhanced physical properties, including better mechanical strength, stability, and controlled degradation rate, while also releasing nontoxic degradation products. Also, hybrid polymers can be tailored to enhance cell attachment and proliferation by incorporating polymers that improve these properties.

2.3.3.1 *PLGA-Collagen*

Composed of PLGA and collagen, this hybrid polymer is used to fabricate porous scaffolds for tissue engineering. The hydrophilic nature of collagen and its RGD motif helps in cell proliferation, and the mechanical strength imparted by PLGA overcomes the mechanical properties of the combined scaffold. Putri et al. (2020) developed a porous PLGA-collagen scaffold by hybridization of PLGA mesh with the collagen sponge by using ice particulates. In this way, the pore structures were developed in a controlled manner. The resultant scaffold was found to have good mechanical strength and promoted neocartilage generation, showing the potential applications of this scaffold in osteochondral tissue engineering.

2.3.3.2 PLGA-Fibrin

Fibrin is a naturally occurring polymer which has various biological activities, such as cell signalling. In addition, fibrin also promotes cell adhesion and transportation, making it an attractive choice for developing hybrid polymers (Rahman et al. 2015). PLGA is a commonly employed synthetic polymer with good biocompatibility, biodegradability and mechanical properties. In a study, Yadegani et al. (2018) concluded that the addition of fibrin to PLGA scaffolds improved the suitability of the scaffold for tissue engineering. This hybrid polymer scaffold had high porosity, better mechanical properties and water absorption, and promoted cell proliferation, having several potential applications in tissue engineering.

2.3.3.3 PCL-Chitosan

In addition to biocompatibility, biodegradability and nontoxicity, chitosan has antimicrobial properties and hydrophilicity. When combined with the mechanical properties and processability of PCL, the resultant hybrid polymer has attractive characteristics suitable for various applications. For example, Tayebi et al. (2021) incorporated chitosan nanoparticles into a PCL-chitosan composite for corneal tissue engineering. The scaffold was used as a carrier for corneal endothelial cells and it was found that the scaffold was biocompatible with excellent transparency.

2.3.3.4 Chitosan-PEG

PEG is a synthetic polymer with unique properties of water solubility and hydrophilicity, in addition to nontoxicity. When combined with the hydrophilic natural polymer chitosan, the water solubility of the resultant hybrid polymer is enhanced, indicating its potential use for developing hydrogels. Chitosan-PEG hydrogels have applications in neural tissue engineering, as shown by Mohrman et al. (2018). The team developed an injectable hydrogel by combining thiolated chitosan with maleimide-terminated PEG to treat spinal cord injury. The mechanical properties of the hydrogel were suitable for the central nervous system and was nontoxic.

2.3.3.5 PU-Gelatin

Gelatin is a widely available natural polymer which is biocompatible and promotes cell adhesion and proliferation with lower immunogenicity. The major drawbacks of gelatin are the faster degradation rate and weak mechanical properties. These properties can be overcome by combining gelatin with PU, which has desirable mechanical properties. For example, Rodrigues et al. (2020) used rotary jet spinning to develop PU-gelatin scaffolds. The scaffolds were developed to reduce the costs involved in their creation. The scaffolds were hydrophilic and promoted cell adhesion, showing that these scaffolds can be further studied and developed for various tissue engineering applications.

2.4 SYNTHETIC POLYMERS IN 3D BIOPRINTING

3D bioprinting is a biofabrication technique in which 3D tissue constructs are created by deposition of bioink, which contains cells and the biomaterial, based on a pre-developed model. While natural polymers are used in 3D bioprinting for promoting cell growth and tissue maturation, synthetic polymers are used to structurally replicate the native tissue or organ, improve the mechanical properties of the construct, and sometimes as sacrificial components (Wang 2019). The main advantages of synthetic polymers in 3D bioprinting applications are that they can be modified such that the rate of degradation of the material matches the rate of regeneration of the tissue and can better mimic the mechanical properties of the target tissue or organ. However, their limitations include the requirement of heat or toxic chemicals during the modification and development processes, as well as the absence of cellular recognition sites. The lack of these motifs calls for the functionalization of synthetic polymers before tissue engineering applications, which is possible by various techniques. Several studies have demonstrated the potential of 3D bioprinting synthetic polymers in advancing regenerative medicine. For instance, Jia et al. (2016) developed a bioink comprising gelatin methacryloyl (GelMA), sodium alginate, and 4-arm poly(ethylene glycol)-tetra-acrylate to create vascular constructs. The covalent photocrosslinking of the synthetic polymer constituents of the bioink greatly enhanced the mechanical strength and morphological integrity of the bioprinted vessels. Hydrogel scaffolds composed of PLA and gellan gum-poly(ethylene glycol) diacrylate (GG-PEGDA) can potentially promote intervertebral disc (IVD) repair and regeneration, as demonstrated by Hu et al. (2018). This synthetic bioink was used because it had the necessary properties for IVD regeneration, combining the mechanical strength of PLA with the biocompatibility and bioprintability of GG-PEGDA. Bendtsen et al. (2017) developed alginate-PVA-hydroxyapatite hydrogel scaffolds with osteoconductive properties, suggesting their potential application in treating bone defects. The osteoconductive properties were due to the incorporation of hydroxyapatite, while PVA increased the viscosity of the hydrogel formulation for easier bioprinting. Additionally, synthetic polymers find applications in dental regeneration, as Nejad et al. (2021) demonstrated. They formulated a PCL/45S5 Bioglass (BG) composite for dentin regeneration and a PCL/hyaluronic acid formulation for pulp regeneration. PCL was used for its biocompatibility, biodegradability and printability, while the incorporation of HG improved the mechanical properties, surface roughness and bioactivity of the scaffolds.

2.5 PROPERTIES OF POLYMERS

Fabrication of bioink is an art in 3D bioprinting and the successful formulation of bioink takes place when the right polymer is chosen as it directly influences the efficiency of bioprinting and cytocompatibility. Different properties of the polymer used in bioink predicts the nature of models during 3D bioprinting, Figure 2.2 depicts visually the different properties of polymers.

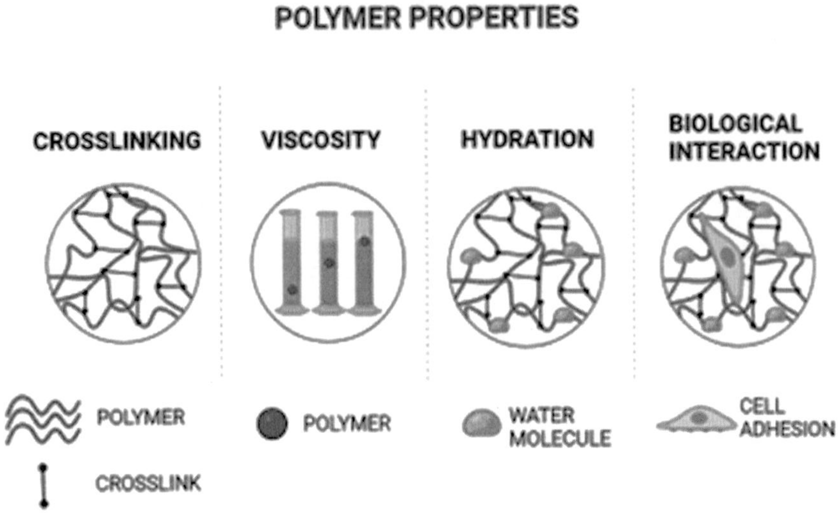

Figure 2.2 Different properties of polymers.

Evaluation of few properties before fabrication of bioinks is done which involves the following:

2.5.1 Mechanism of Cross-Linking

The polymer for 3D bioprinting is selected based on the kind of crosslinking network formed after crosslinking the polymer. The crosslinking property of a bioink is exhibited after bioprinting the 3D model and this can be achieved via various agents. Few of the method to initiate the crosslinking in the polymer is through photo-based, ion-based, electrostatic-based, pH-based and temperature-based cross-linking mechanisms. Considering the kind of cross-linking initiator (like heat and pH) is also crucial as the bioinks contain cells and may affect the cell viability (Lam et al. 2002, Skardal et al. 2010). Many studies have opted for photoinitiator for crosslinking as the polymer used would be chemically modified (via acylate/ methacrylamate groups) which results in network which is covalently cross-linked (Selimović et al. 2012, Billiet et al. 2014).

2.5.2 Viscosity

Majority of the bioink component is the polymer which determines the rheological properties and the shape retention. Viscosity is one such factor where the flow rate of bioink extrusion during the 3D bioprinting is determined (Das et al. 2013). The critical point of the viscosity to affect bioink is during the extrusion while bioprinting as there are high chances of either dispensing too much of bioink or clogging in the syringe. In case of high viscosity, there are shear forces exerted on the cells leading to death of few cells and on the other hand, if the viscosity is too low, then the proper shape formation will not take place. Studies using the semi-interpenetrating polymer networks (elastin-like polypeptides fused with multiple SpyCatcher or SpyTag sequences for polymerization then encapsulated in chitosan) have overcome the issue of viscosity as it can produce a mechanically stable hydrogel and yet be hydrated sufficiently (Dai et al. 2021).

2.5.3 Hydration Properties

The amount of the water constituent in the polymer is also another factor that can affect the bioink extrusion as it directly changes the viscoelasticity and mechanical strength. Another major factor that hydration can directly impact is the transportation of oxygen and nutrition to the laden cells in the hydrogel network. With the increase in the porosity, there is a decrease in hydrogel modulus and this is due to the hydrogel-to- water contact. Synthetic polymers, such as PLGA and PEG, are gaining increasing applicability due to their fluidic properties and, conversely, their ability to maintain good hydration during fabrication, making them suitable for designing complex structures, including blood vessels.

2.5.4 Property of Biological Interaction

Certain physical property like the hydrophilicity and surface energy have effect on the cellular activity and at the same time, on polymer's applicability. Few polymers lack the cell binding moieties, which leads to prevention of cell adhesion and proliferation. To enhance this property in the polymers, cell adhesive molecules are blended with the synthetic polymers. Mostly used synthetic polymers for tissue engineering are PCL and PLLA which have exhibited excellent cytocompatibility along with cell adhesion and proliferation in absence of any blending with cell adhesive substances (Schuurman et al. 2011, Seyednejad et al. 2012, Guo et al. 2013). Few natural polymers (collagen and gelatin) hold the RGD motif (sequence of Arg-Gly-Asp) which itself promotes cell adhesion through receptors like integrin (Liu et al. 2008, Hoch et al. 2013).

2.5.5 Backbone of Polymer

In addition to the aforementioned properties, this particular characteristic holds significant importance, as the polymer backbone plays a critical role in governing the interactions that drive the responses of the fabricated model. Particular factor which plays major role in the response of the polymer backbone is the temperature. Alteration in temperature during 3D bioprinting leads to increase or decrease in the viscosity for easy extrusion for fabrication. The concentration of polymer along with the natural hydrophilicity of backbone constituents is another factor that majorly impacts the swelling and degradation rates (Pataky et al. 2011, Selimović et al. 2012).

The choice of polymer depends on the combination of all properties and shall be chosen based on the application and study. Overall, all the properties contribute to the fabrication of a 3D model and have to be analysed in detail. The application of a specific fabrication technique among emerging technologies, which employ various processes to generate models (such as displacement forces and pressure exerted through the nozzle), is largely influenced by the choice of polymer used in the study. With the increasing development of new complex techniques for 3D bioprinting, there will be a need to increase the intricacy for improving the properties of polymers.

2.6 APPLICATIONS OF THE POLYMERS IN DIFFERENT FIELDS

The most emerging field of additive manufacturing is the 3D bioprinting, which is a very complex process. Just like any other process, 3D bioprinting involves detailed architectural designing, selection of biomaterials which also includes polymers and the understanding of complex equipment. The most important factor in 3D bioprinting involves the support for cellular proliferation and mimicking of the fabricated model as per the study (Rui et al. 2009, Sui et al. 2009, Xu et al. 2009, Wang et al. 2010, Huang et al. 2013, Zhao et al. 2014, Wang et al. 2015b, Wang 2018, Liu et al. 2018). Despite the diversity of the polymers available, their application varies and are sometime unique. Application of polymers in general is in various fields including food industry, packaging application, medical and dental applications, sportswear and materials, industrial and construction materials, appliances for kitchen, smart materials and textile industry. Application of polymers in 3D bioprinting is majorly in the fields like biophysics, medicinal, bio-electronical, and other branches of science and technology (tissue engineering, 3D constructs as implants, synthesis of nanomaterials, drug delivery systems, medical devices, *in vivo* and *in vitro* study models as drug testing and disease models) (van Vlierberghe et al. 2007, Meinel et al. 2009, Guarino et al. 2015, Premkumar et al. 2021, Chen et al. 2022). One of the most commonly used natural polymer is gelatin (porous, easy exchange of solvents and gases) which has established application in medicinal wound healing as dressing patches fabricated by 3D bioprinting (Zhao et al. 2016). Synthetic polymer, which is majorly applied, is PLGA which has established application in tissue engineering as well as a drug-delivery system fabricated by electrospinning (Liao et al. 2011). Hybrid polymers like multiwall carbon nanotubes fabricated by electrospinning has also proven potential application in different aspects of tissue engineering (Reddy et al. 2015).

2.7 APPLICATION OF POLYMERS IN 3D BIOPRINTING

The most revolutionized and emerging technology of this century is the 3D bioprinting. The most exciting development is the 3D bioprinting of living cells, tissues and organs for medical application with a good rate of transition from lab to clinical trials. Polymers have a wide range of applications as mentioned earlier and in the biomedical application, there are multiple studies carried out for development of scaffolds for tissue regeneration, tissue constructs by tissue engineering, *in vitro* drug study models and *in vivo* synthetic organs. This widespread of polymer application is due to their wide range of properties like the biocompatibility, bioprintability, biostability, biodegradability, biostorablitiy and property of biological interactions. In this book chapter, we have highlighted major polymer applications for potential tissue engineering along with regenerative medical application. Different applications of the synthetic polymers used in fabrication of 3D models by 3D bioprinting in for regenerative medicine are illustrated in Figure 2.3.

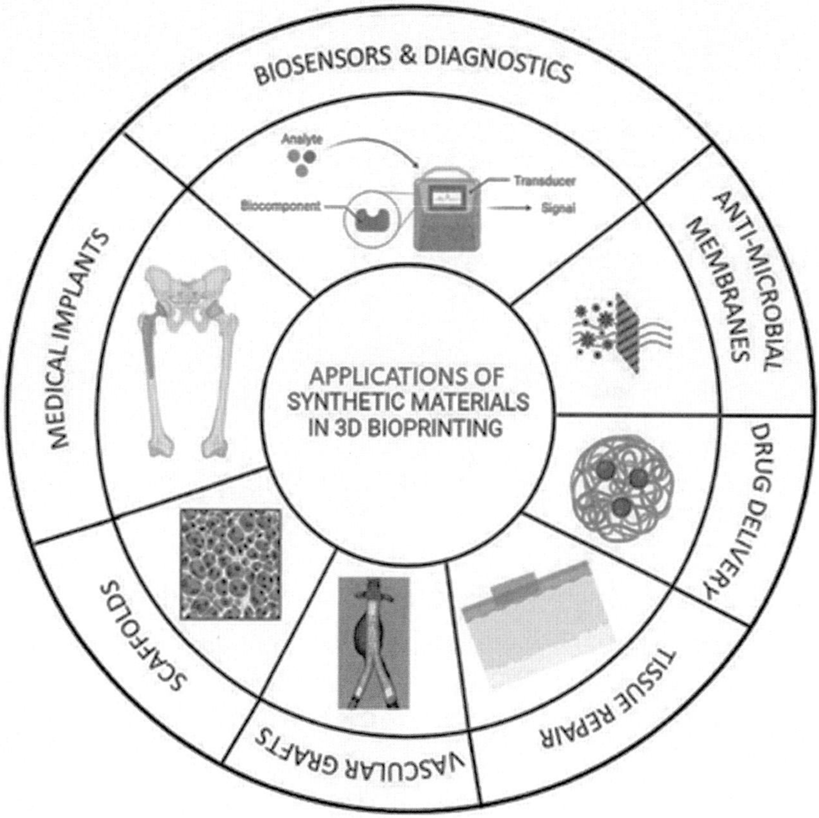

Figure 2.3 Overview of polymers and their application.

2.7.1 3D Bioprinting of Constructs

The base of all the advanced tissue engineering is cell-laden assemblies, which provide a well-defined and organized architecture. Cells are naturally the functional and structural unit of life and this is the main reason for their immense consideration in 3D bioprinting. The main application of the 3D cells and tissue models is not only for the study of the cellular interactions but also the regenerations of the damaged tissues (Ouyang et al. 2015). Initial development of 3D cell-laden bioprinting was with utilization of combination of parenchymal (human neonatal dermal fibroblasts), stromal (human mesenchymal stromal cells) and endothelial cells (human umbilical vein endothelial cells) to form vascularized tissues. This study could provide solid proof for the utilization of multiple bioinks to formulate a construct and long stability with duration of six week (Homan et al. 2016).

2.7.2 3D Bioprinting of Bone and Cartilage Tissue Constructs

Most of the accidental cases where bones are affected, fracture treatment is only done via plastering which takes a long time to recover as well as different factors like age and health conditions delay the process of recovery. In such cases, the 3D bioprinted constructs for the bone and cartilage tissue regeneration is an easy and faster way for recovering the damages. CT (rabbit computed tomography) scans play an important role in providing accurate information for matching the patient's physiology (Sun et al. 2021). A study involving the use of polymer poly(propylene fumarate) for 3D bioprinting showed high porosity along with slow degradation rate (stable for over a period of 7.5 months) which both supported bone tissue engineering applications (Wang et al. 2015a). Synthetic polymers like PCL/PLGA/ß-TCP based 3D bioprinted scaffolds both *in vivo* and *in vitro* showed positive osteogenic properties which included osetoinduction and osteoconduction along with growth of human mesenchymal stromal cells (Fazeli et al. 2021). The application of developing cartilaginous tissues has immense consideration in tissue engineering and one such example is developing ear-shaped models (Das and Basu 2019).

2.7.3 3D Bioprinting of Skin Cells and Tissues

There is a limitation of the 3D bioprinted skin constructs due to the lack of availability of skin donors. These constructs induce a paradigm shift in wound healing management from employment of wide range of passive wound dressing to utilize bioactive skin-based bioprinted constructs. Conventional techniques of skin cells constructs rely majorly on the scaffolds exhibiting cell adhesion and proliferative properties, which leads to the secretion of the extra cellular matrix and remodelling of the microenvironment in the wound vicinity (Shi et al. 2016). Studies involving 3D bioprinting with the use of human fibroblasts and keratinocytes provided promising results for regeneration of human skin tissues. Few well-known products have shown more advantage over hydrogels and the success rates of 3D bioprinted skin constructs composing of collagen and elastin (MatriDerm®) for treatment of facial burns (Abbott and Kaplan 2015).

2.7.4 3D Bioprinting of Blood Vessels

Conventional techniques for bioprinting blood vessels had limitation in the potency of constructs, but with the advancement in the technology of additive manufacturing, the product outcome has shown significant improvement. Technique involving coaxial system in 3D bioprinters marks the potency to 3D bioprint blood vessels (Hoch et al. 2014). A study involving 3D bioprinter with the coaxial bioprinter application successfully bioprinted alginate-based blood vessels (Dolati et al. 2014). Another study used a strategy where encapsulated fibroblasts were added to gelatin methacrylate bioink with human umbilical vein endothelial cells to provide a construct for endothelialisation (Mandrycky et al. 2016). With further advancement in the technology, improvement in the feature of vascularization can further provide base for large vessel construct fabrication.

2.7.5 3D Bioprinting of Neuronal Tissues

Due to the limitation in the availability of potential techniques for regeneration of neurons, 3D bioprinting has immense important breakthrough and hence researchers have been exploring the potential of bioprinting technology for 3D bioprinting the nervous tissues. A study was successful where synthetic nerve graft consisting of rat retinal ganglion cells and glia was bioprinted using inkjet bioprinter (Lorber et al. 2013). Another study where unique technique like micro-stereolithographic was implemented to develop a nerve graft using PEG demonstrated successful nerve repair model (Pateman et al. 2015).

During the fabrication of the 3D bioprinted construct, various factors as discussed above are involved and properties like mechanical stability, shape and design, and biological barriers can be achieved by utilizing natural polymers, imaging clinical data by CT scan, control of motion while fabrication (nozzle), and induction of micro-channels, respectively (Kang et al. 2016). With the increase in the advancement of additive manufacturing techniques, there is increase in the research carried out in the development of varied *in vivo* and *in vitro* implants and models and hence detailed application of the various studies has been discussed in Table 2.4.

2.8 DIFFERENT TECHNIQUES TO CHARACTERIZE CELL INTERACTIONS WITH POLYMERS

Commonly during the process of 3D bioprinting, the term "bio" refers to cells in bioinks and hence it is critical to understand and characterize the interaction of the cells with the polymers. Most of the research carried out by 3D bioprinting is *in vitro* by using cell culture techniques and few of the following characterization techniques have been carried out. Post 3D bioprinting, the cells interact with the polymeric base and after certain duration the adhesion capacity, proliferation, viability, functionality and migration can be analysed. It is a well-known fact that cells are sensitive and their function is affected by the properties of the surface (chemical, physical and morphological) and the microenvironment they are subjected to.

Table 2.4 Application of *in vivo, in vitro* implants and models used in different tissue engineering fields as repair, regeneration and drug testing models.

Field	Sr. No.	Polymer Used	Composition and Method	Outcome	Application	References
Bone tissue engineering	1	PCL[1] and alginate	Encapsulation of MC3T3-E1[2] cells into the polymer	The cell viability was around 84%, PCL provided high mechanical strength and overall supportive framework	Soft and hard tissue regeneration	Lee et al. 2013
	2	PEGDMA[3]–GelMA[4]	Bioprinting with bioink consisting of with hMSCs[5] cells	The model proved to have high degree of differentiation and showcased chondrogenic effects.	Bone and cartilage tissue regeneration	Gao et al. 2015
	3	GelMA-nanosilicates	Encapsulation of MC3T3[6] preosteoblasts hydrogels composed of nanosilicates and GelMA.	The preosteoblast cells grew to maximum extent with almost nil effect on the cell viability.	Bone tissue regeneration and repair	Xavier et al. 2015
	4	Bone-derived methacrylated-dECM[7] and alginate	3D[8] Bioprinting of hASCs[9] and polymers as a cell-laden mesh construct	The osteogenic cells showed high cellular proliferation and differentiation in the bioprinted model.	Polysaccharide-based biomaterial for bone tissue engineering	Lee et al. 2020
	5	Alginate and gelatin	3D Bioprinting of hMSCs and polymers as constructs	Post porous bioprinting that mimic bone-like tissue showed increase in the viability of cells and deposition of mineral.	Test new therapeutics and screen drug in bone tissue engineering.	Zhang et al. 2020
Cartlidge tissue engineering	1	Cellulose–alginate	3D Bioprinting of hNCs[10] and polymers as nano-fibrillated construct	The 3D bioprinted model had strong shape fidelity and good resolution	Generation of cartilage tissue.	Markstedt et al. 2015
	2	PCL, PF127[11], gelatin, fibrinogen, HA[12], and glycerol	3D Bioprinting of Human AFSCs[13] and polymers as human-scale ear constructs	The scaffolds when studied *in vivo* showed successful maturation of vascularized bone tissue.	Ready to use in clinical trial for vascularized tissue regeneration	Kang et al. 2016

Field	Sr. No.	Polymer Used	Composition and Method	Outcome	Application	References
	3	GelMA and chondroitin sulfate aminoethyl methacrylate or GelMA hyaluronic acid methacrylate	Coaxial dispensing of BM-MSCs[14] and polymers as hydrogel scffolds.	GelMA and CS-AEMA[15] model is proven to be the better substitute of other materials for cartilage formation	Cartilaginous tissue application	Costantini et al. 2016
	4	PCL and GelMA	3D Bioprinting of Chondrocyte and polymers as constructs	The 3D bioprinted model had better cellular integrity due to the porous structure	Neocartilage as implants for nose reconstruction	Ruiz-Cantu et al. 2020
	5	Hydroxypropyl methylcellulose to the silk fibrin	3D Bioprinting of BMSCs and polymers as constructs	The model had double network that helped in providing the model required for mechanical strength	Cartilage tissue regeneration	Ni et al. 2020
	6	(Alginate with short PLA[16] fibers), (alginate, GelMA) and (PCL structure modified with acetone and ultrasound)	3D Bioprinting of Chondrocyte, MSCs[17] and polymers as triphasic scaffold (noncalcified cartilage, noncalcified cartilage and subchondral bone)	This bioprinted model had hybrid zones, which helped securing the biological signalling from the cells.	Cartilage tissue repair	Kosik-Kozioł et al. 2020
	7	GelMA with the copolymer of PCL and PEG[18]	3D Bioprinting of MSCs and polymers as triphasic scaffold	The scaffolds when studied *in vivo* showed successful differentiation of the MSCs within all the layers in bone.	Osteochondral regeneration	Qiao et al. 2021
	8	PCL and alginate	3D Bioprinting (multi-head 3D cell printing system) of FCPCs[19] and polymers as heterogeneous scaffold	The scaffolds when studied *in vivo* showed successful stimulation of osteochondrogenesis and repaired osteochondral defects	Engineering multiphase tissue formation	Yu et al. 2020

(*Contd.*)

Field	Sr. No.	Polymer Used	Composition and Method	Outcome	Application	References
	1	PCL	3D Bioprinting (multi-head 3D cell printing system) of rabbit nasal epithelial cells and polymers as co-culture system	The scaffolds when studied *in vitro* showed successful differentiation of the osteochondral tissue, and showed formation of multilayered tissue	Respiratory epithelium regeneration	Park et al. 2019
Trachea tissue engineering	2	PCL and alginate	3D Bioprinting of hMSCs and polymers as tubular structure	The scaffolds when transplanted into rabbits, showed successful formation of respiratory epithelium	Treatment of tracheal stenosis	Ke et al. 2019
	3	Inner layer (nanoscale PCL fibres) and outer layer (microscale PCL fibres)	Electrospinning of hBECs[20], iPSC-MSCs[21], iPSC-Chds[22] and polymers as two-layered hollow structure	The scaffolds when transplanted into rabbits, effective formation of neocartilage at the defected sites	Long-term functional reconstruction of a segmental tracheal defect	Kim et al. 2020
	1	Fibrin-collagen	3D Bioprinting of AFSs[23] and polymers as wound healing	The scaffolds when studied *in vitro* showed successful wound healing and wound closure by promoting angiogenesis	Treatment of large-scale skin wounds and burns	Skardal et al. 2012
Skin tissue engineering	2	Silk and gelatin	3D Bioprinting of Human primary adult dermal fibroblasts cells and polymers as skin constructs	This construct is proven to be a better substitute of collagen-based skin models as it showed higher dimensional stability.	Topical application in cosmetic and pharmaceutical industries.	Admane et al. 2019
	3	Alginate and chitosan–genipin–PEG	3D Bioprinting of Primary Human Epidermal Keratinocytes and Human Dermal Fibroblasts cells and polymers as 3 layered skin constructs	This multi-layered construct is proven to provide excellent self-supportive structures for repair skin wounds	Regeneration of epidermal and dermal skin layers.	Hafezi et al. 2020

(*Contd.*)

Field	Sr. No.	Polymer Used	Composition and Method	Outcome	Application	References
	1	Alginate, agarose, and carboxymethyl-chitosan	3D Bioprinting of hNSCs[24] and polymers as neural constructs	This scaffold when studied *in vitro* showed uniform cellular distribution with good cell viability and most of the neurons showed response of bicuculline-induced increased calcium.	Human neural development	Gu et al. 2016
	2	Fibrin, PVA[25] and thrombin	3D Bioprinting of Schwann cells and polymers as neural constructs	The formulated bioink showed higher viability of cells and better proliferation in the 3D fabricated fibres that were aligned longitudinally.	Nerve regeneration	England et al. 2017
	3	GelMA/PEGDA[26]	3D Bioprinting of BM-MSCs and polymers as bi-layered nerve conduit	The 3D bioprinted construct provides with a great potency of peripheral nervous tissue regeneration and repair.	Repair of peripheral nerve	Liu et al. 2021
Neural tissue engineering	4	Alginate and Matrigel	3D Bioprinting of EMSCs[27] and polymers as neural constructs	The 3D bioprinted scaffolds provided platform for the cells to proliferate, leading to cell differentiation.	Neurons regeneration in spinal cord injury	Li et al. 2020
	5	Gelatin–alginate	3D Bioprinting of Schwann cells and polymers as neural constructs	This 3D bioprinted model when compared to the existing 2D[28] culture had significant difference with increasing exhibition of cellular behavior.	Neural tissue repair	Wu et al. 2020
	6	Alginate-based GelMA-4-arm PEGTA[29]	3D Bioprinting of HUVECs[30], MSCs and polymers as vascular tissue constructs	The hollow 3D bioprinted scaffolds when studied showed higher properties (mechanical and rheological) when compared to other model for neural regeneration.	Regeneration of pre-vascularized tissue	Jia et al. 2016

(*Contd.*)

Field	Sr. No.	Polymer Used	Composition and Method	Outcome	Application	References
	7	Alginate and vascular-derived dECM	3D Bioprinting of EPCs[31] and polymers as bio-blood-vessel	This scaffold when studied *in vivo* shoed enhanced survival and differentiation of EPCs as well as increase in the rate of neovascularization.	The formulated bioink showed results suitable for treating the ischemic disease.	Gao et al. 2017
	8	Gelatin and fibrinogen	3D Bioprinting of HDF-n[32] and polymers as small-diameter vascular constructs	This 3D bioprinted scaffold when studied *in vitro* showed adequate collagen deposition and post 60 d of study, the mechanical properties of the model including the circumferential and axial elastic moduli were significantly higher.	Standard protocol for developing customized and personalized vessel grafts for neural tissue engineering.	Freeman et al. 2019
	9	Alginate and PCL	3D Bioprinting of BM-MSCs and polymers as vascular scaffold	The 3D bioprinted scaffolds when studied *in vivo* showed improved endothelialisation with barely noticeable inflammation	Cell-derived vessels for neural tissue engineering	Jang et al. 2020
Liver tissue engineering	1	PCL and liver-specific dECM	3D Bioprinting of HepG2[33], BM-MSCs and polymers as hybrid construct	The 3D bioprinted scaffolds when studied showed higher potency of stem cell differentiation	Model for promising liver tissue engineering	Lee et al. 2017
	2	Alginate and cellulose nanocrystals	3D Bioprinting of fibroblast and hepatoma cells and polymers as liver-mimetic construct	The bioink showed excellent shear-thinning and post-printing good extrudability, and stability of shape	Bioink is good potential candidate for liver tissue regeneration and repair	Wu et al. 2018
	3	Gelatin and alginate	3D Bioprinting of HepaRG[34] cells and polymers as hepatoganoid	The 3D bioprinted scaffolds when studied *in vivo* showed increase in the survival rates.	Generate human liver tissues for replacement of damaged human liver	Yang et al. 2020

Field	Sr. No.	Polymer Used	Composition and Method	Outcome	Application	References
	4	GelMA and liver-derived dECM	3D Bioprinting of hiHep[35] cells and polymers as inner gear-like construct	The 3D bioprinted scaffolds when studied *in vitro* showed increase in cell viability, higher functional liver activities including albumin and blood urea nitrogen.	Replacement and repair of liver tissues	Mao et al. 2020
	5	PF127[36] and alginate	3D Bioprinting of HepG2 cells and polymers as construct	The 3D bioprinted scaffolds when studied *in vitro* showed increase in functional liver metabolism when compared to traditional 2D models.	Model for study of *in vivo* drug hepatotoxicity.	Gori et al. 2020
Heart tissue engineering	2	Fibronectin-stamped polydimethylsiloxane	3D Bioprinting of cardiomyocytes cells and polymers	The 3D bioprinted scaffolds when studied showed improved tissue functions (beating frequency and force).	Regeneration of laminar cardiac muscle	Grosberg et al. 2011
	3	Fibronectin-stamped polydimethylsiloxane	3D Bioprinting of iPSC-MSCs and polymers as *in vitro* model	Investigated Barth syndrome, suggesting new treatment strategies *in vivo*.	*in vitro* modeling of cardiomyopathy	Wang et al. 2014
	4	SU-8[37] photoresist on silicon substrate	3D model of hiPSC-CMs[38], HUVECs, fibroblast and polymers as a 3D heart model	The 3D bioprinted scaffolds when studied showed higher fidelity of adult cardiac phenotype, with good sarcoplasmic reticulum function as well as contractile properties	Model for studying different treatments of cardiac diseases.	Huang et al. 2020

(Contd.)

Field	Sr. No.	Polymer Used	Composition and Method	Outcome	Application	References
	1	SU-8 2002 on silicon	Photolithography of fetal stem cell and polymers as microfluidic *in vitro* NMJ[39] model	The 3D bioprinted scaffolds when studied showed the contractions of the myotube which shows dose-dependent responses to the toxins	Platform for generating patient-specific NMJ models	Santhanam et al. 2018
Muscle tissue engineering	2	Commercial multielectrode array system	Neuronal cells as opto-genetically engineered NMJ mode	The 3D bioprinted scaffolds when studied showed the generation of the response to calcium transients by NMJs when subjected to the glutamate stimuli.	Platform for screening of drugs affecting the NMJ.	Solomon et al. 2021
	3	Fibrin	3D Bioprinting of iPSCs[40] and polymers as *in vitro* humanized muscular dystrophy model	The 3D bioprinted scaffolds when studied showed abnormal phenotypes, demonstrating muscular dystrophies.	Platform for human skeletal muscle organoid-like model as disease model.	Maffioletti et al. 2018

PCL[1] – polycaprolactone, MC3T3–E1[2] – PEGDMA[3] – Poly(ethylene Glycol Dimethacrylate) , GelMA[4] – gelatin methacrylamide, hMSCs[5] – human mesenchymal stem cells, MC3T3[6] – , dECM[7] – decellularized extra cellular matrix, 3D[8] – three dimensional, hASC[9] – human adipose derived stem, hNC[10] – Human nasoseptal chondrocytes, PF127[11] – Pluronic , HA[12] – hyaluronic acid, AFSCs[13] – Amniotic Fluid Stem Cells, BM–MSCs[14] – bone marrow–derived human mesenchymal stem cells, CS–AEMA[15] – chondroitin sulfate amino ethyl methacrylate, PLA[16] – polylactic acid, MSCs[17] – mesenchymal stem cells, PEG[18] – Polyethylene Glycol, FCPCs[19] – Fetal Cartilage–Derived Cells, hBECs[20] – human bronchial epithelial cells, iPSC–MSCs[21] – induced pluripotent stem cells–derived mesenchymal stem cells, and iPSC–Chds[22] – induced pluripotent stem cells–derived chondrocytes, AFSs[23] – amniotic fluid–derived stem, hNSCs[24] – human neural stem cells, PVA[25] – Polyvinyl alcohol, PEGDA[26] – Poly(ethylene glycol) diacrylate, EMSCs[27] – Ectomesenchymal stem cells, 2D[28] – two dimensional, PEGTA[29] – poly(ethylene glycol)–tetra–acrylate, HUVECs[30] – Human umbilical vein endothelial cells, EPCs[31] – Endothelial progenitor cells, HDF–n[32] – primary neonatal human dermal fibroblasts, HepG2[33] – Human hepatocellular carcinoma, HepaRG[34] – Primary Human Hepatocytes, hiHep[35] – Functional human induced hepatocytes, PF127[36] – Pluronic, SU–8[37] – photoresist, hiPSC–CMs[38] – human induced pluripotent stem cells derived cardiomyocytes, NMJ[39] – neuro–muscular junctions, iPSCs[40] – induced pluripotent stem cells.

2.8.1 Adhesion and Proliferation Rate

3D bioprinted scaffolds require the attachment of the cells onto the matrix for their growth, viability and differentiation. The simplest method to quantify the adhesion of cells onto the scaffolds is by suspending the cells over the matrix, incubation over certain period, followed by detachment of loosely attached and unattached cells. This is again followed by either quantification of the remaining cells adhered on the matrix or the number of cells detached from the matrix by MTT (Saltzman 2004). Alternatively, the adhesion of cells can also be qualitatively analyzed through direct visualization by radiolabeling/ fluorescently labelling cells attached on the scaffolds. Another method is by measuring the concentration of intracellular enzyme released by the cells or via binding of intracellular component like DNA (Gawne et al. 2022). Other technique to quantify the detached cells is by centrifugal detachment assays where the inverted plate is subjected to controlled detachment force via centrifugation and the fluid is collected between two parallel plates. The flow-through fluid is then used to measure the cellular kinetics (Reyes and García 2003). Similarly, an apparatus where spinning desk is used in place of centrifugation for measurement can also be applied (Dumbauld et al. 2010). Lastly, micropipette-based techniques are utilized to measure membrane deformability as well as the forces involved in cell-cell and cell-matrix adhesion (Qin et al. 2005).

2.8.2 Migration Rate

The importance of migration in the 3D bioprinted scaffolds is necessary as the cells in the native form migrate during the tissue formation and are critical for the architecture. Particularly in case of scaffolds for tissue regeneration, this factor plays an important role as the ability of movement of cell in association with the matrix as well as other cells helps in the process of repair. The migration is complicated to measure, particularly in the complex microenvironment. Visually, a small number of movements (~100 cells) can be observed but are tedious to evaluate (Yamada and Sixt 2019). Population techniques like filter chamber assay can be performed to analyse collective movements of large population of cells through a membrane filter (Stoellinger and Alexanian 2022). Another technique involving the blocking of leading front of cell movement on agarose is performed and is termed as agarose assay (Szatmary et al. 2014). Generally, the measuring parameters for cell motility are the random motility coefficient and persistence time.

2.8.3 Cellular Aggregation

Cellular aggregation is an essential tool in tissue engineering research provides critical parameters, such as the correlation between cell-cell interactions and key processes like cell differentiation, migration, and viability. For 3D bioprinted construct to mimic the tissue microenvironment during transplants and repair applications, cells shall aggregate to enhance the functionality and survival rates (Mahoney and Saltzman 2001). Polymers are modified by the addition of cell binding peptides (linking RGD and YIGSR) at both ends to promote the aggregation of cells from the suspension (Dai et al. 1994). Direct visualization to analyse the cellular aggregation is often performed where cells are visualized by histogenic method introduced by Moscona (Moscona 1961). Another technique with the use of specialized aggregometer like small angle light scattering can rapidly measure rate of aggregation (Thomas and Steinberg 1980, Asnaghi et al. 1997).

2.8.4 Cellular Phenotype

Hybrid polymer-based research is carried out in order to enhance the outcome of the model; one such property enhancement is the promotion of cell-specific functioning which is obtained by achieving the perfect morphology of the cell. Many a times the cell specific enzymes and genes are expressed only in presence of perfect morphology and these differ from cells and the kind of study being carried out. The measurement of secreted proteins and enzymatic activity serves as a valuable tool in 3D biological models, such as cytochrome P450 activity in hepatocytes (Bachour-El Azzi et al. 2022), neurotransmitter metabolism assays (e.g., choline acetyltransferase) in neurons (Salvaterra 1987), and extracellular matrix protein expression in skin regeneration models (Rousselle et al. 2019).

2.9 CHALLENGES AND FUTURE PERSPECTIVES

Synthetic polymers are popular choices for 3D bioprinting applications because of their versatility and excellent mechanical properties. Synthetic polymers find applications in 3D bioprinting for a wide range of purposes, from the creation of bioartificial organs to the development of disease models. As with any other emerging technique, there are certain challenges to be overcome before this technology can be utilized to its full potential. Biocompatibility issues such as immunogenic reactions and cytotoxicity constitute a significant challenge for synthetic polymers. While the polymer might not be responsible for the adverse effects, the degradation of byproducts might have immunogenic effects, which can be overcome by chemically modifying the polymer before usage and by copolymerization. Also, the lack of cell-adhesive motifs is a major limitation of synthetic polymers compared to natural polymers. Another challenge concerning synthetic polymers is the slow rate of degradation. Due to this, the native tissue may regenerate faster, and the persisting scaffold can interfere with the natural regeneration processes. A strategy to overcome the challenges of lower biocompatibility of synthetic polymers is functionalizing them by the addition of bioactive materials. Some commonly added additives to synthetic polymers are fibrin gel, hyaluronic acid, polyethylene oxide, calcium phosphate, and polycation polyethyleneimine (Parak et al. 2019). Regarding the slow degradation rate of synthetic polymers, they can be chemically modified so that the rate of degradation can match the rate of regeneration of the native tissues. With further advancements in 3D bioprinting techniques, it will be possible to use synthetic bioink-based bioinks for accurately printing complex and functional tissues and organs with specific cellular components, architecture, and mechanical strength. These developments can transform biomedical science and help solve many problems related to drug testing, disease modelling, tissue and organ shortages, and, potentially, the difficulties of organ transplantation.

2.10 CONCLUSION

The bioink properties are very crucial for 3D bioprinting and since major component of the bioink is the polymer, few major aspects to be seen while formulating bioink have been described in the chapter. To conclude, the criteria to satisfy a property of polymers includes the rheological properties (uniform and smooth extrusion), shift of the state with temperature (liquid to solid phase), maintain the structural integrity, shear thinning modulus and finally the cytocompatibility aspect. Considering all the above aspects, not all polymers can be applied for fabrication of 3D bioprinted models but most of the polymers discussed in the chapter have shown high success rates in the bio-medical applications and have great potential in revolutionizing the tissue engineering field. With the increasing development in the technology of additive manufacturing techniques, the application of polymers will show linearly increasing graphs in terms of outcome. With the development of novel polymers, using multidisciplinary efforts and their application in the field of 3D bioprinting will give rise to virtual organ replacement in case of failed and diseased human organs in the near future.

ACKNOWLEDGEMENTS

The authors wish to acknowledge the Indian Council for Medical Research (ICMR Adhoc – 5/3/8/53/2020 - ITR) for funding the research and Manipal Centre for Biotherapeutics Research, MAHE for infrastructural support.

REFERENCES

Abbott, R.D. and D.L. Kaplan. 2015. Strategies for improving the physiological relevance of human engineered tissues. Trends in Biotechnology. 33: 401–407.

Admane, P., A.C. Gupta, P. Jois, S. Roy, C.C. Lakshmanan, G. Kalsi, et al., 2019. Direct 3D bioprinted full-thickness skin constructs recapitulate regulatory signaling pathways and physiology of human skin. Bioprinting. 15: e00051.

Ahsan, S.M., M. Thomas, K.K. Reddy, S.G. Sooraparaju, A. Asthana and I. Bhatnagar. 2018. Chitosan as biomaterial in drug delivery and tissue engineering. International Journal of Biological Macromolecules. 110: 97–109.

Alicja, K, M. Heljak and W. Święszkowski. 2020. Mechanical properties of hybrid triphasic scaffolds for osteochondral tissue engineering. Materials Letters. 261: 126893–93.

Amoretti, M., C. Amsler, G. Bonomi, A. Bouchta, P. Bowe, C. Carraro, et al., 2002. Production and Detection of Cold Antihydrogen Atoms. Nature. 419: 456–59.

Arefin, A.M.E., N.R. Khatri, N. Kulkarni, and P.F. Egan. 2021. Polymer 3D Printing Review: Materials, Process, and Design Strategies for Medical Applications. Polymers. 13: 1499.

Arslan-Yildiz, A., R.E. Assal, P. Chen, S. Guven, F. Inci and U. Demirci. 2016. Towards Artificial Tissue Models: Past, Present, and Future of 3D Bioprinting. Biofabrication. 8: 014103.

Asnaghi, D., M. Carpineti, M. Giglio and A. Vailati. 1997. Small angle light scattering studies concerning aggregation processes. Current Opinion in Colloid and Interface Science. 2: 246–250.

Bachour-El Azzi, P., Chesné, C. and Uehara, S. 2022. Expression and functional activity of cytochrome P450 enzymes in human hepatocytes with sustainable reproducibility for in vitro phenotyping studies. pp. 285–305. *In*: H. Yamazaki (ed.). Advances in Pharmacology. Academic Press.

Bello, A.B., D. Kim, D. Kim, H. Park and S-H. Lee. 2020. Engineering and functionalization of gelatin biomaterials: from cell culture to medical applications. Tissue Engineering Part B Review. 26: 164–180.

Bendtsen, S.T., S.P. Quinnell and M. Wei. 2017. Development of a novel alginate-polyvinyl alcohol-hydroxyapatite hydrogel for 3D bioprinting bone tissue engineered scaffolds. Journal of Biomedical Materials Research A. 105: 1457–1468.

Boland, T., T. Xu, B. Damon and X. Cui. 2006. Application of inkjet printing to tissue engineering. Journal of Biotechnology. 1: 910–917.

Boparai, K.S., R. Singh and H. Singh. 2016. Development of rapid tooling using fused deposition modeling: a review. Rapid Prototyping Journal. 22(2): 281–299. https://doi.org/10.1108/RPJ-04-2014-0048.

Carothers, W.H. 1929. Studies on polymerization and ring formation. I. An introduction to the general theory of condensation POLYMERS. Journal of the American Chemical Society. 51: 2548–2559.

Chen, W-H., Q-W. Chen, Q. Chen, C. Cui, S. Duan, Y. Kang, et al., 2022. Biomedical polymers: synthesis, properties, and applications. Science China Chemistry. 65: 1010–1075.

Chevallay, B. and D. Herbage. 2000. Collagen-based biomaterials as 3D scaffold for cell cultures: applications for tissue engineering and gene therapy. Medical and Biological Engineering and Computing. 38: 211–218.

Chircov, C., A.M. Grumezescu and L.E. Bejenaru. 2018. Hyaluronic acid-based scaffolds for tissue engineering. Romanian Journal of Morphology and Embryology. 59: 71–76.

Chung, Y-C. and C-Y. Chen. 2008. Antibacterial characteristics and activity of acid-soluble chitosan. Bioresource Technology. 99: 2806–2814.

Chung, J.H.Y., S. Naficy, Z. Yue, R. Kapsa, A. Quigley, S.E. Moulton, et al., 2013. Bioink properties and printability for extrusion printing living cells. Biomaterials Science. 1: 763–773.

Collins, M. N. and C. Birkinshaw. 2007. Comparison of the effectiveness of four different crosslinking agents with hyaluronic acid hydrogel films for tissue-culture applications. Journal of Applied Polymer Science. 104: 3183–3191.

Collins, M.N. and C. Birkinshaw. 2013. Hyaluronic acid based scaffolds for tissue engineering—a review. Carbohydrate Polymers. 92: 1262–1279.

Copes, F., N. Pien, S.V. Vlierberghe, F. Boccafoschi and D. Mantovani. 2019. Collagen-based tissue engineering strategies for vascular medicine. Frontiers in Bioengineering and Biotechnology. 7.

Costantini, M., J. Idaszek, K. Szöke, J. Jaroszewicz, M. Dentini, A. Barbetta, et al., 2016. 3D bioprinting of BM-MSCS-loaded ECM biomimetic hydrogels for *in vitro* neocartilage formation. Biofabrication. 8: 035002.

Crump, S.S. and Wales. 1991. Fast, precise, safe prototypes with FDM. http://hdl.handle.net/2152/64313.

Dai, W., J. Belt and W.M. Saltzman. 1994. Cell-binding peptides conjugated to poly(ethylene glycol) promote neural cell aggregation. Nature Biotechnology. 12: 797–801.

Das, S. and B. Basu. 2019. An overview of hydrogel-based bioinks for 3D bioprinting of soft tissues. Journal of the Indian Institute of Science. 99: 405–428.

Demirtaş, T.T., G. Irmak and M. Gümüşderelioğlu. 2017. A bioprintable form of chitosan hydrogel for bone tissue engineering. Biofabrication. 9: 035003.

Dumbauld, D.W., K.E. Michael, S.K. Hanks and A.J. García. 2010. Focal adhesion kinase-dependent regulation of adhesive forces involves vinculin recruitment to focal adhesions. Biology of the Cell. 102: 203–213.

Elzoghby, A.O. 2013. Gelatin-based nanoparticles as drug and gene delivery systems: reviewing three decades of research. Journal of Controlled Release. 172: 1075–1091.

Emeric, C., S. Marquette, C. Peerboom, K. Amighi and J. Goole. 2021. Development of MAB-loaded 3D-printed (FDM) implantable devices based on PLGA. International Journal of Pharmaceutics. 597: 120337.

England, S., A. Rajaram, D.J. Schreyer and X. Chen. 2017. Bioprinted fibrin-factor XIII-hyaluronate hydrogel scaffolds with encapsulated schwann cells and their *in vitro* characterization for use in nerve regeneration. Bioprinting. 5: 1–9.

Fakhari, A. and C. Berkland. 2013. Applications and emerging trends of hyaluronic acid in tissue engineering, as a dermal filler and in osteoarthritis treatment. Acta Biomaterialia. 9: 7081–7092.

Farzaneh, D., Y. Yu, Y. Zhang, A.M.D. Jesus, E.A. Sander and I.T. Ozbolat. 2014. *In vitro* evaluation of carbon-nanotube-reinforced bioprintable vascular conduits. Nanotechnology. 25: 145101–1.

Fazeli, N., E. Arefian, S. Irani, A. Ardeshirylajimi and E. Seyedjafari. 2021. 3D-printed PCL scaffolds coated with nanobioceramics enhance osteogenic differentiation of stem cells. ACS Omega. 6: 35284–35296.

Flory, P.J. 1953. Principles of Polymer Chemistry. Cornell University Press, New York.

Forget, A., T. Derme, D. Mitterberger, M. Heiny, C. Sweeney, L. Mudili, et al., 2019. Architecture-inspired paradigm for 3D bioprinting of vessel-like structures using extrudable carboxylated agarose hydrogels. Emergent Materials. 2: 233–243.

Freeman, S., R. Ramos, P.A. Chando, L. Zhou, K. Reeser, S. Jin, et al., 2019. A bioink blend for rotary 3D bioprinting tissue engineered small-diameter vascular constructs. Acta Biomaterialia. 95: 152–164.

Fu, Z., L. Ouyang, R. Xu, Y. Yang and W. Sun. 2022. Responsive biomaterials for 3D bioprinting: a review. Materials Today. 52: 112–132.

Gao, G., A.F. Schilling, K. Hubbell, T. Yonezawa, D. Truong, Y. Hong, et al., 2015. Improved properties of bone and cartilage tissue from 3D inkjet-bioprinted human mesenchymal stem cells by simultaneous deposition and photocrosslinking in PEG-GelMA. Biotechnology Letters. 37: 2349–2355.

Gao, G., J.H. Lee, J. Jang, D.H. Lee, J-S. Kong, B.S. Kim, et al., 2017. Tissue engineered bio-blood-vessels constructed using a tissue-specific bioink and 3D coaxial cell printing technique: a novel therapy for ischemic disease. Advanced Functional Materials. 27: 1700798

Gasiorowski, J.Z., C.J. Murphy and P.F. Nealey. 2013. Biophysical cues and cell behavior: the big impact of little things. Annual Review of Biomedical Engineering. 15: 155–176.

Gawne, P.J., F. Man, P.J. Blower and R.T.M. de Rosales. 2022. Direct cell radiolabeling for *in vivo* cell tracking with PET and SPECT imaging. Chemical Reviews. 122: 10266–10318.

Gentile, P., V. Chiono, I. Carmagnola and P.V. Hatton. 2014. An overview of poly(lactic-co-glycolic) acid (PLGA)-based biomaterials for bone tissue engineering. International Journal of Molecular Sciences. 15: 3640–3659.

Glowacki, J. and S. Mizuno. 2008. Collagen scaffolds for tissue engineering. Biopolymers. 89: 338–344.

Gori, M., S.M. Giannitelli, M. Torre, P. Mozetic, F. Abbruzzese, M. Trombetta, et al., 2020. Biofabrication of hepatic constructs by 3D bioprinting of a cell-laden thermogel: an effective tool to assess drug-induced hepatotoxic response. Advanced Healthcare Materials. 9: 2001163.

Grida, I. and J.R.G. Evans. 2003. Extrusion freeforming of ceramics through fine nozzles. Journal of the European Ceramic Society. 23: 629–635.

Grosberg, A., P.W. Alford, M.L. McCain and K.K. Parker. 2011. Ensembles of engineered cardiac tissues for physiological and pharmacological study: heart on a chip. Lab on a Chip. 11: 4165.

Gu, Q., E. Tomaskovic-Crook, R. Lozano, Y. Chen, R.M. Kapsa, Q. Zhou, et al., 2016. Functional 3D neural mini-tissues from printed gel-based bioink and human neural stem cells. Advanced Healthcare Materials. 5: 1429–38.s

Guarino, V., Gloria, A., Santis, R.D. and Ambrosio, L. 2010. Composite hydrogels for scaffold design, tissue engineering, and prostheses. pp. 227–245. *In*: R.M. Ottenbrite, K. Park and T. Okano (eds). Biomedical Applications of Hydrogels Handbook. Springer New York, NY, USA.

Guarino, V., Marrese, M. and Ambrosio, L. 2015. Chemical and physical properties of polymers for biomedical use. pp. 67–90. *In*: F. Puoci (ed.). Advanced Polymers in Medicine. Springer Cham.

Gunatillake, P.A. and R. Adhikari. 2003. Biodegradable synthetic polymers for tissue engineering. European Cells and Materials. 5: 1–16.

Guo, Y. and W.T. Pu. 2020. Cardiomyocyte maturation: new phase in development. Circulation Research. 126: 1086–1106.

Hafezi, F., S. Shorter, A.G. Tabriz, A. Hurt, V. Elmes, J. Boateng, et al., 2020. Bioprinting and preliminary testing of highly reproducible novel bioink for potential skin regeneration. Pharmaceutics. 12: 550–50.

Ho, L. and S-H. Hsu. 2018. Cell reprogramming by 3D bioprinting of human fibroblasts in polyurethane hydrogel for fabrication of neural-like constructs. Acta Biomaterialia. 70: 57–70.

Hoch, E., G.E.M. Tovar and K. Borchers. 2014. Bioprinting of artificial blood vessels: current approaches towards a demanding goal. European Journal of Cardiothoracic Surgery. 46: 767–778.

Hockaday, L.A., K.H. Kang, N.W. Colangelo, P.Y.C. Cheung, B. Duan, E. Malone, et al., 2012. Rapid 3D Printing of anatomically accurate and mechanically heterogeneous aortic valve hydrogel scaffolds. Biofabrication. 4: 035005.

Homan, K.A., D.B. Kolesky, M.A. Skylar-Scott, J. Herrmann, H. Obuobi, A. Moisan, et al., 2016. Bioprinting of 3D convoluted renal proximal tubules on perfusable chips. Scientific Reports. 6: 34845.

Hoskins, J.N. and S.M. Grayson. 2009. Synthesis and degradation behavior of cyclic poly (ε-caprolactone). Macromolecules. 42: 6406–6413.

Hou, R., L. Nie, G. Du, X. Xiong and J. Fu. 2015. Natural polysaccharides promote chondrocyte adhesion and proliferation on magnetic nanoparticle/PVA composite hydrogels. Colloids and Surfaces B: Biointerfaces. 132: 146–154.

Hsieh, C-T. and S-H. Hsu. 2019. Double-network polyurethane-gelatin hydrogel with tunable modulus for high-resolution 3D bioprinting. ACS Applied Materials and Interfaces. 11: 32746–32757.

Hu, D., D. Wu, L. Huang, Y. Jiao, L. Li, L. Lu, et al., 2018. 3D bioprinting of cell-laden scaffolds for intervertebral disc regeneration. Materials Letters. 223: 219–222.

Huang, Y., K. He and X. Wang. 2013. Rapid prototyping of a hybrid hierarchical polyurethane-cell/hydrogel construct for regenerative medicine. Materials Science and Engineering. C. 33: 3220–3229.

Huang, C.Y., R.P.M. Maia-Joca, C.S. Ong, I. Wilson, D. DiSilvestre, G.F. Tomaselli, et al., 2020. Enhancement of human iPSC-derived cardiomyocyte maturation by chemical conditioning in a 3D environment. Journal of Molecular and Cellular Cardiology. 138: 1–11.

Islam, S., M.A.R. Bhuiyan and M.N. Islam. 2016. Chitin and chitosan: structure, properties and applications in biomedical engineering. Journal of Polymers and the Environment. 25: 854–866.

Jang, E.H., J-H. Kim, J.H. Lee, D-H. Kim and Y-N. Youn. 2020. Enhanced biocompatibility of multi-layered, 3D bio-printed artificial vessels composed of autologous mesenchymal stem cells. Polymers. 12: 538.

Jia, W., P.S. Gungo-Ozkerim, Y.S. Zhang, K. Yue, K. Zhu, W. Liu, et al., 2016. Direct 3D bioprinting of perfusable vascular constructs using a blend bioink. Biomaterials. 106: 58–68.

Kang, H-W., S.J. Lee, I.K. Ko, C. Kengla, J.J. Yoo and A. Atala. 2016. A 3D bioprinting system to produce human-scale tissue constructs with structural integrity. Nature Biotechnology. 34: 312–319.

Kasoju, N. and U. Bora. 2012. Silk fibroin in tissue engineering. Advanced Healthcare Materials. 1: 393–412.

Ke, D., H. Yi, S. Est-Witte, S.K. George, C.V. Kengla, S.J. Lee, et al., 2019. Bioprinted trachea constructs with patient matched design, mechanical and biological properties. Biofabrication. 12: 015022.

Kim, H.S., X. Sun, J-H. Lee, H-W. Kim, X. Fu and K.W. Leong. 2019. Advanced drug delivery systems and artificial skin grafts for skin wound healing. Advanced Drug Delivery Reviews. 146: 209–239.

Kim, I.G., S.A. Park, S-H. Lee, J.S. Choi, H. Cho, S.J. Lee, et al., 2020. Transplantation of a 3D-printed tracheal graft combined with IPS cell-derived MSCs and chondrocytes. Science Reports. 10: 4326.

Koh, L-D., Y. Cheng, C-P. Teng, Y-W. Khin, X-J. Loh, S-Y. Tee, et al., 2015. Structures, mechanical properties and applications of silk fibroin materials. Progress in Polymer Science. 46: 86–110.

Kundu, B., R. Rajkhowa, S.C. Kundu and X. Wang. 2013. Silk fibroin biomaterials for tissue regenerations. Advanced Drug Delivery Reviews. 65: 457–470.

Lam, T., T. Dehne, J.P. Krüger, S. Hondke, M. Endres, A. Thomas, et al., 2019. Photopolymerizable gelatin and hyaluronic acid for stereolithographic 3D bioprinting of tissue-engineered cartilage. Journal of Biomedical Materials Research—B Applied Biomaterials. 107: 2649–2657.

Leal Egaña, A. and T. Scheibel. 2010. Silk-based materials for biomedical applications. Applied Biochemistry and Biotechnology. 55: 155–167.

Lee, A., A.R. Hudson, D.J. Shiwarski, J.W. Tashman, T.J. Hinton, S. Yerneni, et al., 2019. 3D bioprinting of collagen to rebuild components of the human heart. Science. 365: 482–487.

Lee, H., S. Ahn, L.J. Bonassar and G. Kim. 2012. Cell(MC3T3-E1)-printed poly (ε-caprolactone)/alginate hybrid scaffolds for tissue regeneration. macromol. Rapid Communications. 34: 142–149.

Lee, H., W. Han, H. Kim, D-H. Ha, J. Jang, B.S. Kim, et al., 2017. Development of liver decellularized extracellular matrix bioink for three-dimensional cell printing-based liver tissue engineering. Biomacromolecules. 18: 1229–1237.

Lee, J., J. Hong, W. Kim and G.H. Kim. 2020. Bone-derived DECM/alginate bioink for fabricating a 3D cell-laden mesh structure for bone tissue engineering. Carbohydrate Polymers. 250: 116914.

Li, S., X. Tian, J. Fan, H. Tong, Q. Ao and X. Wang. 2019. Chitosans for tissue repair and organ three-dimensional (3D) bioprinting. Micromachines. 10: 765.

Li, Y., X. Cao, W. Deng, Q. Yu, C. Sun, P. Ma, et al., 2020. 3D printable sodium alginate-matrigel (SA-MA) hydrogel facilitated ectomesenchymal stem cells (EMSCs) Neuron differentiation. Journal of Biomaterials Applications. 35: 709–719.

Liao, H., R. Qi, M. Shen, X. Cao, R. Guo, Y. Zhang, et al., 2011. Improved cellular response on multiwalled carbon nanotube-incorporated electrospun polyvinyl alcohol/chitosan nanofibrous scaffolds. Colloids Surface B: Biointerfaces. 84: 528–535.

Liu, F., Q. Chen, C. Liu, Q. Ao, X. Tian, J. Fan, et al., 2018. Natural polymers for organ 3D bioprinting. Polymers. 10: 1278.

Liu, F. and X. Wang. 2020. Synthetic polymers for organ 3D printing. Polymers. 12: 1765.

Liu, J., B. Zhang, L. Li, J. Yin and J. Fu. 2021. Additive-lathe 3D bioprinting of bilayered nerve conduits incorporated with supportive cells. Bioactive Materials. 6: 219–229.

Liu, S., T. Wang, S. Li and X. Wang. 2022. Application status of sacrificial biomaterials in 3D bioprinting. Polymers. 14: 2182.

Lorber, B., W-K. Hsiao, I.M. Hutchings and K.R. Martin. 2013. Adult rat retinal ganglion cells and glia can be printed by piezoelectric inkjet printing. Biofabrication. 6: 015001.

Lukin, I., I. Erezuma, L. Maeso, J. Zarate, M.F. Desimone, T.H. Al-Tel, et al., 2022. Progress in gelatin as biomaterial for tissue engineering. Pharmaceutics. 14: 1177.

Maffioletti, S.M., S. Sarcar, A.B.H. Henderson, I. Mannhardt, L. Pinton, L.A. Moyle, et al., 2018. Three-dimensional human iPSC-derived artificial skeletal muscles model muscular dystrophies and enable multilineage tissue engineering. Cell Reports. 23: 899–908.

Mahmoodi, N.M., F. Moghimi, M. Arami and F. Mazaheri. 2010. Silk degumming using microwave irradiation as an environmentally friendly surface modification method. Fibers and Polymers. 11: 234–240.

Mahoney, M.J and W.M. Saltzman. 2001. Transplantation of brain cells assembled around a programmable synthetic microenvironment. Nature Biotechnology. 19: 934–939.

Makadia, H.K. and S.J. Siegel. 2011. Poly lactic-co-glycolic acid (PLGA) as biodegradable controlled drug delivery carrier. Polymers. 3: 1377–1397.

Malafaya, P.B., G.A. Silva and R.L. Reis. 2007. Natural–origin polymers as carriers and scaffolds for biomolecules and cell delivery in tissue engineering applications. Advanced Drug Delivery Reviews 59: 207–233.

Mandrycky, C., Z. Wang, K. Kim and D-H. Kim. 2016. 3D bioprinting for engineering complex tissues. Biotechnology Advances. 34: 422–434.

Mano, J.F., G.A. Silva, H.S. Azevedo, P.B. Malafaya, R.A. Sousa, S.S. Silva, et al., 2007. Natural origin biodegradable systems in tissue engineering and regenerative medicine: present status and some moving trends. Journal of the Royal Society Interface. 4: 999–1030.

Mao, Q., Y. Wang, Y. Li, S. Juengpanich, W. Li, M. Chen, et al., 2020. Fabrication of liver microtissue with liver decellularized extracellular matrix (dECM) bioink by digital light processing (DLP) bioprinting. Materials Science and Engineering C. 109: 110625.

Markstedt, K., A. Mantas, I. Tournier, H.M. Ávila, D. Hägg and P. Gatenholm. 2015. 3D bioprinting human chondrocytes with nanocellulose–alginate bioink for cartilage tissue engineering applications. Biomacromolecules. 16: 1489–1496.

Martínez, A., M.D. Blanco, N. Davidenko and R.E. Cameron. 2015. Tailoring chitosan/collagen scaffolds for tissue engineering: effect of composition and different crosslinking agents on scaffold properties. Carbohydrate Polymers. 132: 606–619.

Matthews, J.A., G.E. Wnek, D.G. Simpson and G.L. Bowlin. 2002. Electrospinning of collagen nanofibers. Biomacromolecules. 3: 232–238.

Meinel, A.J., K.E. Kubow, E. Klotzsch, M. Garcia-Fuentes, M.L. Smith, V. Vogel, et al., 2009. Optimization strategies for electrospun silk fibroin tissue engineering scaffolds. Biomaterials 30: 3058–3067.

Minoura, N., A. Seiichi, M. Higuchi, Y. Gotoh, M. Tsukada and Y. Imai. 1995. Attachment and growth of fibroblast cells on silk fibroin. Biochemical and Biophysical Research Communications. 208: 511–516.

Mishra, N., A.K. Goyal, K. Khatri, B. Vaidya, R. Paliwal, S. Rai, et al., 2008. Biodegradable polymer based particulate carrier(s) for the delivery of proteins and peptides. Anti-Inflammation. Anti-Allergy Agents Medicinal Chemistry. 7: 240–251.

Mohan, D., Z.K. Teong, M.S. Sajab, N.H.N. Kamarudin and H. Kaco. 2021. Intact fibrillated 3D-printed cellulose macrofibrils/$CaCO_3$ for controlled drug delivery. Polymers. 13: 1912.

Mohrman, A.E., M. Farrag, R.K. Grimm and N.D. Leipzig. 2018. Evaluation of *in situ* gelling chitosan-PEG copolymer for use in the spinal cord. Journal of Biomaterials Applications. 33: 435–446.

Moscona, A. 1961. Rotation-mediated histogenetic aggregation of dissociated cells. Experimental Cell Research. 22: 455–475.

Munaz, A., R.K. Vadivelu, J. St. John, M. Barton, H. Kamble and N-T. Nguyen. 2016. Three-dimensional printing of biological matters. Journal of Science: Advanced Materials and Devices. 1: 1–17.

Naghieh, S., M.D. Sarker, E. Abelseth and X. Chen. 2019. Indirect 3D bioprinting and characterization of alginate scaffolds for potential nerve tissue engineering applications. Journal of the Mechanical Behavior of Biomedical Materials. 93: 183–193.

Nathan, K.G., K. Genasan and T. Kamarul. 2023. Polyvinyl alcohol-chitosan scaffold for tissue engineering and regenerative medicine application: a review. Marine Drugs. 21: 304.

Naureen, B., A.S.M.A. Haseeb, W.J. Basirun and F. Muhamad. 2021. Recent advances in tissue engineering scaffolds based on polyurethane and modified polyurethane. Materials Science and Engineering C. 118: 111228.

Nejad, Z.M., A. Zamanian, M. Saeidifar, H.R. Vanaei and M.S. Amoli. 2021. 3D bioprinting of polycaprolactone-based scaffolds for pulp-dentin regeneration: investigation of physicochemical and biological behavior. Polymers. 13: 4442.

Ni, T., M. Liu, Y. Zhang, Y. Cao and R. Pei. 2020. 3D bioprinting of bone marrow mesenchymal stem cell-laden silk fibroin double network scaffolds for cartilage tissue repair. Bioconjugate Chemistry. 31: 1938–1947.

Odde, D.J. and M.J. Renn. 1999. Laser-guided direct writing for applications in biotechnology. Trends In Biotechnology. 17: 385–389.

Ouyang, L., R. Yao, S. Mao, X. Chen, J. Na and W. Sun. 2015. Three-dimensional bioprinting of embryonic stem cells directs highly uniform embryoid body formation. Biofabrications. 7: 044101–1.

Pan, Z. and J. Ding. 2012. Poly(Lactide-Co-Glycolide) Porous scaffolds for tissue engineering and regenerative medicine. Interface Focus. 2: 366–377.

Park, J-H., J-K. Yoon, J.B. Lee, Y.M. Shin, K-W. Lee, S-W. Bae, et al., 2019. Experimental tracheal replacement using 3-dimensional bioprinted artificial trachea with autologous epithelial cells and chondrocytes. Scientific Reports. 9: 2103.

Pateman, C.J., A.J. Harding, A. Glen, C.S. Taylor, C.R. Christmas, P.P. Robinson, et al., 2015. Nerve guides manufactured from photocurable polymers to aid peripheral nerve repair. Biomaterials. 49: 77–89.

Patrício, T., M. Domingos, A. Gloria, U. D'Amora, J.F. Coelho and P.J. Bártolo. 2014. Fabrication and characterisation of PCL and PCL/PLA scaffolds for tissue engineering. Rapid Prototyping Journal. 20: 145–156.

Pranzo, D., P. Larizza, D. Filippini and G. Percoco. 2018. Extrusion-based 3D printing of microfluidic devices for chemical and biomedical applications: a topical review. Micromachines. 9: 374.

Premkumar, J., SonicaSree, K. and Sudhakar, T. 2020. Polymers in biomedical use. pp. 1–28. *In*: C.M. Hussain and S. Thomas (eds). Handbook of Polymer and Ceramic Nanotechnology. Springer Cham.

Putri, N.R.F., X. Wang, Y. Chen, X. Li, N. Kawazoe and G. Chen. 2020. Preparation of PLGA-collagen hybrid scaffolds with controlled pore structures for cartilage tissue engineering. Progress in Natural Science: Materials. 30: 642–650.

Qiao, Z., M. Lian, Y. Han, B. Sun, X. Zhang, W. Jiang, et al., 2021. Bioinspired stratified electrowritten fiber-reinforced hydrogel constructs with layer-specific induction capacity for functional osteochondral regeneration. Biomaterials. 266: 120385.

Qin, T-W., Z-M. Yang, Z-Z. Wu, H-Q. Xie, J. Qin and S-X. Cai. 2005. Adhesion strength of human tenocytes to extracellular matrix component-modified poly(Dl-Lactide-Co-Glycolide) substrates. Biomaterials. 26: 6635–6642.

Rahman, R.A., N.M. Sukri, N.M. Nazir, M.A.A. Radzi, A.H. Zulkifly, A.C. Ahmad, et al., 2015. The potential of 3-dimensional construct engineered from poly(lactic-co-glycolic acid)/fibrin hybrid scaffold seeded with bone marrow mesenchymal stem cells for *in vitro* cartilage tissue engineering. Tissue Cell. 47: 420–430.

Reddy, N., R. Reddy and Q. Jiang. 2015. Crosslinking biopolymers for biomedical applications. Trends Biotechnol. 33: 362–369.

Reyes, C.D. and A.J. García. 2003. A centrifugation cell adhesion assay for high-throughput screening of biomaterial surfaces. Journal of Biomedical Materials Research A. 67A: 328–333.

Rizwana, N., N. Maslekar, K. Chatterjee, Y. Yao, V. Agarwal and M. Nune. 2023. Dual crosslinked antioxidant mixture of poly(vinyl alcohol) and cerium oxide nanoparticles as a bioink for 3D bioprinting. ACS Applied Nano Materials. 7(16): 18177–18188.

Rodrigues, P.C.I., L.F. Woigt, K.D. Pereira, A.D. Luchessi, E.S.N. Lopes, T.J. Webster, et al., 2020. Low-cost hybrid scaffolds based on polyurethane and gelatin. Journal of Materials Research and Technology. 9: 7777–7785.

Rodríguez-Rodríguez, R., H. Espinosa-Andrews, C. Velasquillo-Martínez and Z.Y. García-Carvajal. 2019. Composite hydrogels based on gelatin, chitosan and polyvinyl alcohol to biomedical applications: a review. International Journal of Polymeric Materials and Polymeric Biomaterials. 69: 1–20.

Rose, J.B., S. Pacelli, A.J.E. Haj, H.S. Dua, A. Hopkinson, L.J. White, et al., 2014. Gelatin-based materials in ocular tissue engineering. Materials. 7: 3106–3135.

Rousselle, P., M. Montmasson and C. Garnier. 2019. Extracellular matrix contribution to skin wound re-epithelialization. Matrix Biology. 75–76: 12–26.

Ruiz-Cantu, L., A. Gleadall, C. Faris, J. Segal, K. Shakesheff and J. Yang. 2020. Multi-material 3D bioprinting of porous constructs for cartilage regeneration. Materials Science and Engineering C. 109: 110578.

Saltzman, W.M. 2004. Tissue engineering: engineering principles for the design of replacement organs and tissues. Oxford University Press.

Salvaterra, P.M. 1987. Molecular biology and neurobiology of choline acetyltransferase. Molecular Neurobiology. 1: 247–280.

Santhanam, N., L. Kumanchik, X. Guo, F. Sommerhage, Y. Cai, M. Jackson, et al., 2018. Stem cell derived phenotypic human neuromuscular junction model for dose response evaluation of therapeutics. Biomaterials. 166: 64–78.

Schmitz, C., L.G. Auza, D. Koberidze, S. Rasche, R. Fischer and L. Bortesi. 2019. Conversion of chitin to defined chitosan oligomers: current status and future prospects. Marine Drugs. 17: 452.

Shajani, Y., C. Pan, L. Elomaa and Y. Yang. 2016. Development of a novel bioprinting system for hybrid tissue engineering constructs of biomimetic complexity. Frontiers in Bioengineering and Biotechnology. Conference Abstract: 10th World Biomaterials Congress. doi: 10.3389/conf.FBIOE.2016.01.01464.

Sharma, R., I.P.M. Smits, L.D.L. Vega, C. Lee and S.M. Willerth. 2020. 3D bioprinting pluripotent stem cell derived neural tissues using a novel fibrin bioink containing drug releasing microspheres. Frontiers in Bioengineering and Biotechnology. 8:57.

Shi, Z., X. Gao, M.W. Ullah, S. Li, Q. Wang and G. Yang. 2016. Electroconductive natural polymer-based hydrogels. Biomaterials. 111: 40–54.

Si, H., T. Xing, Y. Ding, H. Zhang, R. Yin and W. Zhang. 2019. 3D bioprinting of the sustained drug release wound dressing with double-crosslinked hyaluronic-acid-based hydrogels. Polymers. 11: 1584.

Singh, S., S. Ramakrishna and R. Singh. 2017. Material issues in additive manufacturing: a review. Journal of Manufacturing Processes. 25: 185–200.

Skardal, A., D. Mack, E. Kapetanovic, A. Atala, J.D. Jackson, J. Yoo, et al., 2012. Bioprinted amniotic fluid-derived stem cells accelerate healing of large skin wounds. Stem Cells Translational Medicine. 1: 792–802.

Solomon, E., A.M. Rooney, A.M. Rodriguez, S. Micheva-Viteva, R. Bashir, R. Iyer, et al., 2021. Neuromuscular junction model optimized for electrical platforms. Tissue Engineering Part C Methods. 27: 242–252.

Sonnenschein, M.F., Z. Lysenko, D.A. Brune, B.L. Wendt and A.K. Schrock. 2005. Enhancing polyurethane properties via soft segment crystallization. Polymer. 46: 10158–10166.

Stansbury, J.W. and M.J. Idacavage. 2016. 3D printing with polymers: challenges among expanding options and opportunities. Dent Materials. 32: 54–64.

Staudinger, H. 1920. Über polymerisation. Berichte der Deutschen Chemischen Gesellschaft (A and B Series) 53: 1073–1085.

Stoellinger, H.M. and A.R. Alexanian. 2022. Modifications to the transwell migration/invasion assay method that eases assay performance and improves the accuracy. ASSAY and Drug Development Technologies. 20(2): 75–82.

Sui, S., X. Wang, P. Liu, Y. Yan and R. Zhang. 2009. Cryopreservation of cells in 3D constructs based on controlled cell assembly processes. Journal of Bioactive and Compatable Polymers. 24: 473–487.

Sun, H., S. Song, X. Xu, J. Dai, J. Yu, W. Zhou, et al., 2021. Recent progress on structurally ordered materials for electrocatalysis. Advanced Energy Materials. 11: 2101937.

Szatmary, A.C., C.H. Stuelten and R. Nossal. 2014. Improving the design of the agarose spot assay for eukaryotic cell chemotaxis. RSC Advances. 4: 57343–57349.

Tappa, K. and U. Jammalamadaka. 2018. Novel biomaterials used in medical 3D printing techniques. Journal of Functional Biomaterials. 9: 17.

Tayebi, T., A. Baradaran-Rafii, A. Hajifathali, A. Rahimpour, H. Zali, A. Shaabani, et al., 2021. Biofabrication of chitosan/chitosan nanoparticles/polycaprolactone transparent membrane for corneal endothelial tissue engineering. Scientific Reports. 11: 7060.

Teixeira, M.A., M.T.P. Amorim and H.P. Felgueiras. 2019. Poly(Vinyl Alcohol)-based nanofibrous electrospun scaffolds for tissue engineering applications. Polymers. 12: 7.

Tessmar, J.K. and A.M. Göpferich. 2007. Customized PEG-derived copolymers for tissue-engineering applications. Macromolecular Bioscience. 7: 23–39.

Thomas, W.A. and M.S. Steinberg. 1980. A twelve-channel automatic device for continuous recording of cell aggregation by measurement of small-angle light-scattering. Journal of Cell Science. 41: 1–18.

Toprakhisar, B., A. Nadernezhad, E. Bakirci, N. Khani, G.A. Skvortsov and B. Koc. 2018. Development of bioink from decellularized tendon extracellular matrix for 3D bioprinting. Macromolecular Bioscience. 18: 1800024.

Tsai, M-C., K-C. Hung, S-C. Hung and S-H. Hsu. 2015. Evaluation of biodegradable elastic scaffolds made of anionic polyurethane for cartilage tissue engineering. Colloids and Surfaces B: Biointerfaces. 125: 34–44.

Tsai, K.J., S. Dixon, L.R. Hale, A. Darbyshire, D. Martin and A. de Mel. 2017. Biomimetic heterogenous elastic tissue development. NPJ Regenerative Medicine. 2: 16.

Vanaei, H., M. Shirinbayan, M. Deligant, K. Raissi, J. Fitoussi, S. Khelladi, et al., 2020a. Influence of process parameters on thermal and mechanical properties of polylactic acid fabricated by fused filament fabrication. Polymer Engineering Science. 60: 1822–1831.

Vanaei, H.R., K. Raissi, M. Deligant, M. Shirinbayan, J. Fitoussi, S. Khelladi, et al., 2020b. Toward the understanding of temperature effect on bonding strength, dimensions and geometry of 3D-printed parts. Journal of Materials Science. 55: 14677–14689.

Vanaei, H.R., M. Shirinbayan, S.F. Costa, F.M. Duarte, J.A. Covas, M. Deligant, et al., 2020c. Experimental study of PLA thermal behavior during fused filament fabrication. Journal of Applied Polymer Science. 138: 49747.

Vanaei, R.H., M. Deligant, M. Shirinbayan, K. Raïssi, J. Fitoussi, S. Khelladi, et al., 2020d. A comparative in-process monitoring of temperature profile in fused filament fabrication. Polymer Engineering and Science. 61: 68–76.

Vanaei, R.H., M. Shirinbayan, S. Vanaei, J. Fitoussi, S. Khelladi and A. Tcharkhtchi. 2021. Multi-scale damage analysis and fatigue behavior of PLA manufactured by fused deposition modeling (FDM). Rapid Prototyping Journal. 27: 371–378.

Vasconcelos, E.V., F.B. da Luz, S.P.A. da Paz, M.A.L. dos Reis, A.C.R. da Silva, M.F. Passos, et al., 2023. Nanostructured 3D bioprinting of PLA with bioglass-CNT Scaffolds for osseus tissue graft manufacturing. Journal of Materials Research and Technology. 23: 5923–5938.

Vlierberghe, V.S., V. Cnudde, P. Dubruel, B. Masschaele, A. Cosijns, I.D. Paepe, et al., 2007. Porous Gelatin hydrogels: 1. Cryogenic Formation and Structure Analysis. Biomacromolecules. 8: 331–337.

Volkov, A.V., A.A. Muraev, I.I. Zharkova, V.V. Voinova, E.A. Akoulina, V.A. Zhuikov, et al., 2020. Poly(3-Hydroxybutyrate)/hydroxyapatite/alginate scaffolds seeded with mesenchymal stem cells enhance the regeneration of critical-sized bone defect. Materials Science and Engineering C. 114: 110991.

Wang, X., Y. Yan, F. Lin, Z. Xiong, R. Wu, R. Zhang, et al., 2005. Preparation and characterization of a collagen/chitosan/heparin matrix for an implantable bioartificial liver. Journal of Biomaterials Science. 16: 1063–1080.

Wang, X., Y. Yan, Y. Pan, Z. Xiong, H. Liu, J. Cheng, et al., 2006. Generation of three-dimensional hepatocyte/gelatin structures with rapid prototyping system. Tissue Engineering. 12: 83–90.

Wang, H., O.C. Boerman, K. Sariibrahimoglu, Y. Li, J.A. Jansen and S.C.G. Leeuwenburgh. 2012. Comparison of micro- vs. nanostructured colloidal gelatin gels for sustained delivery of osteogenic proteins: bone morphogenetic protein-2 and alkaline phosphatase. Biomaterials. 33: 8695–8703.

Wang, X., Y. Huang and C. Liu. 2015a. A Combined rotational mold for manufacturing a functional liver system. Journal of Bioactive and Compatable Polymers. 30: 436–451.

Wang, M.O., C.M. Piard, A. Melchiorri, M.L. Dreher and J.P. Fisher. 2015b. Evaluating changes in structure and cytotoxicity during *in vitro* degradation of three-dimensional printed scaffolds. Tissue Engineering Part A. 21: 1642–1653.

Wang, X., Q. Ao, X. Tian, J. Fan, H. Tong, W. Hou, et al., 2017. Gelatin-based hydrogels for organ 3D bioprinting. Polymers. 9: 401.

Wang, X. 2018. Bioartificial organ manufacturing technologies. Cell Transplant. 28: 5–17.

Wang, X. 2019. Advanced Polymers for three-dimensional (3D) organ bioprinting. Micromachines. 10: 814.

Wohlers, T., and R.I. Campbell. 2017. Wohlers Report 2017: 3D Printing and Additive Manufacturing State of the Industry: Annual Worldwide Progress Report. Fort Collins, Colo.: Wohlers Associates, Inc.

Wohlers Associates. 2019. Wohlers Report 2019: 3D Printing and Additive Manufacturing State of the Industry. Fort Collins, Colorado: Wohlers Associates.

Woodfield, T.B.F., J. Malda, J. de Wijn, F. Péters, J. Riesle and C.A. van Blitterswijk. 2004. Design of porous scaffolds for cartilage tissue engineering using a three-dimensional fiber-deposition technique. Biomaterials. 25: 4149–4161.

Wu, Y., Z.Y.(William) Lin, A.C. Wenger, K.C. Tam and X.(Shirley) Tang. 2018. 3D bioprinting of liver-mimetic construct with alginate/cellulose nanocrystal hybrid bioink. Bioprinting. 9: 1–6.

Wu, Z., Q. Li, S. Xie, X. Shan and Z. Cai. 2020. *In vitro* and *in vivo* biocompatibility evaluation of a 3D bioprinted gelatin-sodium alginate/rat Schwann-cell scaffold. Materials Science and Engineering. C. 109: 110530.

Xavier, J.R., T. Thakur, P. Desai, M.K. Jaiswal, N. Sears, E. Cosgriff-Hernandez, et al., 2015. Bioactive nanoengineered hydrogels for bone tissue engineering: a growth-factor-free approach. ACS Nano. 9: 3109–3118.

Xiang, Y., K. Miller, J. Guan, W. Kiratitanaporn, M. Tang and S. Chen. 2022. 3D bioprinting of complex tissues *in vitro*: state-of-the-art and future perspectives. Archives of Toxicology. 96: 691–710.

Xiaohong, W., S. Sui, Y. Yan and R. Zhang. 2010. Design and fabrication of PLGA sandwiched cell/fibrin constructs for complex organ regeneration. Journal of Bioactive and Compatable Polymers. 25: 229–240.

Xin, S., D. Chimene, J.E. Garza, A.K. Gaharwar and D.L. Alge. 2019. Clickable PEG hydrogel microspheres as building blocks for 3D bioprinting. Biomaterials Science. 7: 1179–1187.

Xu, M., Y. Van, H. Liu, R. Yag and X. Wang. 2009. Controlled adipose-derived stromal cells differentiation into adipose and endothelial cells in a 3D structure established by cell-assembly technique. Journal of Bioactive and Compatable Polymers. 24: 31–47.

Yadegari, M., B. Hashemibeni, M. Pourentezari, E. Tavakoli, M. Anvari, A. Valiani, et al., 2019. Evaluation of the mechanical characteristics of PLGA and PLGA/fibrin scaffolds in providing an appropriate environment for viability and growth of human adipose-derived stem cells. Eurasian Journal of Analytical Chemistry. 13: 151–159.

Yamada, K.M. and M. Sixt. 2019. Mechanisms of 3D cell migration. Nature Reviews Molecular Cell Biology. 20: 738–752.

Yang, X., Z. Lu, H. Wu, W. Li, L. Zheng and J. Zhao. 2018. Collagen-alginate as bioink for three-dimensional (3D) cell printing based cartilage tissue engineering. Materials Science and Engineering C. 83: 195–201.

Yang, H., L. Sun, Y. Pang, D. Hu, H. Xu, S. Mao, et al., 2021. Three-dimensional bioprinted hepatorganoids prolong survival of mice with liver failure. Gut. 70: 567–574.

Yang, P., Y. Ju, Y. Hu, X. Xie, B. Fang and L. Lei. 2023. Emerging 3D bioprinting applications in plastic surgery. Biomaterials Research. 27.

Yao, R., R. Zhang, Y. Yan and X. Wang. 2009. *In vitro* angiogenesis of 3D tissue engineered adipose tissue. Journal of Bioactive and Compatable Polymers. 24: 5–24.

Yin, J., M. Yan, Y. Wang, J. Fu and H. Suo. 2018. 3D bioprinting of low-concentration cell-laden gelatin methacrylate (GelMA) bioinks with a two-step cross-linking strategy. ACS Applied Materials and Interfaces. 10: 6849–6857.

Yu, J., S. Lee, S. Choi, K.K. Kim, B. Ryu, C-Y. Kim, et al., 2020. Fabrication of a polycaprolactone/alginate bipartite hybrid scaffold for osteochondral tissue using a three-dimensional bioprinting system. Polymers. 12: 2203.

Zarif, M-E. 2018. A review of chitosan-, alginate-, and gelatin-based biocomposites for bone tissue engineering. Biomater. Tissue Engineering Bulletin. 5: 97–109.

Zeltinger, J., J.K. Sherwood, D.A. Graham, R. Müeller and L.G. Griffith. 2001. Effect of pore size and void fraction on cellular adhesion, proliferation, and matrix deposition. Tissue Engineering. 7: 557–572.

Zhang, K., Q. Fu, J. Yoo, X. Chen, P. Chandra, X. Mo, et al., 2017. 3D Bioprinting of urethra with PCL/PLCL blend and dual autologous cells in fibrin hydrogel: an *in vitro* evaluation of biomimetic mechanical property and cell growth environment. Acta Biomaterialia. 50: 154–164.

Zhang, D., X. Wu, J. Chen and K. Lin. 2018. The Development of collagen based composite scaffolds for bone regeneration. Bioactive Materials. 3: 129–138.

Zhang, J., E. Wehrle, P. Adamek, G.R. Paul, X-H. Qin, M. Rubert, et al., 2020. Optimization of mechanical stiffness and cell density of 3D bioprinted cell-laden scaffolds improves extracellular matrix mineralization and cellular organization for bone tissue engineering. Acta Biomaterialia. 114: 307–322.

Zhao, X., L. Liu, J. Wang, Y. Xu, W. Zhang, G. Khang, et al., 2014. *In vitro* vascularization of a combined system based on a 3D printing technique. Journal of Tissue Engineering and Regenerative Medicine. 10: 833–842.

Zhao, W., J. Li, K. Jin, W. Liu, X. Qiu and C. Li. 2016. Fabrication of functional PLGA-based electrospun scaffolds and their applications in biomedical engineering. Materials Science and Engineering C. 59: 1181–1194.

Zhu, J. 2010. Bioactive modification of poly(ethylene glycol) hydrogels for tissue engineering. Biomaterials. 31: 4639–4656.

Chapter **3**

Natural Polymers for Fabrication of Tissue Engineering Scaffold

Naznin Sultana*[1], Anisa Cole[1], Francine Strachan[2], Max W. Fontus[1], Tony Grady[2], Quyen Huynh[1] and Dennis E. Daniels[1]

[1]Undergraduate Medical Academy

[2]Department of Chemistry,

Prairie View A&M University, Prairie View, TX77446, USA

3.1 INTRODUCTION

Tissues play an essential role in organisms, and any damage to the tissues impairs the body's normal functioning and the quality of life of an organism. Although tissues such as bones and skeletal muscles can heal themselves after incurring some form of injury, clinical interventions are required to address severe injuries (Jahani and Brooks 2020, Eltom et al. 2019). Under these circumstances, conventional treatment methods, including autografts or donor grafts, may not work because of unavailability, and there is an increased risk of infection or immunological rejection (Aldana and Abraham 2017). As a result, Tissue Engineering (TE) is the most effective alternative to guarantee a patient's full recovery in case of severe tissue damage (Gregor et al. 2017). TE restores, improves, and maintains tissues that have been damaged due to injury or disability that arises from congenital conditions. This is achieved using natural and synthetic polymers.

*For Correspondence: nasultana@pvamu.edu

Developing biocomposite scaffolds with appropriate properties is of great importance in TE. This chapter explores the materials, fabrication techniques, and future directions of biocompatible scaffolds. Key findings highlight the significance of scaffold properties such as porosity, mechanical strength, degradation rate, and bioactivity in promoting effective tissue regeneration. The chapter underscores the importance of innovative scaffold development in advancing regenerative medicine, addressing critical challenges in tissue engineering, and meeting the growing demand for biocompatible tissues and organs.

Regenerative medicine is an interdisciplinary field that applies life sciences and engineering principles to promote regeneration or restoration of injured tissues and organs. Using regenerative medicine, tissues and organs damaged through aging, trauma, or disease can be healed or even replaced (Mao and Mooney 2015). Central to regenerative medicine is TE. TE is a field in which biological and engineering principles are employed to create new tissues and organs (Dzobo et al. 2018). TE is required because it provides an alternative source of biocompatible and biodegradable tissues and organs, which addresses the crucial gap where there is a high demand for tissues and organs but a limited supply.

For TE to be functional, a strategy of regenerating tissues and organs, a triad of cells, scaffolds, and growth factors are usually employed (Figure 3.1) (Sultana et al. 2024). The goal is to either stimulate the original tissue to regenerate or replace the original tissue. A scaffold is used not only to provide a structure similar to that provided by an extracellular matrix but also to provide a framework where biological cues can be used, added, or infused to promote regeneration and mimic an environment found naturally in healthy tissues and organs.

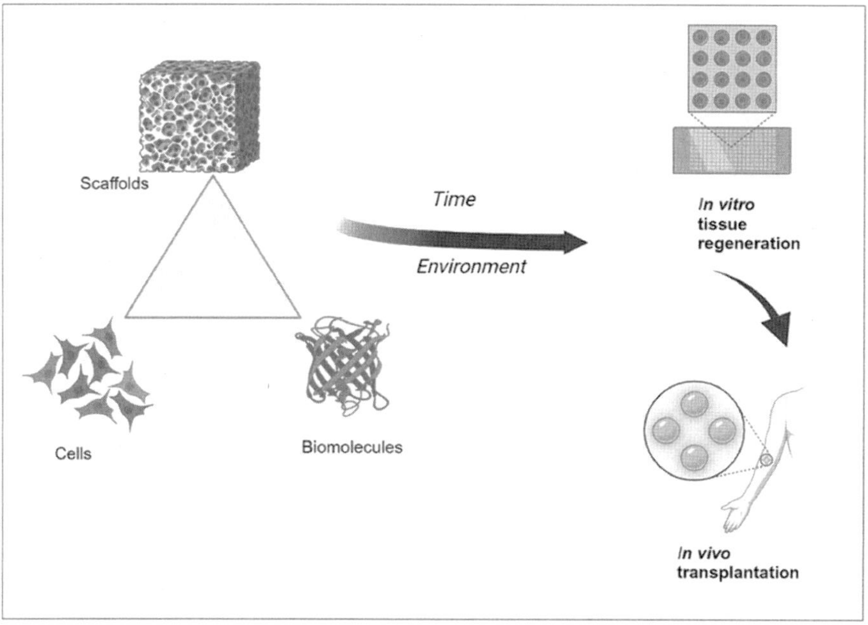

Figure 3.1 Tissue Engineering strategy (Created with BioRender.com).

The use of scaffolds in TE has evolved throughout history. Despite being constructed from varying materials, such as linen in Egypt, catgut in Europe, and ant heads in India and South Africa, these scaffolds provided a framework for a tissue to heal or regenerate. One common barrier faced by many pioneers in scaffold creation was constructing scaffolds that could not only provide a structural framework but also facilitate cell attachment, differentiation, and proliferation. The ideal scaffold must also allow for the crucial cell-to-cell and cell-to-matrix interactions demonstrated *in vivo* (Kaul and Ventikos 2015).

Scaffolds in TE are three-dimensional and porous, providing a favorable environment for tissue reparation and regeneration (Krishani et al. 2023). There are several requirements for scaffolds. The ideal scaffold should be biodegradable, biocompatible, possess mechanical properties, have a highly porous architecture, and be cost-effective (Krishani et al. 2023). Cell adhesion and proliferation rely on the scaffold's biocompatibility. Furthermore, another measure of how biocompatible a scaffold is an immune response. A negligible immune response is ideal for preventing chronic inflammatory reactions that impede healing or may even cause rejection (Krishani et al. 2023).

This chapter comprehensively analyzes natural polymer-based biocomposite scaffolds used in TE, focusing on their materials, fabrication techniques, and future directions. This chapter also aims to synthesize current knowledge on scaffold properties and their impact on tissue regeneration. By examining various materials and methods, the review seeks to identify the strengths and limitations of each approach, offering insights into potential advancements in scaffold technology. The scope of the chapter spans from foundational principles to the latest innovations to ensure a thorough understanding of the field's current state and future possibilities.

3.2 MATERIALS USED IN SCAFFOLDS

The primary natural polymers in TE scaffolds include chitosan, collagen, and chitin. Collagen forms part of the extracellular matrix in most connective tissues within the body of mammals. Generally, it is composed of fibrous protein (Marque et al. 2019). Since it has a fibrillar structure, collagen plays an essential role in maintaining the biological structure and integrity of the extracellular matrix, thus providing support for tissues (Silver et al. 2018). Furthermore, collagen is less immunogenic, more permeable, biocompatible, biodegradable, and has a structural porosity (Silver et al. 2018). Unfortunately, collagen does not have the mechanical strength ideal for complex tissue engineering (Fan et al. 2022). As a result, bioceramics, such as hydroxyapatite (HA) and beta-tricalcium phosphate (β-TCP) are used with collagen to fabricate composite scaffolds to improve strength (Fan et al. 2022). Furthermore, β-TCP is essential in providing better osteoconductivity and accelerating the scaffold degradation rate for replacement with a newly formed tissue (Fan et al. 2022).

Gelatin is another natural polymer that is used in tissue engineering scaffolds. It is mainly a protein derived from collagen through hydrolysis (Afewerki et al. 2018).

Since it contains Arginylglycylaspartic acid (RGD), gelatin is highly biocompatible. Furthermore, the Arginylglycylaspartic acid found within the gelatin structure makes it easy for cells to attach, spread, and proliferate (Kuttappan et al. 2016). Since gelatin has poor mechanical properties, it cannot be directly used to treat bone defects. Instead, gelatin-based scaffolds are combined with other materials, including silica nanoparticles and polymer microparticles, to improve their mechanical strength (Kuttappan et al. 2016).

Another crucial natural polymer employed in tissue engineering scaffolds is chitosan. Chitosan is made up of a linear polysaccharide, usually found in crustaceans, mollusks, and cuticles of insects (Lavanya et al. 2020). It is biologically compatible, degradable, adhesive, and renewable (Islam et al. 2017). Furthermore, it contains free amino groups, which can be involved in protonation, thus making chitosan easy to modify using biochemical groups (Lavanya et al. 2020). The protonation of amino groups makes it easy for chitosan to electrostatically interact with DNA, proteins, lipids, or any other negatively charged synthetic polymers (Ahmed et al. 2018). Mechanical enhancers are used together with chitosan to improve its mechanical strength. Table 3.1 shows the commonly used natural polymers used to fabricate tissue engineering scaffolds.

Recently, another natural polymer, pectin, developed with chitosan, was explored in tissue engineering as a scaffolding material to regenerate bone and skin tissues (Table 3.2). Blending chitosan and pectin produces a polyelectrolyte complex that results in a new scaffold that has enhanced mechanical resistance, swelling capacity, porous microstructures, stabilized crosslinking, and biocompatibility (Sultana 2023).

Biodegradable polymer matrix nanocomposites are highly effective for bone tissue engineering (Surav et al. 2024). These materials have regenerative capabilities as they were shown to stimulate the regeneration of bones and tissues. Surav et al. emphasized natural biopolymers because they are cost-effective, have high availability, cause low immune response, have negligible toxicity, and are biocompatible. Natural biopolymers such as hyaluronic acid, elastin, alginate, collagen, gelatin, chitosan, and GAGs (glycosaminoglycans) are some of the natural polymers used in scaffolds for bone tissue engineering, and they are also the most researched (Surav et al. 2024).

Scaffolds can also be derived from synthetic origins. Synthetic scaffolds meet the inconsistency found in natural scaffolds, which tend to differ from batch to batch. Furthermore, the same ethical concerns about animal-derived scaffolds do not apply to synthetic ones. With synthetic scaffolds, there is room for modification, higher availability, and low costs. Polyglycolic acid (PGA) and polylactic acid (PLA) have been investigated extensively in bone tissue engineering (Wong et al. 2023). After implantation, PGA scaffolds, made from a material with high crystallinity, also demonstrated controlled degradation with decreased mechanical strength. Osteoblasts have been shown to proliferate and calcify on PGA scaffolds. PLA scaffolds are known for their mechanical and thermal properties. In addition to being biocompatible, pure PGA scaffolds also allow for the expression of osteogenic bone markers and the formation of calcium nodules (Wong et al. 2023).

Table 3.1 Applications of some natural polymers in tissue engineering (Sultana et al. 2024)

Natural polymer	Structure/Method of Production	Biological Properties	References
Collagen	Fibrillar structure, which contributes to the extracellular scaffolding	Promotes regeneration and angiogenesis of the bone through monocyte immunomodulation	(Correa-Araujo et al. 2024)
Chitosan	It contains an amine group, vital in pH sensitivity and functionality	Induces biological activity by showing excellent antimicrobial activity against bacteria	(Gholap et al. 2024)
Gelatin	Composed of a freeze-dried fiber scaffold	Produces a scaffold that is enzymatically crosslinked to enhance bone regeneration.	(Echave et al. 2019)
Collagen	Consists of a triple helix chain formed by α chains	Offers low immunogenicity, a porous structure, permeability, good biocompatibility, and biodegradability	(Correa-Araujo et al. 2024)
Collagen	Classical fibril-forming collagens, including types I, II, and III collagens	Crosslink formation can shield or modify major antigenic sites and, thus, reduce their capacity to interact with antibodies.	(Dong and Lv 2016)
Chitosan	Semicrystalline biopolymer containing several hydrogen bonds forming functional groups, including amino and hydroxyl groups.	The hydrogels are pH-sensitive in aqueous media, so these stimuli-responsive hydrogels are the best choice for drug delivery.	(Eivazzadeh-Keihan et al. 2022)
Chitosan	Scaffolds, cells, and bio-signals, all together to minimize artificial and cellular environment	Cardiovascular tissue engineering	(Ahmed et al. 2018)
Gelatin	In its structure, amino acid sequences such as the arginine-glycine-aspartic acid (RGD) motif improve cell adhesion, differentiation, and proliferation.	Obtain different isometric points.	(Echave et al. 2019)

Table 3.2 A summary of several pectin systems in tissue engineering approach (Sultana 2023)

Pectin Systems	Method	Application
Low-methoxyl citrus pectin	UV photocrosslinking with peptide crosslinkers (cell-degradable) and adhesive ligands (integrin-specific); lyophilization	Skin tissue engineering
Sugar beet pectin (SBP) crosslinked by visible light	Applying 405 nm visible light in the presence of tris(bipyridine)ruthenium (II) chloride hexahydrate and sodium persulfate, rapid hydrogenation of SBP was obtained; 3D hydrogel constructs were obtained using 3D bioprinting	Promising for liver and other soft tissue engineering
Citrus peel's pectin crosslinked with (3glycidyloxypropyl) trimethoxysilane (GPTMS)	Freeze-drying or 3D bioprinting	Various tissue regeneration
Pectin/chitin/nano $CaCO_3$	Lyophilization	Bone regeneration
Pectin/chitosan	Freeze-drying	Bone tissue engineering
Pectin/strontium/hydroxyapatite	Solution-based chemical technique	Bone regeneration
Collagen/polyurethane/pectin	Semi-interpenetration process	Bone regeneration
Pectin/PVA	Freezing–thawing	Bone regeneration
Poly(L-lactide-co-ε-caprolactone) (PLCA)/pectin	Scaffolds functionalized with pectin	*In vitro* and *in vivo* bone regeneration

Composite scaffolds contain a combination of biomaterials: biopolymers, bioceramics, degradable metals or natural products. When creating scaffolds for TE naturally, a composite material, such as bone, is needed, and it is logical to utilize a composite scaffold for increased biomimicry and bioactivity (Turnbull et al. 2018). Many materials are used in these composite scaffolds; Turnbull et al. outline polymers, hydrogels, metals, ceramics, and bio-glasses, to name a few (Turnbull et al. 2018). These composite scaffolds demonstrate desirable mechanical properties and cell interactions. Furthermore, it has been concluded that ceramic and polymer composite scaffolds have the most success.

Each type of scaffold material exhibits distinct advantages and disadvantages in TE. Natural materials are ideal because of their biocompatibility, low toxicity, and ability to stimulate regeneration in a tissue (Krishani et al. 2023). However, a disadvantage is their inconsistency from batch to batch. Synthetic materials are more consistent and accessible to modify and may lack that intrinsic bioactivity seen in natural scaffolds. Composite scaffolds take the best features of both natural and synthetic scaffolds. However, manufacturing biocomposite scaffolds is complex and expensive (Gang et al. 2022).

3.3 METHODS OF SCAFFOLD FABRICATION

In addition to the material used in the scaffold, the scaffold fabrication method is also fundamental in tissue engineering (Kaviani and Geramizadeh 2023). Scaffold fabrication is the process 'by which scaffolds with a continuous, uninterrupted pore structure can be made' (Kundu et al. 2014, Sultana et al. 2024).

A traditional scaffold fabrication method is solvent casting, where a polymer solution is poured into a mold, and the solvent is evaporated to form a scaffold. This method allows the fabrication of the desired architecture (Joseph et al. 2023). Another traditional method is gas foaming, a method of fabrication of synthetic scaffolds that avoids solvents. In this method, a polymer is compressed under high temperatures and molded into a solid disc. This disc will be placed in a carbon dioxide chamber for several days. This process allows gas to infiltrate the polymer disc and create a porous structure. Thermally-induced phase separation (TIPS) and freeze drying are other promising techniques for fabricating scaffolds from natural or synthetic polymers (Sultana and Wang 2008). Highly porous and interconnected composite scaffolds were fabricated using a freeze-drying technique. This technique is effective in producing composite scaffolds due to the ability to incorporate bioactive materials or drugs (Figure 3.2).

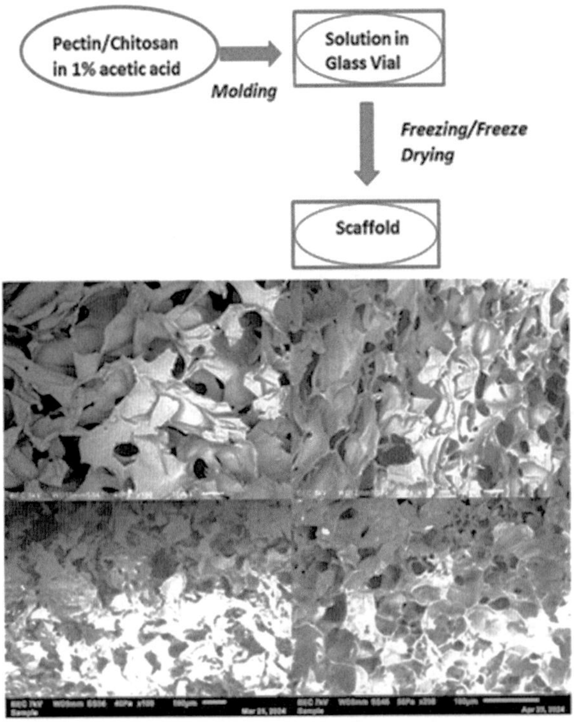

Figure 3.2 Schematic diagram of scaffold fabrications using freeze-drying technique and scanning electron micrograph of a typical pectin/chitosan and gelatin scaffold fabricated using the technique.

3D printing and electrospinning are advanced methods of tissue scaffold fabrication. These techniques are believed to craft intricate scaffolds with enhanced characteristics compared to traditional methods (Ejiohuo 2023). 3D printing is an additive manufacturing (AM) technique for fabricating structures and complex geometries from three-dimensional (3D) model data. The process consists of printing successive layers of materials formed on each other. Despite allowing for a broader range of geometries, 3D-printed polymer scaffolds tend to have inferior mechanical properties compared to scaffolds fabricated through different methods. Electrospinning is a fabrication technique that uses an electric field to propel and spin a polymer solution into nanofibers on a substrate (Ejiohuo 2023). Electrospinning is valued as it creates structures with higher surface area-to-volume ratios, which is helpful in tissue engineering (Xue et al. 2019). Combining 3D printing and electrospinning when creating a scaffold can make structures superior to those when the methods are used alone. Figure 3.3 shows the morphology of the electrospun nanofibrous scaffold and the growth of human skin fibroblast (HSF) cells on the scaffold (Lim 2015, Sultana et al. 2024).

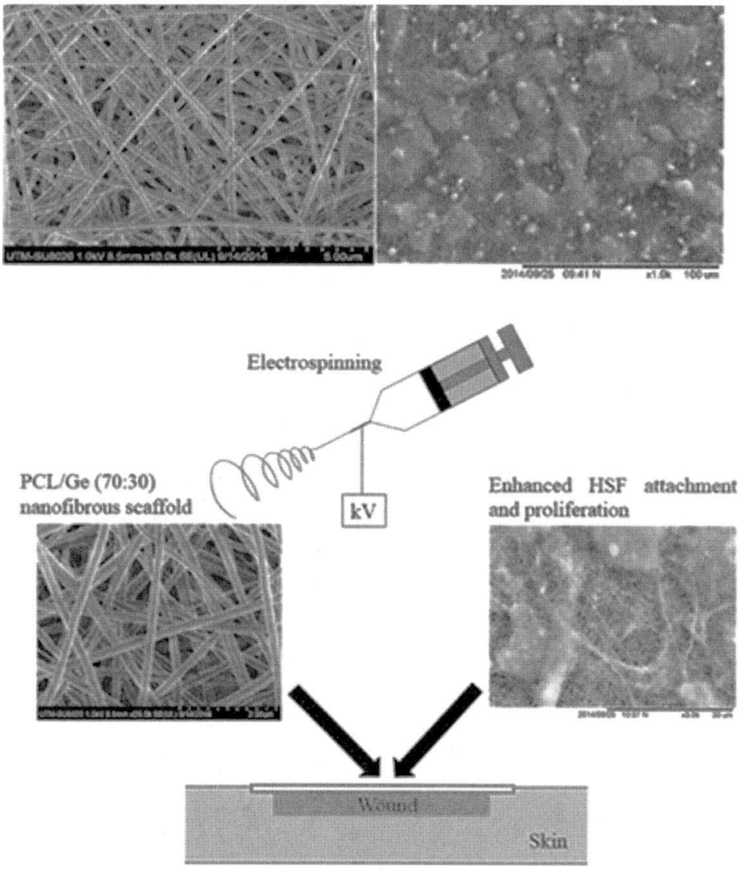

Figure 3.3 Morphology of electrospun scaffold and HSF cell growth after three days of culture on Polycaprolactone/Gelatin scaffold (reproduced from Lim 2015).

Considering the function of the scaffold to direct cellular behaviors such as proliferation, migration, and differentiation, there are a few requirements of the scaffold that must be met. These requirements discussed earlier, such as a porous structure, large pore sizes, and a uniform interconnectedness of the pores, can be met using nanotechnology. Nanotechnology can manipulate a scaffold at the atomic, molecular, and macromolecular levels (Wen et al. 2005). In addition, the scaffold can be constructed using the specific geometrical and topological structures found in natural tissue. The use of nanotechnology in scaffolds has been connected to higher mechanical strength, porosity, and enhanced biocompatibility (Wen et al. 2005).

Advanced scaffold fabrication methods, such as 3D printing and electrospinning, have demonstrated higher efficacy and precision in replicating the geometries seen in tissue, while traditional methods, although quicker and more straightforward, lack the complexities required in a scaffold for biomimicry. Conventional methods, such as solvent casting and gas foaming, are still the most cost-effective since advanced methods require expensive equipment. Therefore, even though advanced methods for scaffolding are more capable of mimicking the complex structures found *in vivo,* the cost problem remains a significant barrier to their widespread adoption in tissue engineering (Ejiohuo 2023).

3.4 TISSUE ENGINEERED PRODUCT APPLICATIONS

Researchers addressed the limitations of collagen (Col) hydrogels, which tend to have poor physicochemical and mechanical properties, by adding mesoporous bioactive glass nanoparticles (mBGns) (El-Fiqi et al. 2013). These Col–mBGn nanocomposite hydrogel scaffolds demonstrated excellent stability due to increased chemical bonding between collagen molecules. Furthermore, mechanical strength was increased, as the Col–mBGn hydrogel scaffolds had improved loading and stiffness compared to those without mBGns (El-Fiqi et al. 2013).

The influential role of mBGns in collagen scaffolds is further demonstrated by the seeding of mesenchymal stem cells. (MSCs) MSCs cultivated within the Col–mBGn hydrogels demonstrated viability and enhanced cytoskeleton extensions in hydrogels with mBGn added. In addition, while Col hydrogen demonstrated extreme shrinkage within a few days of culture, Col–mBGn hydrogels demonstrated no observable shrinkage over 21 days (El-Fiqi et al. 2013).

In one study, Alshammari et al. (2023) explored the relationship of scaffold pore architecture to bone tissue formation using mechanobiological modeling. Two scaffold designs were used: IREG1 and REG. IREG1 (irregular pore architecture) had larger pores, measuring 0.4 × 0.2 mm, while REG (regular pore architecture) had smaller pores, measuring 0.3 × 0.2 mm. In the earlier period of the study, the IREG1 scaffold resulted in more considerable bone growth, while in the later period, the REG scaffold resulted in more considerable bone growth than IREG1 (Alshammari et al. 2023). These results suggest that scaffolds' design and pore architecture can significantly influence bone growth rates over time, highlighting the importance of tailoring scaffold designs for different phases of bone regeneration.

Researchers have used regenerative medicine strategies for decades to develop tissue-engineered vascular grafts (TEVG) (Breuer et al. 2023). The TEVG was created by seeding tubular biodegradable polymeric scaffolds with autologous bone marrow-derived mononuclear cells. In the first-in-human study, the feasibility of TEVGs was demonstrated as the TEVG transformed into a living vascular graft that could grow. However, this initial clinical trial revealed a high incidence of early TEVG stenosis, preventing this promising technology's widespread use (Breuer et al. 2023). Researchers have used mouse and computational models to address this complication and design strategies to inhibit stenosis. Recently, these strategies have been validated in large animal models and clinics via an FDA-approved clinical trial (Breuer et al. 2023). This case highlights the challenges faced in using tissue-engineered vascular grafts. It demonstrates the effective solutions developed, underscoring this innovative technology's continued promise and potential.

The quantitative analysis assessed the biocompatibility of freeze-cast tissue scaffolds for several materials tested in mice. The freeze-cast method is versatile in creating porous and structured scaffolds. In the quantitative analysis, measurements were taken for encapsulation thickness, leukocyte cell counts, and density (of the scaffolds). Furthermore, researchers did lymphocyte assays, counting the number of capillaries and histological scoring. All of these methods were used to grade the biocompatibility of the scaffolds based on the cell-matrix and cell-cell interactions (Divakar et al. 2020). All of these scaffolds were derived from natural sources, and some had chemical modifications on top of them. *In vivo* biocompatibility was tested using mice, and all showed to be biocompatible. However, some differed in histopathological responses. This study outlines future directions, suggesting a focus on a systematic investigation of the effects of composition and processing parameters on the response to the scaffold (Divakar et al. 2020).

3.5 IMPORTANCE OF CONTINUED INNOVATION IN SCAFFOLD DEVELOPMENT

The case studies reviewed underscore the ongoing need for innovation in scaffold development (Breuer et al. 2023). The TEVG example demonstrates that overcoming clinical challenges such as stenosis requires a multidisciplinary approach, integrating insights from small and large animal models, computational simulations, and clinical trials (Breuer et al. 2023). The continued promise of tissue-engineered scaffolds in regenerative medicine highlights the importance of sustained research and development efforts. By addressing current limitations and exploring new materials and techniques, the field can advance toward creating effective, biocompatible scaffolds for various tissue engineering applications.

The field of TE has made substantial progress in developing scaffolds that support tissue regeneration. Advances in materials science and fabrication technologies have enabled the creation of scaffolds that closely mimic the complex structures of natural tissues. Despite these advancements, challenges remain, particularly in achieving consistent biocompatibility and mechanical properties while managing cost and scalability.

3.6 FUTURE RESEARCH DIRECTIONS

Future research should refine scaffold materials to enhance biocompatibility and mechanical properties while reducing immune responses (Antmen et al. 2021). Continued exploration of advanced fabrication techniques, such as integrating 3D printing and electrospinning, will be crucial for creating more intricate and functional scaffolds. Additionally, understanding the biological mechanisms underlying scaffold-tissue interactions will be essential for optimizing scaffold designs. Research should also aim to develop scalable and cost-effective production methods to facilitate widespread clinical application.

3.7 CONCLUSIONS

This chapter highlights the significant contribution of natural biopolymer scaffolds in TE and regenerative medicine by providing an overview of existing literature, key findings, and a review of the highlights and critical importance of biopolymer scaffold compatibility in regenerative improvement to damaged organs. This chapter identifies the essential research findings and contextualizes them within a broader aspect of the research. Innovations in scaffold materials, including natural, synthetic, and composite options, offer distinct advantages and challenges. The choice of fabrication techniques, from traditional methods like solvent casting and gas foaming to advanced techniques such as 3D printing and electrospinning, significantly impacts scaffolds' efficiency, cost, and suitability for various tissues. From the diverse array of biopolymers utilized in scaffold fabrication to exploring their essential qualities, such as structural integrity and biological properties, this chapter provides an understanding of the vital role of natural biopolymer scaffolds in tissue engineering.

ACKNOWLEDGEMENT

Authors acknowledge Mim Mim Lim's scholarly contributions, Undergraduate Medical Academy, and Prairie View A&M University's support.

REFERENCES

Afewerki, S., A. Sheikhi, S. Kannan, S. Ahadian and A. Khademhosseini. 2018. Gelatinpolysaccharide composite scaffolds for 3D cell culture and tissue engineering: towards natural therapeutics. Bioengineering and Translational Medicine. 4(1): 96–115.

Ahmed, S., A. Ali and J. Sheikh. 2018. A review on chitosan-centered scaffolds and their applications in tissue engineering. International Journal of Biological Macromolecules. 116: 849–862.

Al-Abduljabbar, A. and I. Farooq. 2022. Electrospun polymer nanofibers: processing, properties, and applications. Polymers. 15(1): 65. https://doi.org/10.3390/polym15010065.

Aldana, A.A. and G.A. Abraham. 2017. Current advances in electrospun gelatin-based scaffolds for tissue engineering applications. International Journal of Pharmaceutics. 523(2): 441453.

Alshammari, A., F. Alabdah, W. Wang and G. Cooper. 2023. Virtual design of 3D-printed bone tissue engineered scaffold shape using mechanobiological modeling: relationship of scaffold pore architecture to bone tissue formation. Polymers. 15(19): 3918.

Amza, C.G., A. Zapciu, G. Constantin, F. Baciu and M.I. Vasile. 2021. Enhancing mechanical properties of polymer 3D printed parts. Polymers. 13(4): 562.

Antmen, E., N.E. Vrana and V. Hasirci. 2021. The role of biomaterials and scaffolds in immune responses in regenerative medicine: macrophage phenotype modulation by biomaterial properties and scaffold architectures. Biomaterials Science. 9(24): 8090–8110.

Breuer, T., M. Jimenez, J.D. Humphrey, T. Shinoka and C.K. Breuer. 2023. Tissue engineering of vascular grafts: A case report from bench to bedside and back. Arteriosclerosis. Thrombosis, and Vascular Biology. 43(3): 399–409.

Correa-Araujo, L., A. Lara-Bertrand and I. Silva-Cote. 2024. Tissue Engineering Scaffolds: The Importance of Collagen.

Dai, C., Y. Li, W. Pan, G. Wang, R. Huang, Y. Bu, et al., 2019. Three-dimensional high-porosity chitosan/honeycomb porous carbon/hydroxyapatite scaffold with enhanced osteoinductivity for bone regeneration. ACS Biomaterials Science and Engineering. 6(1): 575–586.

Divakar, P., K.L. Moodie, E. Demidenko, P.J. Hoopes and U.G. Wegst. 2020. Quantitative evaluation of the *in vivo* biocompatibility and performance of freeze-cast tissue scaffolds. Biomedical Materials. 15(5): 055003.

Dong, C. and Y. Lv. 2016. Application of collagen scaffold in tissue engineering: recent advances and new perspectives. Polymers. 8(2): 42.

Dzobo, K., N.E. Thomford, D.A. Senthebane, H. Shipanga, A. Rowe, C. Dandara, et al., 2018. Advances in regenerative medicine and tissue engineering: innovation and transformation of medicine. Stem Cells International. 2018: 2495848. https://doi: 10.1155/2018/2495848

Echave, M.C., C. Pimenta-Lopes, J.L. Pedraz, M. Mehrali, A. Dolatshahi-Pirouz, F. Ventura, et al., 2019. Enzymatic crosslinked gelatin 3D scaffolds for bone tissue engineering. International Journal of Pharmaceutics. 562: 151–161.

Ejiohuo, O. 2023. A perspective on the synergistic use of 3D printing and electrospinning to improve nanomaterials for biomedical applications. Nano Trends. 100025.

Eivazzadeh-Keihan, R., E.B. Noruzi, S.F. Mehrban, H.A.M. Aliabadi, M. Karimi, A. Mohammadi, et al., 2022. Review: the latest advances in biomedical applications of chitosan hydrogel as a powerful natural structure with eye-catching biological properties. Journal of Materials Science. 57(6): 3855. https://doi.org/10.1007/s10853-021-06757-6.

El-Fiqi, A., J.H. Lee, E.J. Lee and H.W. Kim. 2013. Collagen hydrogels incorporated with surface-aminated mesoporous nano bioactive glass: improved physicochemical stability and mechanical properties are adequate for complex tissue engineering. Acta Biomaterialia. 9(12): 9508–9521.

Eltom, A., G. Zhong and A. Muhammad. 2019. Scaffold techniques and designs in tissue engineering functions and purposes: a review. Advances in Materials Science and Engineering. 2019: 3429527. https://doi.org/10.1155/2019/3429527

Fan, J., K. Abedi-Dorcheh, A. Sadat Vaziri, F. Kazemi-Aghdam, S. Rafieyan, M. Sohrabinejad, et al., 2022. A review of recent advances in natural polymer-based scaffolds for musculoskeletal tissue engineering. Polymers. 14(10): 2097.

Gang, F., W. Ye, C. Ma, W. Wang, Y. Xiao, C. Liu, et al., 2022. 3D printing of PLLA/ biomineral composite bone tissue engineering scaffolds. Materials. 15(12): 4280. https://doi.org/10.3390/ma15124280.

Gholap, A.D., S. Rojekar, H.S. Kapare, N. Vishwakarma, S. Raikwar, A. Garkal, et al., 2023. Chitosan scaffolds: expanding horizons in biomedical applications. Carbohydrate Polymers. 121394.

Gregor, A., E. Filová, M. Novák, J. Kronek, H. Chlup, M. Buzgo, et al., 2017. Designing of PLA scaffolds for bone tissue replacement fabricated by ordinary commercial 3D printer. Journal of Biological Engineering. 11: 1–21.

Islam, S., M.R. Bhuiyan and M.N. Islam. 2017. Chitin and chitosan: structure, properties, and applications in biomedical engineering. Journal of Polymers and the Environment. 25: 854–866.

Jahani, B. and A. Brooks. 2020. Additive manufacturing techniques for fabrication of bone scaffolds for tissue engineering applications. Recent Progress in Materials. 2(3): 1–41.

Joseph, B., C. Jose, S.V. Kavil, N. Kalarikkal and S. Thomas. 2023. Solvent-casting approach for design of polymer scaffolds and their multifunctional applications. Functional Biomaterials: Design and Development for Biotechnology, Pharmacology, and Biomedicine. 2: 371–394.

Kaul, H. and Y. Ventikos. 2015. On the genealogy of tissue engineering and regenerative medicine. Tissue Engineering Part B: Reviews. 21(2): 203–217.

Kaviani, M. and B. Geramizadeh. 2023. Fundamental aspects of skin tissue engineering: Cells, biomaterials, scaffold fabrication techniques, and signaling factors. Journal of Medical and Biological Engineering. 43: 508–521. https://doi.org/10.1007/s40846-023-00822-y

Krishani M., W.Y. Shin, H. Suhaimi and N.S. Sambudi. 2023. Development of scaffolds from bio-based natural materials for tissue regeneration applications: a review. Gels. 9(2): 100. doi: 10.3390/gels9020100. PMID: 36826270; PMCID: PMC9957409.

Kundu, J., F. Pati, J-H. Shim and D-W. Cho. 2014. Rapid prototyping technology for bone regeneration. pp. 254–284. *In*: R. Narayan (ed.). Rapid Prototyping of Biomaterials: Principles and Applications, Woodhead Publishing.

Kuttappan S., D. Mathew and M.B. Nair. 2016. Biomimetic composite scaffolds containing bioceramics and collagen/gelatin for bone tissue engineering—A mini-review. International Journal of Biological Macromolecules. 93(Pt B): 1390–1401. doi: 10.1016/j.ijbiomac.2016.06.043. Epub 2016 Jun 15. PMID: 27316767.

Lavanya, K., S.V., Chandran, K. Balagangadharan and N.J.M.S. Selvamurugan. 2020. Temperature-and pH-responsive chitosan-based injectable hydrogels for bone tissue engineering. Materials Science and Engineering: C. 111: 110862.

Lim, M.M., T. Sun and N. Sultana. 2015. *In vitro* biological evaluation of electrospun polycaprolactone/gelatine nanofibrous scaffold for tissue engineering. Journal of Nanomaterials. 2015: 303426. https://doi.org/10.1155/2015/303426.

Mao, A.S. and D.J. Mooney. 2015. Regenerative medicine: current therapies and future directions. Proceeding of the National Academy of Sciences. 112(47): 14452–14459.

Marques, C.F., G.S. Diogo, S. Pina, J.M. Oliveira, T.H. Silva and R.L. Reis. 2019. Collagen-based bio-inks for complex tissue engineering applications: a comprehensive review. Journal of Materials Science: Materials in Medicine. 30: 1–12.

Sánchez, P., J.L. Pedraz and G. Orive. 2017. Biologically active and biomimetic dual gelatin scaffolds for tissue engineering. International Journal of Biological Macromolecules. 98: 486–494.

Shukla, A., N., Dasgupta, S. Ranjan, S. Singh and R. Chidambram. 2017. Nanotechnology towards prevention of anemia and osteoporosis: from concept to market. Biotechnology and Biotechnological Equipment. 31(5): 863–879.

Silver, F.H., M. Jaffe and R.G. Shah. 2018. Structure and behavior of collagen fibers. pp. 345–365. *In*: A.R. Bunsell (ed.). Handbook of Properties of Textile and Technical Fibres, 2nd Ed., Woodhead Publishing. Elsevier. https://doi.org/10.1016/B978-0-08-101272-7.00011-0.

Sultana, N. and M. Wang. 2008. Fabrication of HA/PHBV composite scaffolds through the emulsion freezing/freeze-drying process and characterization of the scaffolds. Journal of Materials Science: Materials in Medicine. 19: 2555–2561. https://doi.org/10.1007/s10856-007-3214-3:

Sultana, N. 2023. Biological properties and biomedical applications of pectin and pectin-based composites: a review. Molecules. 28: 7974.

Sultana, N., A. Cole and F. Strachan. 2024. Biocomposite scaffolds for tissue engineering: Materials, fabrication techniques and future directions. Materials. 17(22): 5577. https://doi.org/10.3390/ma17225577

Sun, J.L., K. Jiao, L.N. Niu, Y. Jiao, Q. Song, L.J. Shen, et al., 2017. Intrafibrillar silicified collagen scaffold modulates monocytes to promote cell homing, angiogenesis, and bone regeneration. Biomaterials. 113: 203–216.

Surav, S., P. Sharma, A. Kumar, Z. Tabassum, M. Girdhar, N. Mamidi, et al., 2024. Harnessing natural polymers for nano-scaffolds in bone tissue engineering: a comprehensive overview of bone disease treatment. Current Issues in Molecular Biology. 46(1): 585–611. https://doi.org/10.3390/cimb46010038.

Turnbull, G., J. Clarke, F. Picard, P. Riches, L. Jia, F. Han, et al., 2018. 3D bioactive composite scaffolds for bone tissue engineering. Bioactive Materials. 3(3): 278–314. https://doi.org/10.1016/j.bioactmat.2017.10.001.

Wang, H., R. Sun, S. Huang, H. Wu and D. Zhang. 2024. Fabrication and properties of hydroxyapatite/chitosan composite scaffolds loaded with periostin for bone regeneration. Heliyon. 10: e25832.

Wen, X., D. Shi and N. Zhang. 2005. Applications of nanotechnology in tissue engineering. *In*: Handbook of Nanostructured Biomaterials and Their Applications in Nanobiotechnology. 1: 1–23.

Wong, S.K., M.M.F. Yee, K.-Y. Chin and S. Ima-Nirwana. 2023. A review of the application of natural and synthetic scaffolds in bone regeneration. Journal of Functional Biomaterials. 14(5): 286. https://doi.org/10.3390/jfb14050286.

Xue, J., T. Wu, Y. Dai and Y. Xia. 2019. Electrospinning and electrospun nanofibers: Methods, materials, and applications. Chemical Reviews. 119(8): 5298–5415. https://doi.org/10.1021/acs.chemrev.8b00593.

Chapter **4**

Bioprinting the Bone

Tanmay Gupta[1], Amirjalal Jalali[1], Farnaz Azarnia[1], Subrata Bandhu Ghosh[2], Mohini Sain[1] and Sanchita Bandyopadhyay- Ghosh*[2]

[1]Department of Mechanical and Industrial Engineering, University of Toronto, Toronto, ON, Canada

[2] Engineered Biomedical Materials Research and Innovation Centre (EnBioMatRIC), Department of Mechanical Engineering, Manipal University Jaipur, Jaipur, Rajasthan, India

4.1 INTRODUCTION

The need for reconstructive procedures to address bone defects arising from a variety of causes including accidental fractures, osteoporosis, hereditary diseases, and cancer is substantial (Sun et al. 2022, Sheen et al. 2023). Although bone possesses an inherent ability to heal itself, the complete regeneration of critically sized defects often necessitates medical intervention. For minor defects, immobilization with casts or splints may suffice, but complex fractures or significant bone defects typically require the surgical implantation of bone substitutes to provide stabilization and promote regeneration (Park et al. 2021, Kim et al. 2020, Einhorn and Gerstenfeld 2015). Furthermore, research on the global incidence of bone fractures in 2019 indicated over 190 million new cases, with projections suggesting a continual rise in the future (Wu et al. 2021, Dare and Hu 2017, Gupta et al. 2019, Farr et al. 2017).

Traditional approaches to bone defect repair involve the use of implants, such as autografts and allografts. However, these methods face challenges including limited supply, the risk of complications, and the possibility of immune rejection (Sanz-Sánchez et al. 2022, Moy and Aghaloo 2019, Reznikov et al. 2018). On the other hand, fixation devices made from non-biodegradable materials like titanium and zirconia are not constrained by supply issues. Yet, their inert nature and lack

*For Correspondence: sanchitab.ghosh@jaipur.manipal.edu

of a structure that emulates native tissue can lead to difficulties integrating into the host site. Additional concerns include complications during implant removal, patient discomfort due to the weight of the materials, and the challenge of fabricating patient-specific shapes (Raffa et al. 2021, Huang et al. 2023, Dapunt et al. 2014).

In response, promising bone tissue engineering (BTE) techniques like three-dimensional bioprinting (3DBP) have emerged in the past decade as a potential means for creating constructs that closely mimic native bone anatomy, thus enhancing implant success rates (Crowley et al. 2013, Zeng et al. 2018, Ambikaipalan et al. 2013).

Recent advancements in 3D printing (3DP) science have enabled the production of scaffolds seeded with cells, facilitating precise control over construct architecture and bioactive cues. 3DBP further benefits from uniform cell distribution, precise control over porosity, and accurate deposition of biomaterials and cells in predefined structures. This technology facilitates the creation of tailor-made constructs for personalized therapy, real-time production during surgery, and the repair of intricate defects, such as craniomaxillofacial defects (Yu et al. 2019, Li et al. 2022).

Researchers have explored various biomaterials, including metals, polymers, ceramics, and natural materials, as bone-like biomaterials to optimize tissue regeneration. Key factors for scaffold biomaterials in bone tissue engineering encompass biocompatibility, degradability, and mechanical characteristics (Holzapfel et al. 2017).

Bone is a sophisticated tissue characterized by unique structural and mechanical properties, featuring hierarchical organizational levels. It comprises of an organic phase comprising primarily of collagenous proteins (like Col-1) along with a smaller fraction of non-collagenous proteins (proteoglycans, glycosaminoglycans etc.), in conjunction to an inorganic phase having calcium phosphate based chemistry (Fontcuberta-Rigo et al. 2023, Delmas and Malaval 1993, Feng 2009, Florencio-Silva et al. 2015). During our lifetime bone continually undergoes remodeling in response to diverse stimuli and is intricately associated with blood vessels in the periosteum. Both the processes of bone remodeling and healing involve a complex interplay of modulatory signals orchestrated by resident immune cells and extracellular matrix coordinating activities such as cellular migration, transdifferentiating, protein synthesis, biomineralization etc. (Szwed-Georgiou et al. 2023, Huang et al. 2021, Longoni et al. 2018, Löffler et al. 2023, Schlundt et al. 2018, Reznikov et al. 2018)

Bone repair scaffolds play a critical role in BTE by providing an environment conducive to cell adhesion, proliferation, and differentiation. Ideal bone repair scaffolds possess characteristics such as excellent biocompatibility, appropriate degradation rates, interconnected pore structures, and adequate mechanical strength (Karageorgiou and Kaplan 2005).

As such, designing biomaterials for bone tissue engineering is a complex endeavor necessitating a comprehensive understanding of the constituents of normal bone tissue. Additionally, continuous advancements in biomaterial scaffold development, including innovations in 3D printing technology, offer promising solutions for more effective bone tissue engineering. Despite the array of available methods, the overarching objective remains to print bone tissue effectively, aiming to alleviate patient suffering and enhance treatment outcomes (Rezwan et al. 2006). The current article briefly discusses the anatomy of bone along with the key factors involved in bone healing and remodeling. We then introduce the state of the art of the bioprinting techniques and materials reported for BTE, followed by a discussion on the expansion of the conventional 3DBP towards development of load bearing BTE scaffolds.

4.2 BONE ANATOMY, STRUCTURE AND HEALING

Bones are essential in the body's musculoskeletal system, serving key roles like protecting organs, generating blood cells, storing minerals, and providing structural support (Eliaz and Metoki 2017). They come in two macrostructural types: cortical (dense) and trabecular (porous) bone, with varying strengths. Cortical bone comprises about 80% of total bone mass and is denser and stronger than trabecular bone (Zioupos and Currey 1998). Trabecular bone, with its porous structure, facilitates interactions between bone and blood cells, crucial for haematopoiesis and homeostasis (Koester et al. 2008). Within bones, microscopic structures like bone lamellae and collagen fibers contribute to toughness and prevent crack propagation. Bones mainly consist of hydroxyapatite and collagen, along with other substances like sodium, magnesium, potassium, and various proteins. Cells like osteoblasts, osteocytes, osteoclasts, and bone lining cells play vital roles in bone maintenance, repair, and remodelling (Ke et al. 2014).

4.3 BONE TYPES

There are four main categories of bones, each with distinct characteristics. Long bones, such as the femur, humerus, and radius, have a long, tubular channel (diaphysis), a conoid-shaped metaphyseal segment near the epiphyseal plategrowth, beyond the growth plate. The diaphysis is primarily comprised of solid bone material, the metaphysis and epiphyses primarily contain trabecular bone, whereas long bones accommodate both yellow and red bone marrow, serving as a central location for blood cell production (Clarke 2008). In contrast, short bones, such as carpal and tarsal bones, exhibit a cube-like shape with comparable horizontal and vertical dimensions. They are primarily composed of trabecular bone, enveloped by a slender cortical bone layer. Flat bones like the mandible and calvaria are characterized by their thinness and structure, comprising a central layer of trabecular bone encased between two outer layers of cortical bone. Irregular bones, like the vertebrae and the hyoid bone, don't fit neatly into the previous three categories. They are primarily composed of trabecular bone and are covered by a thin layer of cortical bone. Regarding bone composition, bones are heterogeneously made up of inorganic minerals and other organic components (Wu et al. 2014). Composition of the bones exhibits a heterogeneous composition, comprising both inorganic minerals and an organic scaffold. The primary inorganic mineral in bone is hydroxyapatite ($Ca_{10}[PO_4]_6[OH]_2$), constituting most of the bone material. Hydroxyapatite contributes to bone rigidity and strength, playing a crucial role in defining its physical properties. The regulation of calcium, a key component of hydroxyapatite, is carefully managed by various hormones (Yazdanpanah et al. 2015, Florencio-Silva et al. 2015). Parathyroid hormone (PTH) stimulates bone resorption, releasing calcium from bone into the bloodstream. Conversely, calcitonin inhibits bone breakdown by deactivating osteoclasts. Calcitriol, another hormone, has multifaceted functions, including enhancing the uptake of calcium and phosphorus—crucial constituents of the inorganic structure within bones. Conversely, the organic matrix predominantly consists of type I collagen.

Comprising about 90% of the organic matrix, collagen type I is crucial for bone structure maintenance and influences various bone functions like apoptosis, cell proliferation, and differentiation. Additionally, approximately 5% of the composition includes glycosylated proteins, particularly small leucine-rich proteins, which contribute to cell proliferation, bone remodeling, and mineral deposition. Furthermore, growth factors such as transforming growth factor-beta (TGF-beta), growth hormone (GH), and insulin-like growth factors (IGF) serve as vital regulators and signalling markers, impacting bone homeostasis, remodeling, and matrix protein synthesis (CURREY 2002, Betts et al. 2021). Bone modeling is the process by which bones alter their shape or structure in response to physiological stress or mechanical forces. First proposed by German Surgeon Julius Wolff in 1892, this theory suggests that long bones adapt their shape to suit the stresses they encounter. Currently, it is recognized that strains exerted on bone tissue generate signals that regulate bone deposition and subsequent formation. Alterations in bone size or orientation entail the addition and removal of bone, facilitated by the activity of osteoblasts and osteoclasts, respectively (Plochocki et al. 2008, Einhorn 1998, Feng 2009).

4.4 BONE REMODELING

Bone remodeling is an ongoing, systematically orchestrated process focused on renewing healthy bone, preserving strength, maintaining mineral content, and regulating calcium and phosphate levels. This cyclic process involves the breakdown of old bone tissue and the deposition of a freshly produced protein-rich matrix, which subsequently undergoes calcification, as shown in Figure 4.1. The four primary phases of bone remodeling include:

1. **Cell Activation:** Involves recruiting and activating monocyte macrophages and osteoclast precursors from circulation. While turnover of bone is a continuous process in healthy individuals, variety of external and internal factors may affect osteoclastic activity and imbalance of which may result in increased bone loss and even osteoporosis (Yin et al. 2019). Moreover, paracrine signalling from activated T-cells and hypoxia-inducible factors (HIF) have been reported to regulate osteoclastic activity (Lorenzo 2017, Meng et al. 2021).

2. **Resorption:** Osteoclasts, under the influence of inflammatory markers such as interleukin (IL)-1 and IL-6, initiate bone resorption by reducing the pH within the bone microenvironment and releasing enzymes to degrade the organic matrix.
3. **Reversal:** Preosteoblasts are recruited, signalling the cessation of bone resorption and the onset of bone formation.
4. **New bone formation:** Recruited osteoblasts are responsible for producing a fresh collagen-rich organic matrix. As osteoblasts secrete this matrix, they become embedded within it, transitioning into osteocytes. These osteocytes maintain communication through a network, forming a functional syncytium. The culmination of bone remodeling is the formation of a new osteon, which enhances overall bone health and strength (Allen and Burr 2013, Porter et al. 2009, Toosi and Behravan 2020).

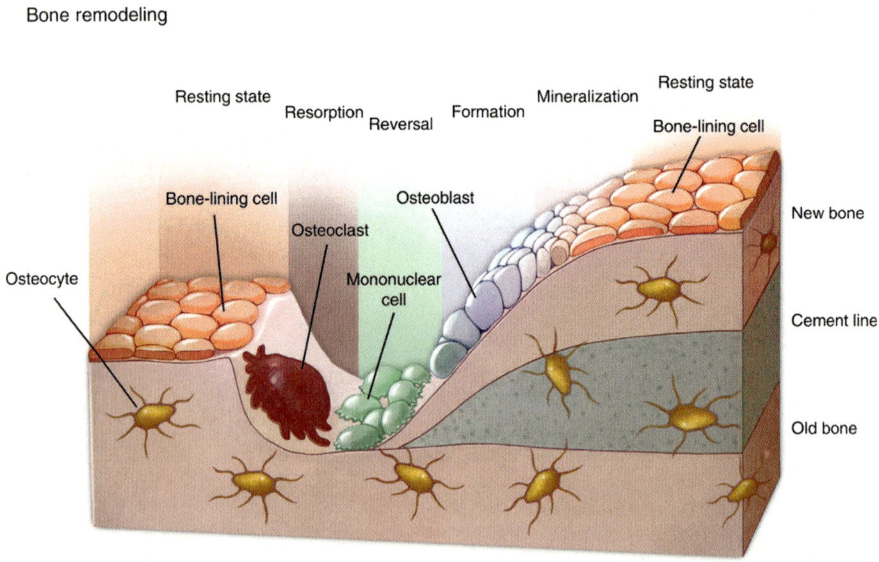

Figure 4.1 Overview of the various stages/process involved in bone remodeling (Yin et al. 2019) (Reproduced under Creative Commons Attribution 4.0 International License).

4.5 BONE HEALING

Depending on the nature of the fracture, the healing mechanism may be classified into two categories:

4.5.1 Primary Healing

Also known as direct healing, it involves the precise alignment and immobilization of fractured bone fragments under compression, thereby minimizing movement at the fracture site. Throughout this process, bone healing ensues through the direct remodeling of lamellar bone and Haversian canals. At the fracture site, cutting cones emerge at the edges of osteons, traversing the fracture line and generating longitudinal voids because of osteoclastic activity. Subsequently, osteoblasts fill these voids with bone matrix, culminating in the fusion of bones and the restoration of Haversian systems. This mode of healing facilitates the maturation and remodeling of osteons into lamellar bone, devoid of callus formation or inflammatory responses (Einhorn 1998).

4.5.2 Secondary Healing

Secondary or indirect healing is the more prevalent mechanism than primary healing and involves both intramembranous and endochondral ossification. This process unfolds through three main stages as represented in Figure 4.2.

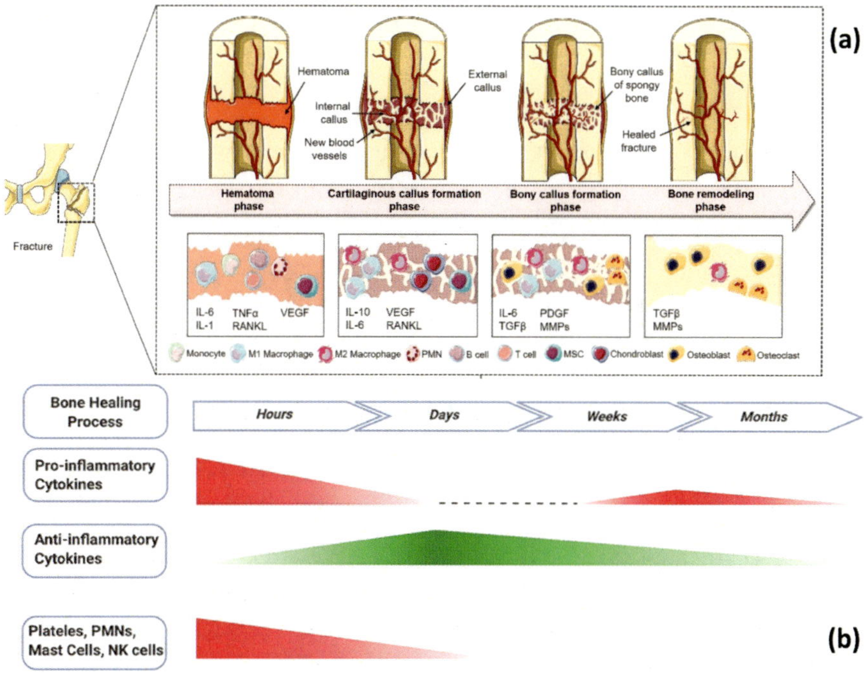

Figure 4.2 The key steps in endochondral ossification. (a) Three phases in bone repair, cartilaginous callus and bony callus formation collectively represent the repair phase (Pfeiffenberger et al. 2021) (Reproduced under Creative Commons CC-BY licence (CC-BY 4.0), (b) the timeline of the relative expression of inflammatory and anti-inflammatory cytokines along with the activation of various immune cells (Maruyama et al. 2020) (Reproduced under Creative Commons CC-BY licence (CC-BY 4.0)).

Acute Inflammatory Phase

This phase is initiated in response to trauma by hematoma formation. The damage to the peripheral nerves further triggers a cascade of cellular/molecular events including the initiation of neuronal damage repair, activation of resident macrophages and the upregulation of inflammatory markers such as TNF-α, IL-6 and IL-1β (Huang et al. 2021b, Nazzal et al. 2024, Tao et al. 2023). The inflammatory phase is further accompanied by vasodilation and recruitment of immune cells like monocytes, lymphocytes, and macrophages which facilitate the elimination of necrotic tissue.

After a period of a few days, gradual transition of pro-inflammatory signals to anti-inflammatory signals commences with the upregulation of cytokines like IL-4, IL-10, TGF-β and VEGF-α, (Lavorato et al. 2021, Hellwinkel et al. 2020) along with transition macrophages to M2 phenotype (Frade et al. 2023, Mass et al. 2023). Overall, these changes support angiogenesis. A robust vasculature within the defect site is not only crucial for resolving the hypoxic conditions, but also for recruitment of haematopoietic cells (like MSCs) necessary for bone healing (Keramaris et al. 2008, Marenzana and Arnett 2013). Consequently, a significant effort has been invested in enhancing angiogenesis in BTE scaffolds (Simunovic and Finkenzeller 2021).

Repair Phase

The repair phase entails the transition of the haematoma towards callus formation. Depending on the factors, such as location and extent of hypoxia, callus formation may occur from either intramembranous ossification or endochondral ossification as represented in Figure 4.3 (Kenkre and Bassett 2018).

Reparative activities near the periosteum are dominated intramembranous ossification processes. Herein, the recruited periosteal MSCs differentiate to osteochondral progenitor cells, followed by differentiation directly into osteoblastic lineage that directly form immature (or woven) bone. Incidentally, intramembranous ossification closely resembles primary healing mechanism.

Endochondral ossification, in contrast, occurs at the bone marrow and endosteum regions of the bone, via initial formation of a soft, cartilaginous callus followed by its mineralization to form a hard callus. Herein, the recruited MSCs differentiate into chondrocytes which deposit a collagen-dense fibrocartilaginous network that acts as a template for subsequent ossification. Next, the chondrocytes undergo hypertrophic differentiation, which initiate mineralization of the callus. At this stage, the chondrocytes undergo apoptosis, and the pro-inflammatory markers are eliminated, which triggers vascular invasion and migration, proliferation of osteoblasts through a complex interplay of growth factors and other paracrine signals mediated by osteoblasts, immune cells and osteoclasts (Maruyama et al. 2020, Hellwinkel et al. 2020). Progression of periosteal MSCs during the repair phase of bone healing is further depicted in Figure 4.3.

Bone Remodeling

In the ultimate phase, the osteoclasts and osteoblasts facilitate restructuring of the bony callus into mature bone. This is crucial since the hard callus only provides a rigid structure contributing to biomechanical stability, it does not fully restore the biomechanical properties of normal bone. Typically, trabecular bone develops at the centre, while cortical bone forms at the periphery. This remodeling stage may span from months to years.

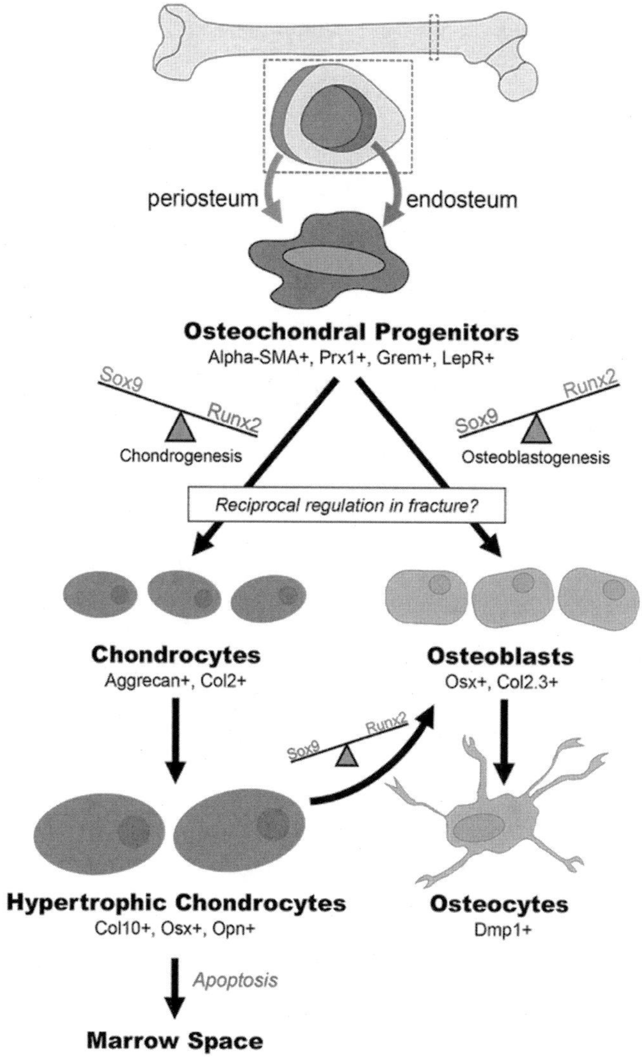

Figure 4.3 Progression of periosteal MSCs during the repair phase of bone healing (Bahney et al. 2019) (reproduced with permission from John Wiley & Sons, Inc.)

Understanding the different phases of bone healing and their timelines offers valuable insights into treatment strategies, which may prove instrumental for the overall success of the BTE scaffolds. Consequently, driving the research interest in studies aimed at improving vasculature, and resolution of inflammation (Budharaju et al. 2024, Kolesky et al. 2016). Moreover, BTE strategies employing developmentally inspired scaffolds may be particularly advantageous (Daly et al. 2016).

4.6 BONE BIOPRINTING: FABRICATION TECHNIQUES

Various methods are used in bioprinting to create scaffolds. Inkjet, laser-assisted, and microextrusion techniques have become the main methods in use (Li et al. 2021).

Inkjet bioprinting operates by using either thermal or piezoelectric mechanisms to propel tiny droplets of the bioink through the printer nozzle (Groll et al. 2019). Thermal inkjet bioprinting is advantageous due to its rapid printing speeds and low operational costs. Although there are concerns about the possible negative effects of heat and mechanical stress on cells, studies have demonstrated high cell viability rates. This is attributed to the short exposure of the bioink to heat, usually less than 2 μs (Ashammakhi et al. 2019, Gao et al. 2015, Li et al. 2021).

As an example, inkjet bioprinting of peptide-conjugated PEG bone scaffolds containing human mesenchymal stem cells (hMSCs) showed remarkable cell viability levels, reaching approximately 88% (refer to Table 4.2) (Gao et al. 2015). In a similar vein, thermal inkjet bioprinting of scaffolds loaded with human mesenchymal stem cells (hMSCs), consisting of poly(ethylene glycol) di-methacrylate (PEGDMA) combined with bioactive glass (BG) and hydroxyapatite (HAp) nanoparticles (nHAp), yielded a commendable cell viability rate of 86.7% (Gao et al. 2014). In a separate study focusing on BTE, thermal inkjet bioprinting of various collagen I/agarose hydrogels incorporating mesenchymal stem cells (MSCs) exhibited exceptional cell viability, measuring around 98% after 21 days of culture (Duarte Campos et al. 2016). Moreover, inkjet printing has been validated as a safe method for handling delicate substances like growth factors. This was exemplified by the successful immobilization of BMP-2 using inkjet printing, facilitating the creation of stem cell microenvironments with spatially defined patterns (Phillippi et al. 2008). Thermal inkjet bioprinters encounter limitations in terms of precision concerning droplet size and shape, along with constraints on using biomaterials that are not resistant to heat or mechanical stress. Conversely, piezoelectric-driven inkjet bioprinters present potential solutions to these limitations. Nonetheless, concerns have been raised regarding potential cell damage at frequencies of 15–25 kHz specific to piezoelectric-driven inkjet bioprinters (Li et al. 2021).

Laser-assisted bioprinting (LAB) stands out for its nozzle-less method and comprises three core elements: a laser emitter, a ribbon coated with an absorbing layer (such as gold or titanium) holding the bioink, and a collector located beyond the ribbon (Bishop et al. 2017). The procedure starts with the bioink being suspended beneath the ribbon. Laser-induced evaporation is then initiated as the laser beam moves across the ribbon. This leads to the propulsion of individual droplets onto the collector, driven by the formation of vapor bubbles due to increased pressure (Bishop et al. 2017). LAB presents numerous benefits, such as the capacity to print high cell densities, preserve excellent cell viability, accomplish swift printing speeds, and achieve a superior level of printing resolution (Li et al. 2021). This innovative technology promises precise manipulation of cell density at the individual cell level, empowering the modulation of cell functionality

(Keriquel et al. 2017). Recent demonstrations have showcased successful *in-situ* bioprinting of mesenchymal stromal cells encapsulated within a collagen/nHAp matrix onto a mouse calvarial bone defect using LAB (refer to Table 4.2) (Keriquel et al. 2017). Additionally, research has shown that LAB is an efficient technique for producing bioprinted scaffolds consisting of nHAp and osteoblastic cells. Notably, no changes were observed in the physicochemical properties of nHAp or in the viability, proliferation, and phenotype of osteoblastic cells over a period of up to 15 days (Catros et al. 2011). Although offering promise for tissue engineering applications, LAB has seen limited utilization in BTE research. This minimal application might be attributed to various challenges, including the intricate ribbon preparation process, potential residue concerns related to metallic components in the scaffold, and elevated production expenses (Mandrycky et al. 2016, Bishop et al. 2017).

Within microextrusion technique, the bioink resides within a cartridge. It is then propelled onto a platform using either pneumatic or mechanical forces, including screw- or piston-based mechanisms (Naghieh and Chen 2021, Ashammakhi et al. 2019). Recognized as the predominant method for constructing bioprinted bone scaffolds, microextrusion offers manifold benefits. It boasts versatility in accommodating diverse biomaterials, ranging from soft hydrogels to synthetic polymers and polymer/ceramic composites. Moreover, microextrusion excels in achieving high cell deposition densities (Li et al. 2021, Mandrycky et al. 2016). This process is capable of handling biomaterials with viscosities ranging from 30 mPa.s to over 6×10^7 mPa.s (Mandrycky et al. 2016, Ashammakhi et al. 2019) and can employ multiple nozzles for separate biomaterial deposition (Izadifar et al. 2016), or co-axial nozzles for simultaneous deposition, such as in core/shell designs (Raja and Yun 2016). Microextrusion has been extensively utilized for bioprinting hydrogels (Table 4.2), including alginate, chitosan, gelatin, collagen, carboxymethyl chitosan, Methacrylamide Gelatin (MG), Methacrylated Hyaluronic Acid (MeHA), and cell-laden composite scaffolds (Bendtsen et al. 2017). Various cell types, including MC3T3-E1, MG-63, BMSCs, Dental Pulp Stem Cells (DPSCs), human Amniotic Fluid Stem Cells (hAFSCs), and human Tooth-derived Mesenchymal Stem Cells (hTMSCs), have been successfully bioprinted using this technique (Demirtaş et al. 2017, Du et al. 2015). However, microextrusion bioprinters typically offer moderate resolution, ranging between 50 and 500 μm [52], [59], and may subject cells to shear stress during the printing process, potentially affecting cell structure and viability. Nevertheless, strategies such as reducing extrusion pressure or increasing needle size can mitigate these issues. Figure 4.4 summarizes the types of bioprinting techniques most commonly used.

4.7 BONE BIOPRINTING: MATERIALS

The process of selecting biomaterials plays a vital role in the development of bone-engineered constructs, as the characteristics of these materials directly impact scaffold properties. While a wide range of both natural and synthetic biomaterials have been utilized in scaffold fabrication for bone regeneration, not all are conducive to cell embedding (Salgado et al. 2004, Zhang et al. 2017). As a result, only a limited number of biomaterials are presently available for effectively encapsulating cells.

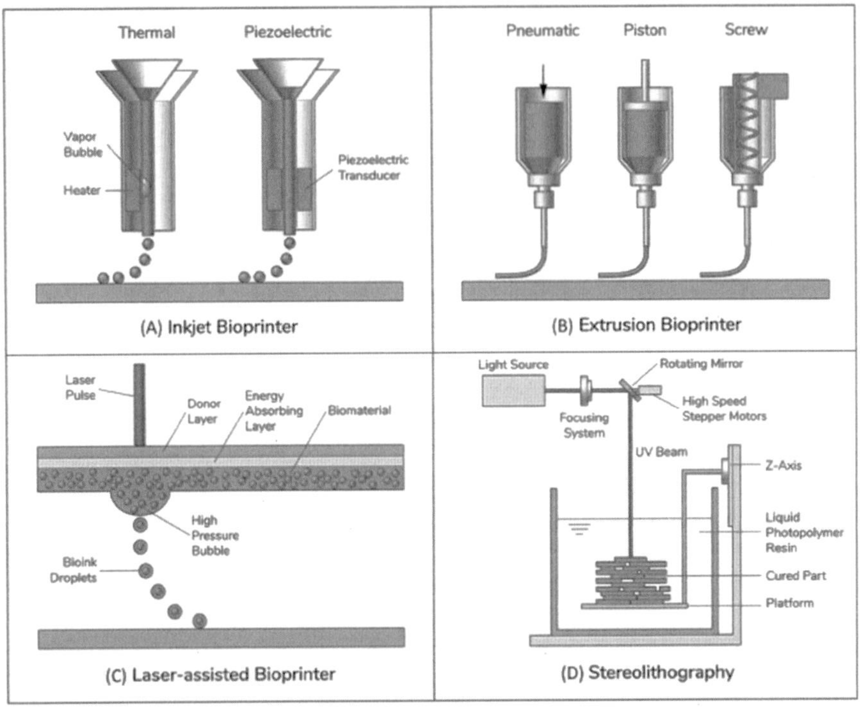

Figure 4.4 Prevalent types of bioprinting techniques used (A) inkjet bioprinting, (B) extrusion or microextrusion bioprinting, (C) laser-assisted bioprinting, and (D) stereolithography bioprinting (Reproduced with permission from Elsevier Inc. (Teixeira et al. 2022)).

Within the domain of bioprinting, hydrogels have risen to prominence as a favored option for serving as matrices for cell encapsulation. This is primarily due to their biocompatible and biodegradable nature, resemblance to the native extracellular matrix (ECM), and ability to create a hydrated environment. Such characteristics are conducive to facilitating the diffusion of vital elements such as nutrients, oxygen, and waste products necessary for cellular proliferation (Salgado et al. 2004, Bishop et al. 2017, Naghieh and Chen 2021). In hydrogels designed for bioprinting applications, meeting several crucial criteria is imperative (Du et al. 2015, Huang et al. 2016, Salgado et al. 2004): (1) they should exhibit suitable viscoelastic properties favorable to the formation of durable strands. (2) They should maintain structural integrity throughout the entirety of the 3D printing process, guaranteeing that open pores persist and do not fail. (3) They should display nontoxicity towards cells, thus facilitating the proliferation and differentiation of encapsulated cells.

A diverse array of natural hydrogels, including gelatin, chitosan, alginate, hyaluronic acid (HA) and collagen are commonly utilized in bioprinted products

(Huang et al. 2016). Table 4.1 provides an overview of the benefits and potential limitations associated with these natural hydrogels, specifically focusing on their applicability for bioprinting bone scaffolds. Alginate is notably favored among these options for bioprinting bone scaffolds, primarily attributed to its adjustable degradation rate and crosslinking capabilities, which allow for the creation of diverse shapes such as microspheres and fibers (Penolazzi et al. 2010, Loozen et al. 2013, Kim and Kim 2015, Raja and Yun 2016).

Table 4.1 Bioprinted bone scaffolds fabricated with various bioprinting techniques

Bioink	Printing procedure	Cell	Reference
Alginate, HAp, PVA, and Collagen	Microextrusion	MC3T3-E1	(Bendtsen and Wei 2017)
Alginate, PVA and HAp	Microextrusion	MC3T3-E1	(Bendtsen, Quinnell and Wei 2017)
Alginate, chitosan, Alginate, HAp,	Microextrusion	MC3T3-E1	(Demirtaş et al. 2017)
Alginate	Microextrusion	MC3T3-E1	(Raja and Yun 2016)
Alginate, gelatin, carboxymethyl and chitosan	Microextrusion	BMSCs	(Huang et al. 2016)
MG hydrogel	Microextrusion	BMSCs	(Du et al. 2015)
Matrigel and alginate	Microextrusion	EPCs	(Poldervaart et al. 2014)
Matrigel	Microextrusion	ASCs	(Murphy et al. 2017)
Gelatin, silk fibroin	Microextrusion	hTMSCs	(Das et al. 2015)
Alginate	Microextrusion	MG63	(Y. B. Kim et al. 2016)
Alginate, gelatin, and collagen	Microextrusion	DPSCs	(Park et al. 2015)
Alginate	Microextrusion	MC3T3-E1	(Ahn et al. 2013)
Alginate, gelatin, and MSCs	Microextrusion	HUVECs MSCs	(Chen et al. 2018)
Acrylated PEG and acrylated peptide	Inkjet	hMSCs	(Gao et al. 2015)
PEGDMA with nHAp and BGs	Inkjet	hMSCs	(Gao et al. 2014)
Collagen type I and agarose	Inkjet	MSCs	(Duarte Campos et al. 2016)
Collagen and nHAp	LAB	MSCs	(Keriquel et al. 2017)
Human osteoprogenitor cells and nHAp	LAB	Human osteoprogenitor	(Catros et al. 2011)
Alginate and gelatin	Microextrusion	SaOS-2	(Neufurth et al. 2014)
Alginate and gelatin	Microextrusion	HUVECs	(Akkineni et al. 2016)

Despite its advantages, challenges persist in Bone Tissue Engineering (BTE) with the use of alginate. These include the rapid decline in mechanical strength during *in-vitro* culture and the limited control over internal microstructure because of its inherent hydrophilicity (Kim and Kim 2015). Research indicates that alginate undergoes a decrease in mechanical strength by approximately

40% following 9 days of *in-vitro* culture (Shoichet et al. 1996). Furthermore, alginate presents restrained bioactivity owing to its deficiency in cell-binding sites essential for facilitating cell adhesion.

To tackle this challenge, attempts have been made to enhance alginate by incorporating cell-adhesive ligands or blending it with other biomaterials. This approach aims to bolster cell responses within bone scaffolds (Luo et al. 2018, Sarker et al. 2019, Giuseppe et al. 2018). Surface modification techniques, such as coating with arginine-glycine-aspartate peptides, have been employed to enhance the bioactivity of alginate (Sarker et al. 2019). Moreover, the combination of alginate with other hydrogels such as chitosan and gelatin has fostered an environment conducive to the bioprinting of Bone Marrow-derived Mesenchymal Stem Cells (BMSCs) and Mesenchymal Dental Pulp-derived Stem Cells (DPSCs) (Park et al. 2015, Huang et al. 2016). The combination of alginate with Matrigel, a gelatinous protein mixture, has been also reported for the encapsulation of Endothelial Progenitor Cells (EPCs) (Poldervaart et al. 2014).

Gelatin emerges as another promising choice for bolstering cellular functions, such as proliferation, attachment, and differentiation, ascribed to its possession of cell-ligand patterns (Luo et al. 2015, Milazzo et al. 2019). Furthermore, the thermos-responsive characteristics of gelatin have rendered it a well-established macromolecular choice for bioinks formulation (Milazzo et al. 2019). However, gelatin is marked by a high degradation kinetics and insufficient mechanical strength, posing challenges for BTE (Gautam et al. 2014). Consequently, gelatin is frequently utilized in conjunction with other biomaterials such as chitosan, alginate and similar substances to function as a cell carrier in bioprinting systems particularly designed for BTE applications (Kang et al. 2016, Das et al. 2015).

Methacrylamide gelatin (MG), a modified form of gelatin, has exhibited notable promise in cell printing applications, boasting exceptional cell viability rates surpassing 97% (Billiet et al. 2014) and this underscores its effectiveness as a synthetic hydrogel in the realm of BTE (Du et al. 2015). Moreover, MG has proven to be a suitable substrate for BMSCs and the growth factor bone morphogenetic protein-2 (BMP-2) (Du et al. 2015). Another derivative of gelatin, known as gelatin methacryloyl (GelMA), has been employed in coating titanium implants, resulting in improved osteointegration. This application underscores its potential suitability as a bioink for Bone Tissue Engineering (BTE) endeavors (McBeth et al. 2017). A variety of crosslinking techniques have been utilized to enhance and fine-tune the mechanical characteristics of gelatin. These methods include the use of chemical agents such as 1-ethyl-3-(3-dimethylaminopropyl) carbodiimide (EDC), glutaraldehyde, or genipin (Huh et al. 2018, Martínez-Vázquez et al. 2015, Akkineni et al. 2016).

Chitosan is recognized as an exceptionally favorable biomaterial for a wide array of biomedical applications. This cationic polymer is characterized by a hydrophilic surface, which promotes cell adhesion, proliferation, differentiation, and viability. The notable charge density of chitosan in solution facilitates the creation of insoluble ionic complexes when combined with water-soluble anionic polymers like alginate (Zhu et al. 2011, Bakhsheshi-Rad et al. 2019, Francis Suh and Matthew 2000, Muzzarelli et al. 2012). Chitosan has also exhibited

Table 4.2 Benefits and potential limitations of natural hydrogels commonly used in bioprinted bone scaffolds

Hydrogel Material	Advantages	Limitations	References
Alginate	• Cost-effectiveness • Processability • Excellent biocompatibility • Rapid gelation • Robust biodegradability and minimal • Immunological activation • Fulfilling cell viability and osmolar needs of cells • Great crosslink-ability • Great rheological properties	• Biologically inert • Limited extended stability • Rapid decline in mechanical features during *in-vitro* culture • Restricted 3D shape adaptability	(Luo et al. 2017, Park et al. 2017, Aldaadaa et al. 2018, Chimene et al. 2016, You et al. 2016, Narayanan et al. 2016, Giuseppe et al. 2018, Milazzo et al. 2019, Neufurth et al. 2014)
Gelatin	• Faster gelation time • Effective in reversible thermal gelation • Excellent biocompatibility • Great biodegradability	• Poor mechanical performance • Accelerated degradation rate	(Luo et al. 2015)
Chitosan	• Mimicking native tissue ECM • Non-hazardous by-products • Great cell adhesion proliferation, differentiation, and viability	• Prolonged gelation rate • Poor mechanical performance • Potentially interfering with printing of cells and pH-sensitive substances	(Muzzarelli et al. 2012, Francis Suh and Matthew 2000, Demirtaş et al. 2017)
Collagen	• Minimal immunogenic response • Excellent biocompatibility • Great biodegradability • Modulating cell adhesion and differentiation	• Poor mechanical performance • Structural deterioration • Reduced fluidity and prolonged gelation	(Rhee et al. 2016, Turnbull et al. 2018, Mohseni et al. 2018)
Hyaluronic acid	• Good biocompatibility • Non-hazardous by-products • Viscoelastic features • Hydrophilic • Antimicrobial characteristics	• Poor mechanical performance	(Zhu et al. 2017, Zhai et al. 2020, Kim et al. 2007, Ashammakhi et al. 2019)

considerable promise as a carrier for rabbit BMSCs and the growth factor BMP-2 within hybrid scaffolds composed of polycaprolactone (PCL) and chitosan. This blend has resulted in the creation of a biomimetic microenvironment characterized by improved cell retention, growth, and distribution (Dong et al. 2017).

Collagen possesses a plethora of remarkable characteristics, including low immunogenicity, permeability, enhanced biocompatibility, and biodegradability. These attributes endow collagen with the potential to regulate cellular morphology, adhesion, migration, and differentiation (Mohseni et al. 2018, Turnbull et al. 2018). Of the various types of collagen, Collagen type I predominates as the preferred form for integration into bioprinted bone scaffolds (Semba et al. 2020). Incorporating Collagen type I into alginate-polyvinyl alcohol (PVA)-hydroxyapatite (HAp) matrices has been demonstrated to offer essential adhesion ligands, fostering the attachment and proliferation of preosteoblast cells (MC3T3-El) (Bendtsen et al. 2017). Coating scaffolds with collagen has also proven effective in enhancing the adhesion, proliferation, and osteogenic differentiation of osteoblast-like cells (MG63), along with facilitating the osteogenic differentiation of human adipose-derived stem cells (hADSCs) (Vandrovcová et al. 2011, Linh et al. 2020).

While Collagen type I exhibits slow gelation kinetics and low viscosity, these characteristics render it less ideal for bioprinting bone scaffolds (Ashammakhi et al. 2019). Efforts have been undertaken to develop a bio-printable form of Collagen type I by incorporating agarose into its matrix. This innovative approach enables the printing of mesenchymal stem cells (MSCs), overcoming the challenges associated with aforementioned challenges of Collagen type I (Duarte Campos et al. 2016). Moreover, the combination of Collagen type I with hyaluronic acid (HA) has resulted in the development of osteochondral scaffolds that harbor an extracellular matrix (ECM) conducive to osteoblast activity (Park et al. 2014).

Hyaluronic acid (HA) is a hydrophilic natural polymer that serves as a fundamental component of the extracellular matrix (ECM) in mammalian connective tissues. (Bae et al. 2011). The HA's potential for biomedical applications is highlighted by its biocompatibility, production of nontoxic degradation by-products, viscoelastic properties, and ability to retain water, thus ensuring tissue hydration (Ashammakhi et al. 2019, Duarte Campos et al. 2016). Furthermore, HA's antimicrobial properties make it a promising candidate for implantation in bone defects (Zhai et al. 2020). Despite its advantages, the mechanical fragility of HA hydrogel renders it susceptible to degradation in aqueous environments. To address this issue, modified forms of the HA network, such as acrylated HA followed by ultraviolet (UV) light crosslinking, have been employed. This approach allows for the fabrication of a matrix with customizable mechanical and degradation properties, ideal for encapsulating cells and growth factors in BTE endeavors (Kim et al. 2007). Another promising option in BTE is photo-crosslinked methacrylated hyaluronic acid (MeHA), a chemically modified variant of HA. *In-vitro* studies have demonstrated favorable levels of cell viability of around 65% and osteogenic differentiation when utilizing MeHA as a matrix for bioprinting. Furthermore, combining HA with methylcellulose has shown promising capability for bioprinting MSCs (Law et al. 2018, Du et al. 2015).

Numerous natural hydrogels share a common limitation characterized by insufficient mechanical properties that are unable to mimic those of native bone (Baino et al. 2015). To address this challenge, endeavors have been focused on improving the mechanical performance of bioprinted scaffolds by incorporating synthetic polymers (Gao et al. 2014). As an illustration, bioprinted bone scaffolds comprising poly(ethylene glycol) dimethacrylate (PEGDMA) and hydroxyapatite (HAp) showcased an elastic modulus of approximately 359 kPa. Although this figure falls short of replicating the mechanical properties of natural bone, it surpasses the strength of natural polymers, which typically exhibit moduli below 5 kPa. Another notable polymer blend used in bioprinting to create mechanically durable constructs involves poly(lactic-co-glycolic acid) (PLGA) and poly(ethylene glycol) (PEG). These constructs possess an elasticity modulus of approximately 57 MPa. When incorporating immortalized human mesenchymal stem cells (hMSCs), these constructs demonstrate promise for applications in bone repair (Sawkins et al. 2015).

4.8 LOAD-BEARING 3D BIOPRINTING

Load-bearing 3D bioprinting (LB-3DP) stands out as a specialized approach in bioprinting, focusing on the fabrication of constructs capable of withstanding mechanical loads, particularly for applications in load-bearing bone implants (Gupta et al. 2023). This technique differs from conventional bioprinting methods by prioritizing the development of scaffolds with enhanced mechanical properties to provide structural support in load-bearing environments.

In conventional bioprinting, the emphasis lies on replicating biological structures using hydrogel-based bioinks that lack the necessary strength for load-bearing applications (Zhang et al. 2021). LB-3DP, on the other hand, addresses this limitation by incorporating specialized load-bearing inks or materials designed to mimic the mechanical properties of native tissues, such as bones, ensuring adequate support and functionality. This is realized by employing multi-material 3DBP, such that the material properties are decoupled. For example, LB-3DBP may involve at least one cell-laden bioink co-printed with a load-bearing or stiff phase (Gupta et al. 2023). Typical materials reported for load-bearing phase have been based on phosphate cement-based chemistry or biocompatible thermoplastics (such as, PLA and PLGA) (Krujatz et al. 2022, Yang et al. 2015, Romanazzo et al. 2021, Cunniffe et al. 2017).

One of the key challenges in LB-3DP is overcoming limitations in material properties. Traditional bioprinted constructs may exhibit poor mechanical strength, hindering their effectiveness in load-bearing scenarios (Fang et al. 2021). To address this, LB-3DP integrates materials specifically tailored to offer superior mechanical properties, such as high stiffness and toughness, essential for supporting physiological loads (Gupta et al. 2023). The load-bearing materials in LB-3DBP must allow printing as well as hardening/curing in 'mild' conditions, thereby limiting the library of load-bearing inks feasible. For, example thermoplastics with lower melting point (such as PCL and PLGA) are preferred to those with higher melting thermoplastics like PLA which may require print temperatures >200°C. Similarly, conventional bioceramics 3DP using selective laser sintering (SLS) or binder jetting (followed by sintering) may not be feasible owing to high temperature and/or use of acidic binders (Lin et al. 2019, Dorozhkin 2009, Xu et al. 2017).

LB-3DP opens a plethora of potential applications in orthopedic regenerative medicine by enabling the creation of personalized load-bearing bone implants with tailored mechanical properties. These implants have the potential to significantly improve outcomes for bone fractures, defects, and other orthopedic conditions by closely matching the patient's anatomy and biomechanical requirements. A qualitative comparison on conventional 3DP, bioprinting and LB-3DBP (designed as biphasic bioprint here) is presented in Figure 4.5.

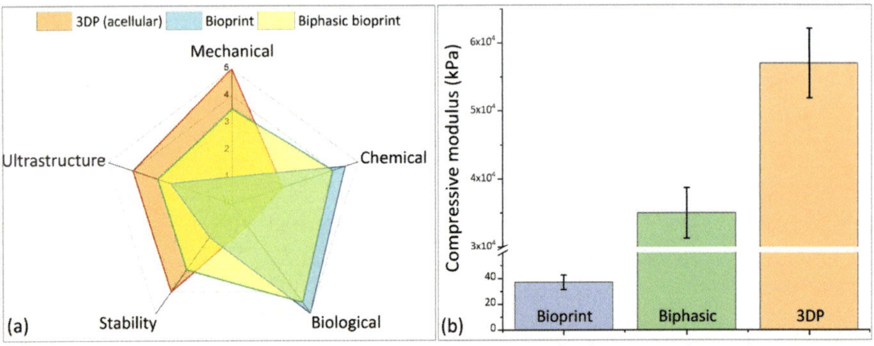

Figure 4.5 (a) Qualitative comparison of mechanical properties, ultrastructure (representative of morphological mimicry and print resolution), stability (in physiological medium), biological and chemical biomimicry capabilities of acellular 3DP, bioprint and biphasic scaffolds. b) Comparison of compressive strength of conventional acellular 3DP, bioprint and biphasic scaffolds (Gupta et al. 2023) (reproduced under Creative Commons Attribution 4.0 license).

By surpassing the limitations of conventional bioprinting methods and incorporating specialized materials, LB-3DP offers innovative solutions for creating robust constructs capable of withstanding physiological stresses in load-bearing environments. Subsequent discussions introduce the different class of materials used for such applications, while comparing the opportunities and pitfalls using them.

4.9 THERMOPLASTIC MATERIALS FOR LB-3DBP

Thermoplastic biomaterials play a significant role in load-bearing 3D bioprinting (LB-3DBP) due to their biodegradability, mechanical properties, and 3D printability. Specifically, the use of thermoplastics like poly(lactic-co-glycolic acid) (PLGA), polylactic acid (PLA), and polycaprolactone (PCL) has been explored for creating load-bearing implants through various printing methods (Tavafoghi et al. 2021, Borkar et al. 2021, Kim et al. 2016a, Christau et al. 2022, Xu et al. 2022). The

3DP inks from these biopolymers have two modalities i.e. those based on melt extrusion or solution-based formulations printed at room temperature (in organic solvents such as dichloromethane or 2,2,2-Trifluoroethanol) (Cao et al. 2022, Gharibshahian et al. 2023, Adarkwa et al. 2023).

Even so, pristine biopolymers are hydrophobic and exhibit slow degradation rate (Woodruff and Hutmacher 2010, Lee et al. 2013), in addition to PLGA reportedly resulting in prolonged inflammatory response (Chor et al. 2022), thereby compromising their biocompatibility and osteogenic behavior. Incorporation of biomaterials, such as bioceramics (tricalcium phosphate, TCP; hydroxyapatite, HA; bioglass; fluorcanasite etc.) (Thamma et al. 2021, Vyas et al. 2022, 2024, Jodati et al. 2020, Verma et al. 2023) or biopolymers (like gelatin, alginate) (Chimene et al. 2020, Chen et al. 2023a), aid in partially addressing issues related to hydrophilicity and slow degradation rate.

Another concern in studies involving melt extrusion is heat buildup; it is essential to ensure rapid heat dissipation towards preventing cell necrosis (Kang et al. 2016). Although initial study involving osteoblasts reported post-print cell viability of 90% (approx.) (Shim et al. 2016), follow up studies employed use of ultrasonic humidifier for thermal management and improvement of cellular viability (Lee et al. 2013). In another study, Kundu et al. demonstrated significant reduction in viability of chondrocytes (within alginate bioinks) for multilayer PCL/alginate scaffolds when compared to single layer scaffolds (~85% versus 95% in single layer prints) (Kundu et al. 2015). It is anticipated that the issue of heat build would intensify with increase in size and complexity of the print.

4.9.1 Extrusion-based Bioprinting with Thermoplastics

Among thermoplastics, PCL stands out for its low melting temperature (~60°C), making it suitable for 3D printing applications where minimizing thermal shock on printed cells is crucial. Extrusion-based 3D printing of PCL and its composites involve printing temperatures ranging from 80 to 100°C, ensuring cell viability during the process (Borkar et al. 2021). Bone scaffolds with bioceramics, such HA and TCP, incorporated in PCL have been widely explored for bone tissue engineering with favorable osteogenic response along with improved cellular adhesion, proliferation, and faster biodegradation (Neufurth et al. 2017, Zhang et al. 2022). Similar results were reported when PCL biocomposites were co-printed with appropriate bioinks 3DBP (Yao et al. 2015, Kang et al. 2016). Another study focussed on LB-3DBP reported blends of alginate and PCL as the load-bearing phase (Kim et al. 2016b). Besides improving the hydrophilicity of stiff phase, this resulted in an increased porosity on account of the dissolution of alginate, thereby improving the overall biological performance of the scaffold without significantly compromising its mechanical properties.

Despite their dynamic capabilities, extrusion-based bioprinting is limited by the rheological property requirement of the bioinks, high-shear stresses during printing, low cellular density, slow print speed and poor print resolution (Tavafoghi et al. 2021), consequently driving the research interest towards other forms of bioprinting and LB-3DBP.

4.9.2 Hybrid Bioprinting Methods with Thermoplastics

Hybrid bioprinting methods represent an alternative strategy to conventional multi-head extrusion bioprinters. Herein, an extrusion head is combined with other 3DP approaches like laser-assisted or inkjet-based printing to enhance the capabilities of the bioprinter. These hybrid methods enable multi-material 3D printing by integrating extrusion with SLA or inkjet-based techniques. For instance, hybrid printers can mechanically reinforce bioinks with PCL filaments, expanding the range of printable materials and bioinks available for both SLA and extrusion bioprinting (Shanjani et al. 2015, Cui et al. 2016, Daly and Kelly 2019).

Such bioprinters offer several advantages over extrusion-based methods when working with thermoplastics. It allows for higher resolution and cellular density compared to traditional extrusion printing. Hybrid printers can achieve z-axis resolutions as low as ~35 μm while maintaining cellular viability around 90%. Additionally, hybrid bioprinters facilitate the creation of complex structures like lumen-like channels within scaffolds, essential for vascularization and fluid exchange in implants (Shanjani et al. 2015).

Hybrid bioprinting stands out for its ability to combine the strengths of different printing techniques to overcome limitations associated with individual methods when working with thermoplastics in LB-3DBP applications. Nevertheless, these techniques are still in their early stages of development and issues, such as inter-material compatibility and poor print speed, need to be addressed before they can finder wider acceptance.

4.10 BIOCERAMIC MATERIALS FOR LB-3DBP

Bioceramics, such as hydroxyapatite (HA) and tricalcium phosphate (TCP), have emerged as attractive materials for bone tissue engineering applications due to their excellent biocompatibility, osteoconductivity, and ability to promote bone regeneration (Zhang et al. 2022). These ceramic materials closely mimic the mineral composition of natural bone, making them well-suited for the development of bone grafts and implants (Dorozhkin 2016, 2009).

Use of bioceramics into 3D bioprinting has been a significant area of research, as it allows for the fabrication of complex, patient-specific bone tissue constructs. By incorporating these bioceramics (as micro or nanoparticles) in hydrogel-based bioinks, researchers have been able to create 3DBP that actively encourages the growth and differentiation of bone cells (Qin and Wu 2022, Heid and Boccaccini 2020). Nevertheless, such scaffolds suffer poor mechanical properties and are ill suited for bone tissue engineering applications.

As an alternative, use of bulk bioceramics in the form of phosphate cement inks have also garnered attention in the field of 3D bioprinting for bone tissue engineering (Zhou et al. 2021, Ostrowski et al. 2016, Lin et al. 2019, Fang et al. 2021). Phosphate cements, such as calcium phosphate cement (CaPC) and magnesium phosphate cement (MgPC), are self-setting materials that can be extruded and printed into complex shapes (Götz et al. 2021, Xu et al. 2017).

These cement-based inks offer several advantages for bioprinting applications. Firstly, they can be formulated to have suitable rheological properties for extrusion-based printing, allowing for the fabrication of intricate, patient-specific bone tissue constructs. Secondly, these cement-based inks harden or set in 'mild' conditions (i.e. at physiological conditions, without production of toxic byproducts), thereby making them suitable for printing along bioinks. Finally, phosphate cements also possess excellent biocompatibility and osteoconductivity, making them well-suited for bone regeneration (Gupta et al. 2023). Phosphate cement inks can be further enhanced by the incorporation of bioactive additives, such as growth factors or specific drugs (Lee et al. 2014, Ahlfeld et al. 2017) to further promote bone regeneration capability and add multifunctionality.

A key design criterion in cement-based inks is the provision of sufficiently long biofabrication window. This is in contrast to conventional aqueous CaPC formulation which is typically combined immediately prior to application. Consequently, such cement inks incorporate hardening retardants or employ non-aqueous carrier formulations to prevent change in ink viscosity during printing and possibly enable finite shelf-life (Ahlfeld et al. 2018, Götz et al. 2021). The primary bioceramics used in LB-3DBP are calcium phosphate cements (CaPC) and magnesium phosphate cements (MgPC).

4.10.1 Calcium Phosphate Cements (CaPCs)

CaPCs are based on the solution precipitation of apatitic phases from water-soluble calcium phosphates, such as tetra-calcium phosphate (TTCP) and dicalcium phosphate (DCPA). CaPCs can form two major end-products after setting: calcium deficient hydroxyapatite (CDHA) at $pH > 4.2$ and dicalcium phosphate dihydrate (DCPD) at $pH < 4.2$. CDHA, owing to its physiological pH requirement and ability for osteoclastic as well as hydrolytic biodegradation, is the preferred phase for LB-3DBP (Zeng et al. 2012, Lin et al. 2019).

An interesting approach for enabling prolonged biofabrication window using CaPCs is the use of water-miscible, non-aqueous carriers such as oligosaccharides (Ahlfeld et al. 2021). The authors demonstrate that such formulations can potentially have long shelf-life without significant degradation and can be set post-printing using an aqueous bath. The solvent exchange, thereof, results in the formation of a complex surface with high porosity which may be potentially beneficial for cellular adhesion and proliferation. Coextrusion of such inks in conjunction with host derived bioinks resulted in LB-3DBP with stiffness >30 MPa, and significantly higher rate of bone regeneration *in-vivo* (Ahlfeld et al. 2020).

In an alternative approach, Naren et al. reported LB-3DBP scaffold fabrication via co-axial extrusion printing (Raja and Yun 2016). The study involved multi-material bioprinting of a-TCP ink (in aqueous hydroxypropyl methyl cellulose carrier) as core, along with alginate-based bioink as sheath material. The authors report synergistic improvement in mechanical properties owing to stabilization of ceramic core by the bioink sheath. Moreover, the method results in reduction of overall print time in comparison to multi-nozzle printing techniques.

4.10.2 Magnesium Phosphate Cements (MgPCs)

MgPCs are based on a solid magnesium precursor (such as MgO and tri-magnesium phosphate anhydrate, TMPA) cured using a soluble phosphate salt (such as diammonium hydrogen phosphate, DAHP and dipotassium hydrogen phosphate, DKHP). They have emerged as a potential substitute to CaPCs due to their higher mechanical strength and faster curing and degradation rate *in-vivo* (Kanter et al. 2018, Götz et al. 2021). The setting reaction yields in insoluble MgP compound, with the struvite phase being of primary interest due to its significant bioactivity and biodegradability (Eugen et al. 2023).

Acellular prints using MgPCs have shown to have favorable properties including upregulation of osteogenic biomarkers and mechanical stiffness >56 MPa (Golafshan et al. 2020, Kanter et al. 2018). Nevertheless, no study till-date employing MgPCs for LB-3DBP has been reported thus far. It is anticipated that DKHP base formulations may be particularly suitable for such applications (Gong et al. 2021, Le Rouzic et al. 2017).

4.10.3 Magnesium versus Calcium Phosphate Cements

CaPCs have been relatively more studied both in terms of acellular scaffolds as well as for LB-3DBP application owing to their bioactivity and similarity to the mineral phase in native bone. CaPCs have received regulatory approval for clinical use in rehabilitating skeletal and craniofacial tissues. Some CaPCs are available under brand names like HydroSet™, Norian SRS, and Biobon. These cements are bioresorbable and have shown potential for bone regeneration through topographical and chemical cues, facilitating new bone formation over the ceramic implant (Gupta et al. 2023). Biodegradation of CaPCs may be driven through simple dissolution or osteoclastic resorption (Sheikh et al. 2015); either of these processes further supports bone growth through osteoclastogenesis and immunomodulatory processes like M2 polarization of macrophages.

In comparison, MgPCs support even faster bone regeneration through a combination of topographical and chemical cues. Moreover, MgPCs exhibit a higher mechanical strength and stiffness than CaPCs. Even though MgPCs do not exhibit osteoclastic resorption, their rate of dissolution is relatively faster. In a study involving large animal models over a period of 10 months, MgPCs resulted in complete replacement of the defect site with new bone as opposed to CaPCs which demonstrated significantly less new bone formation (Kanter et al. 2014, 2018). Table 4.3 compares the properties of various inks used in LB-3DBP.

4.11 TECHNIQUES FOR LB-3DBP

Conventional 3DBP employs hydrogel based bioinks, which allows for controlled distribution of one or more cell types, thereby replicating the biological and chemical niche of the host tissue extracellular matrix (ECM), while controlling the pore architecture and geometry of the overall scaffold. Nevertheless, such scaffolds neither possess sufficient mechanical properties for use in connective tissue (including bone), nor do they exhibit adequate degradation stability to support rehabilitation of large defects (Gu et al. 2022, Murphy et al. 2020). Moreover, virtually all study reports employ a single type of bioprinting method (viz. extrusion, or stereolithography alone).

In contrast, the techniques used for load-bearing 3D bioprinting (LB-3DBP) encompass a variety of innovative approaches that aim to fabricate scaffolds capable of withstanding mechanical loads for applications in load-bearing bone implants (Muskan et al. 2022, Gupta et al. 2023). These techniques leverage different printing technologies to create complex structures with enhanced mechanical properties and stability, suitable for orthopedic applications. Understanding the types of printing technologies employed in LB-3DBP is crucial for advancing regenerative medicine and tissue engineering.

4.11.1 Extrusion-based Bioprinting

This technique involves the deposition of materials through extruders, enabling the printing of multi-material scaffolds with at least one load-bearing phase. Extrusion-based bioprinting is versatile and straightforward to implement, allowing for the use of multiple printheads and high cell density. However, challenges such as high shear stress affecting cell viability, low print speed, and resolution are associated with this technique (Jorgensen et al. 2020, Zhang et al. 2021).

In certain reports, extrusion 3DBP has been extended to incorporate variations of the conventional co-extrusion 3DP to further expand the capabilities of the technique.

Co-extrusion LB-3DBP: This involves sequential printing of individual materials in a layer. Commonplace multi-head extrusion bioprinters can be used to implement these types of prints. Owing to its simplicity, a vast majority of the research efforts on LB-3DBP employ co-extrusion 3DBP (Yang et al. 2021, Gharibshahian et al. 2023, Raja and Yun 2016). Nevertheless, the overall print time increases significantly with the increase in part complexity and size owing to rapid tool changes. These may also affect the overall quality of the 3DP construct.

Co-axial LB-3DBP: This technique involves printing of multi-material scaffolds using a co-axial extrusion nozzle. The core-sheath type of print has a larger inter-material interface thereby implying higher adhesion. Core-sheath type print may also act to improve toughness of the otherwise brittle phosphate cement inks (Raja and Yun 2016). Moreover, this technique reduces printing time, as there is no need to switch between print heads.

COBICS LB-3DBP: This type of bioprinting is essentially an extension of FRESH bioprinting, wherein bioinks are printed in a support bath instead of printing in air. Consequently, the part geometry is not limited to planer 3D printing, where parts are printed as stacks of 2D layers. COBICS is an acronym for ceramic omnidirectional bioprinting in cell-suspension. In a recent work, this method was employed to print a-TCP ceramic ink in a cell-laden gelatin bath with promising

Table 4.3 Summary of the biomaterials and their properties, used for 3D bioprinting of load-bearing orthopedic scaffolds (Gupta et al. 2023) (reproduced under Creative Commons Attribution 4.0 license)

Biomaterial	Bulk compressive strength	Bulk compressive modulus	Curing/setting method	Degradation behavior	Advantage	Limitation	Reference
PCL	~38 MPa	~297–400 MPa	Solidification below melting point (59–64°C)	• No biodegradation *in-vivo* • Hydrolytic degradation • Degradation in 1–4 years	• 3D printability well established. • Lowest melting biopolymer • Easy to blend additives	• Cytotoxic melting and 3DP temperature • No bioactivity • Hydrophobicity • Slow degradation rate	Cameron and Kamvari-Moghaddam 2012, Manoukian et al. 2019, Eshraghi and Das 2010, Borkar et al. 2021
PLGA	~57 MPa	~2GPa	Solidification 57°C	• Degradation in 3–6 months	• High print temperature (95– 145°C) • Lower hydrophobicity than PCL • Faster degradation rate than PCL	• Poor bioactivity • High 3DP temperature • Acidic degradation by-products	Kang et al. 2011, Gentile et al. 2014, Guo et al. 2017, Cameron and Kamvari-Moghaddam 2012, Yang et al. 2015
CaP cement ink	25–45 MPa	2–8 GPa	Setting in aqueous environment at room temperature	• Dissolution in physiological media • Osteoclastic resorption • Degradation in 12–18 months	• High stiffness and strength • Biomimetic chemistry • Complex surface topography • Osteo-inductivity • Room temperature setting	• Brittle behaviour • Slow degradation rate • Complex 3dp ink synthesis	Panzavolta et al. 2011, Zhang et al. 2011, Heinemann et al. 2013, Raja and Yun 2016, Ahlfeld et al. 2018

MgP cement ink	6–91 MPa	>1.2 GPa	Setting in aqueous di-ammonium hydrogen phosphate (DAHP) solution at room temperature	• Dissolution in physiological media • Degradation in 10 months	• High stiffness and strength • Rapid resorption • Complex surface topography • Osteoinductivity • Room temperature setting	• Use of DAHP for crosslinking • Brittle behavior	Zhou et al. 2013, Vorndran et al. 2011, Mestres and Ginebra 2011, Ostrowski et al. 2016, Lee et al. 2014

outcomes, including favorable osteogenic differentiation of MSCs (Lin et al. 2019, Romanazzo et al. 2021).

4.11.2 SLA/Extrusion Hybrid Bioprinting

In this approach, load-bearing ink is 3D printed via melt extrusion, while bioink is 3D printed using stereolithography. SLA/extrusion hybrid bioprinting offers high bioink print resolution, high cell viability due to low shear stress during printing (Levato et al. 2023, Schwab et al. 2020), and the ability to tailor physical properties through light exposure control. However, it requires large volumes of bioink for successful implementation (Kumawat et al. 2020, Vyas et al. 2022, 2024).

4.11.3 Inkjet/Extrusion Hybrid Bioprinting

Inkjet/extrusion hybrid bioprinting combines inkjet technology with extrusion methods to create multi-material scaffolds. This technique provides high bioink print resolution and cellular viability but may face challenges related to limited z-direction control during printing (Li et al. 2020, Angelopoulos et al. 2020, Wu et al. 2024). Inkjet/extrusion hybrid bioprinting allows for the deposition of thermoplastic load-bearing inks alongside bioinks, offering unique advantages and challenges in LB-3DBP (Gupta et al. 2023).

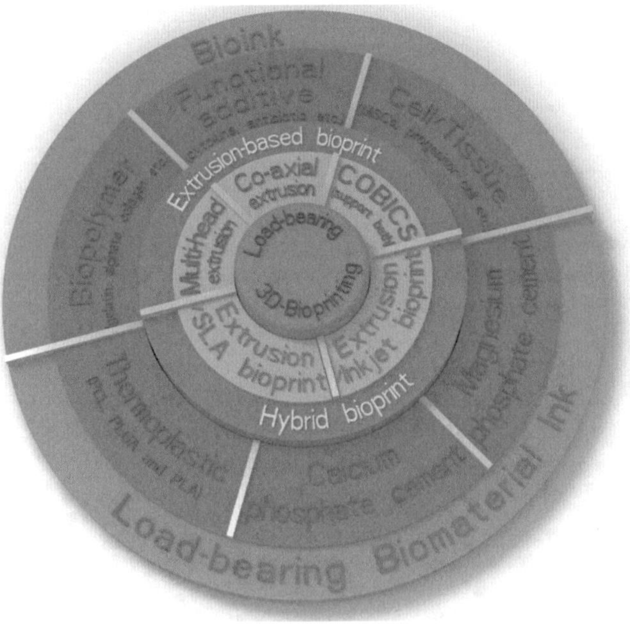

Figure 4.6. Overview of the materials and bioprinting methods reported for load-bearing 3D bioprinting (Gupta et al. 2023) (reproduced under Creative Commons Attribution 4.0 license).

Table 4.4 Techniques used for 3D bioprinting of multi-material orthopedic scaffolds for load-bearing application (Gupta et al. 2023) (reproduced under Creative Commons Attribution 4.0 license).

Description	Modality	Materials for Load-Bearing Phase	Cell Lines	Advantage	Limitation	Reference
Extrusion bioprinting	3DP via ink deposition via • Multiple extruders, or • Coaxial extruder • COBICS	CaP cement ink, MgP cement ink, PCL, PLGA	hTERT-MSC, HUVEC, rat MSC, BM-MSC, MC3T3-E1 preosteoblasts	• Versatile and straightforward implementation • Capability to use multiple printheads. • High cell density • Ability to print high viscosity inks	High shear stress affecting cell viability. • Low print speed • Low resolution • Cannot be used for low viscosity ink	(Borkar et al. 2021, Ahlfeld et al. 2018, Raja and Yun 2016, Cunniffe et al. 2017, Romanazzo et al. 2021)
SLA/Extrusion bioprinting	• Load-bearing ink 3DP via melt extrusion. • Bioink 3DP via stereolithography	PCL, PLA	HUVEC, mouse MSC, hMSC	• High bioink print resolution. • High cell viability (low shear stress and visible light curing) • No need of supports • Tailor-ability of physical properties via light exposure control	• Large volume of bioink required • Switching between two modes time consuming and cumbersome • Bioink prepolymer entrapment may deteriorate interlayer adhesion of PCL or PLA	(Cui et al. 2016, Shanjani et al. 2015)
Inkjet/ Extrusion bioprinting	• Load-bearing ink and support ink 3DP via extrusion. • Bioink 3DP via inkjet printing	PCL	BM-MSC	• High bioink print resolution. • High cellular viability	• Bioink can only be deposited in PCL channels • No z-direction control during bioprinting	(Daly and Kelly 2019)

Each of these techniques presents distinct advantages and challenges in the realm of LB-3DBP. Extrusion-based bioprinting offers versatility but may suffer from low resolution and cell viability issues due to shear stress. SLA/extrusion hybrid bioprinting provides high resolution and cell viability but requires large bioink volumes. Inkjet/extrusion hybrid bioprinting offers high resolution and cellular viability but may have limitations in achieving full z-direction control during printing.

A summary of the state of the art of LB-3DBP is represented in Figure 4.6. These techniques represent cutting-edge advancements in the field of LB-3DBP, paving the way for the development of functional orthopedic scaffolds with enhanced mechanical properties and have been compared in Table 4.4. Overcoming the challenges associated with each technique through ongoing research and technological innovations will be crucial for realizing the full potential of LB-3DBP in revolutionizing load-bearing bone.

4.12 CONCLUDING REMARKS AND OUTLOOK

Bone bioprinting is a promising field in regenerative medicine, aiming to create functional bone tissue for clinical applications. However, it faces several challenges that must be addressed to achieve successful translation from the laboratory to the clinic. This is on account of the fascinating yet complex, hierarchical structure of bone, having a highly optimised arrangement of organic and inorganic phases that enable high fracture toughness and load-bearing capabilities (Rodriguez-Palomo et al. 2024). Furthermore, this structure is interspersed with an even more complex network of vascular and neuronal tissue. Faithful replication of this complexity necessitates development of biomaterials and printing technologies that allow deposition of heterogeneous biomaterials as well as cellular tissues with high fidelity.

Recent advances in microextrusion technology have enabled the patterning of multicomponent constructs containing both synthetic and natural materials with resolutions down to 2 μm for biomaterials alone and down to 50 μm for encapsulated cells (Kang et al. 2016). Moreover, multi-material bioprinting especially load-bearing 3D bioprinting techniques have emerged in the recent years towards enabling heterogeneous deposition biomaterial and diverse cellular tissue (Chen et al. 2023b, Ravanbakhsh et al. 2021), while simultaneously improving the stiffness of the scaffold. Nevertheless, such processes allow only deposition of 2D layers typically having a flat bottom face, thereby limiting the geometries that can be printed. In addition, multilateral printing results in further compromise of the print resolution.

Recent advances in bioprinting technology, such as omnidirectional printing (ODP), have enabled the high-resolution patterning of matrix materials and the use of fugitive bioinks within a photocurable gel reservoir, allowing for truly omnidirectional free-form fabrication (Wu et al. 2011, Budharaju et al. 2024). Early studies using load-bearing inks (such as COBICS) have also demonstrated significant promise in allowing free-form printing for LB-3DBP (Romanazzo et al. 2021).

Vascularization is another performance criterion for bone scaffolds. Ensuring the integration of vasculature within bioprinted bone tissue is crucial for their functionality. Current bioprinting approaches primarily rely on endogenous vascularization, which can be improved by embedding micro-channels to enhance nutrient and oxygen diffusion or patterning angiogenic growth factors or cells within the construct to facilitate vascular development (Chen et al. 2021, Barrs et al. 2020). However, biofabrication of hollow microchannels as vascular network suitable for the provision of oxygen and nutrients into clinically relevant sized 3D-printed tissues remains a significant challenge. Incubation of these coarsely printed channels in a bioreactor may support maturation of vascular network into a multi-scale perfusable network (Ozbolat 2015). Bioprinting of tissue spheroids comprising of distinct cell lines such as fibroblasts and MSCs is another upcoming strategy that is being explored for enhancing vascularization (Kolesky et al. 2016, Ong et al. 2018).

Next, procurement of a sufficient number of functional cells for bioprinting is a challenge, particularly for older patients who may have limited bone marrow-derived mesenchymal stem cells (MSCs) available for chondrogenesis induction (Ong et al. 2018). Developing methods to derive and expand functional osteogenic cells from stem cells is crucial for the success of bone bioprinting.

Furthermore, overall efficacy of the bone tissue engineering construct is largely governed by the biomaterial. Optimization of biomaterials is not only crucial for ensuring sufficient printability but also for modulating the biological response. Bioinks must allow high print resolution without adversely affecting the cellular viability. Additives such as use of viscosity modifiers, biomaterial blends and/or micro/nano fillers have been employed to improve printability (Chen et al. 2023a). Recent advances in bioprinting technology, such as omnidirectional printing (ODP), have also been reported to support the high-resolution patterning of matrix materials and the use of fugitive bioinks within a photocurable gel reservoir (Budharaju et al. 2024).

While additives support bioink printability, effect of such additives on the biological performance must be evaluated rigorously. Developing biocompatible bioinks that can support cell function, tissue structure, and biomechanical properties is essential for successful bone bioprinting. These bioinks should be able to support cell viability, proliferation, and differentiation into osteogenic lineages. Studies employing cell-laden hydrogels, such as those based on collagen and hyaluronic acid, have been shown to direct cellular proliferation and osteogenic differentiation of stem cells (Osidak et al. 2020).

Despite their numerous benefits, 3DBP scaffolds are mechanically weak and cannot be used for bearing even a small amount of mechanical load. Research efforts towards improving stiffness of such hydrogel bioinks (either at a microscopic or macroscopic scales) have been successful in further improving their biological response through topographical and mechanical cues (Nudelman et al. 2013, Vyas et al. 2024, Ahlfeld et al. 2020). Although bioinspired mineralization of collagen hydrogel has been able to replicate the staggered structure of mineralised bone, they exhibit poor overall strength and toughness. In contrast, LB-3DBP scaffolds show significantly higher strength, yet fail to exhibit mechanical strength and fracture toughness expected of load-bearing bone.

In conclusion, bone bioprinting holds great promise for the future of regenerative medicine, but significant challenges remain in replicating the complex architecture and functionality of native bone tissue, achieving vascularization, developing suitable bioinks, sourcing functional cells, and ensuring sufficient mechanical properties. Addressing these challenges through continued research and technological innovation will be crucial for the successful translation of bone bioprinting from the laboratory to clinical applications.

REFERENCES

Adarkwa, E., A. Roy, J. Ohodnicki, B. Lee, P.N. Kumta and S. Desai. 2023. 3D printing of drug-eluting bioactive multifunctional coatings for orthopedic applications. International Journal of Bioprinting. 9(2): 661.

Ahlfeld, T., A.R. Akkineni, Y. Förster, T. Köhler, S. Knaack, M. Gelinsky, et al., 2017. Design and fabrication of complex scaffolds for bone defect healing: combined 3D plotting of a calcium phosphate cement and a growth factor-loaded hydrogel. Annals of Biomedical Engineering. 45(1): 224–236.

Ahlfeld, T., F. Doberenz, D. Kilian, C. Vater, P. Korn, G. Lauer, et al., 2018. Bioprinting of mineralized constructs utilizing multichannel plotting of a self-setting calcium phosphate cement and a cell-laden bioink. Biofabrication. 10(4): 045002.

Ahlfeld, T., N. Cubo-Mateo, S. Cometta, V. Guduric, C. Vater, A. Bernhardt, et al., 2020. A novel plasma-based bioink stimulates cell proliferation and differentiation in bioprinted, mineralized constructs. ACS Applied Materials and Interfaces. 12(11): 12557– 12572.

Ahlfeld, T., A. Lode, R.F. Richter, W. Pradel, A. Franke, M. Rauner, et al., 2021. Toward biofabrication of resorbable implants consisting of a calcium phosphate cement and fibrin—a characterization *in vitro* and *in vivo*. International Journal of Molecular Sciences. 22(3): 1218.

Ahn, S., H. Lee and G. Kim. 2013. Functional cell-laden alginate scaffolds consisting of core/shell struts for tissue regeneration. Carbohydrate Polymers. 98(1): 936–942.

Akkineni, A.R., T. Ahlfeld, A. Lode and M. Gelinsky. 2016. A versatile method for combining different biopolymers in a core/shell fashion by 3D plotting to achieve mechanically robust constructs. Biofabrication. 8(4): 45001.

Aldaadaa, A., N. Owji and J. Knowles. 2018. Three-dimensional printing in maxillofacial surgery: hype versus reality. Journal of Tissue Engineering. 9: 2041731418770909.

Allen, M.R. and D.B. Burr. 2013. Bone modeling and remodeling. pp. 76–90. *In*: Allen, M.R. and D.B. Burr (eds). Basic and Applied Bone Biology. Cambridge: Academic Press.

Ambikaipalan, A., J. Min-Leong Wong and W.S. Khan. 2013. Preclinical and clinical studies on the use of stem cells for bone repair: a systematic review. Current Stem Cell Research and Therapy. 8(3): 210–216.

Angelopoulos, I., M.C. Allenby, M. Lim and M. Zamorano. 2020. Engineering inkjet bioprinting processes toward translational therapies. Biotechnology and Bioengineering. 117(1): 272–284.

Ashammakhi, N., A. Hasan, O. Kaarela, B. Byambaa, A. Sheikhi, A.K. Gaharwar, et al., 2019. Advancing frontiers in bone bioprinting. Advanced Healthcare Materials. 8: 1801048. https://doi.org/10.1002/adhm.201801048.

Bae, M.S., D.H. Yang, J.B. Lee, D.N. Heo, Y.D. Kwon, I.C. Youn, et al. 2011. Photo-cured hyaluronic acid-based hydrogels containing simvastatin as a bone tissue regeneration scaffold. Biomaterials. 32(32): 8161–8171.

Bahney, C.S., R.L. Zondervan, P. Allison, A. Theologis, J.W. Ashley, J. Ahn, et al., 2019. Cellular biology of fracture healing. Journal of Orthopaedic Research. 37(1): 35–50.

Baino, F., G. Novajra and C. Vitale-Brovarone. 2015. Bioceramics and scaffolds: a winning combination for tissue engineering. Frontiers in Bioengineering and Biotechnology. 3: 202. doi: 10.3389/fbioe.2015.00202

Bakhsheshi-Rad, H.R., X. Chen, A.F. Ismail, M. Aziz, E. Abdolahi and F. Mahmoodiyan. 2019. Improved antibacterial properties of an Mg-Zn-Ca alloy coated with chitosan nanofibers incorporating silver sulfadiazine multiwall carbon nanotubes for bone implants. Polymers for Advanced Technologies. 30(5): 1333–1339.

Barrs, R.W., J. Jia, S.E. Silver, M. Yost and Y. Mei. 2020. Biomaterials for bioprinting microvasculature. Chemical Reviews. 120(19): 10887–10949.

Bendtsen, S.T., S.P. Quinnell and M. Wei. 2017. Development of a novel alginate-polyvinyl alcohol-hydroxyapatite hydrogel for 3D bioprinting bone tissue engineered scaffolds. Journal of Biomedical Materials Research—Part A 105(5): 1457–1468.

Bendtsen, S.T. and M. Wei. 2017. *In vitro* evaluation of 3D bioprinted tri-polymer network scaffolds for bone tissue regeneration. Journal of Biomedical Materials Research—Part A. Vol. 105.

Betts, J.G., K.A. Young and J.A. Wise (eds). 2021. Anatomy and Physiology 2nd Ed. OpenStax.

Billiet, T., E. Gevaert, T.D. Schryver, M. Cornelissen and P. Dubruel. 2014. The 3D printing of gelatin methacrylamide cell-laden tissue-engineered constructs with high cell viability. Biomaterials. 35(1): 49–62.

Bishop, E.S., S. Mostafa, M. Pakvasa, H.H. Luu, M.J. Lee, J.M. Wolf, et al., 2017. 3-D bioprinting technologies in tissue engineering and regenerative medicine: current and future trends. Genes and Diseases. 4(4): 185–195.

Borkar, T., V. Goenka and A.K. Jaiswal. 2021. Application of poly-ε-caprolactone in extrusion-based bioprinting. Bioprinting. 21: e00111.

Budharaju, H., D. Sundaramurthi and S. Sethuraman. 2024. Embedded 3D bioprinting—an emerging strategy to fabricate biomimetic and large vascularized tissue constructs. Bioactive Materials. 32 (February): 356–384.

Cameron, R.E. and A. Kamvari-Moghaddam. 2012. Synthetic bioresorbable polymers. pp. 96–118. *In*: M. Jenkins and A.Stamboulis (eds). Durability and Reliability of Medical Polymers. Amsterdam: Elsevier.

Cao, C., P. Huang, A. Prasopthum, A.J. Parsons, F. Ai and J. Yang. 2022. Characterisation of bone regeneration in 3D printed ductile PCL/PEG/hydroxyapatite scaffolds with high ceramic microparticle concentrations. Biomaterials Science. 10(1): 138–152.

Catros, S., J.C. Fricain, B. Guillotin, B. Pippenger, R. Bareille, M. Remy, et al., 2011. Laser-assisted bioprinting for creating on-demand patterns of human osteoprogenitor cells and nano-hydroxyapatite. Biofabrication. Vol. 3.

Chen, Y.W., Y.F. Shen, C.C. Ho, J. Yu, Y.H.A. Wu, K. Wang, et al., 2018. Osteogenic and angiogenic potentials of the cell-laden hydrogel/mussel-inspired calcium silicate complex hierarchical porous scaffold fabricated by 3D bioprinting. Materials Science and Engineering C. 91: 679–687.

Chen, E.P., Z. Toksoy, B.A. Davis and J.P. Geibel. 2021. 3D bioprinting of vascularized tissues for *in vitro* and *in vivo* applications. Frontiers in Bioengineering and Biotechnology. 9 (May).

Chen, X.B., A. Fazel Anvari-Yazdi, X. Duan, A. Zimmerling, R. Gharraei, N.K. Sharma, et al., 2023a. Biomaterials/bioinks and extrusion bioprinting. Bioactive Materials. 28: 511–536.

Chen, A., W. Wang, Z. Mao, Y. He, S. Chen, G. Liu, et al. 2023b. Multimaterial 3D and 4D bioprinting of heterogenous constructs for tissue engineering. Advanced Materials. December.

Chimene, D., K.K. Lennox, R.R. Kaunas and A.K. Gaharwar. 2016. Advanced bioinks for 3D printing: a materials science perspective. Annals of Biomedical Engineering. Vol. 44.

Chimene, D., R. Kaunas and A.K. Gaharwar. 2020. Hydrogel bioink reinforcement for additive manufacturing: a focused review of emerging strategies. Advanced Materials. 32(1).

Chor, A., C.M. Takiya, M.L. Dias, R.P. Gonçalves, T. Petithory, J. Cypriano, et al., 2022. *In vitro* and *in vivo* cell-interactions with electrospun poly (Lactic-Co-Glycolic Acid) (PLGA): morphological and immune response analysis. Polymers. 14(20): 4460.

Christau, S., A. López Ruiz, N. Habibi, J. Witte, M.S. Bannon, K. McEnnis, et al., 2022. Macrophage-targeting poly(lactide-co-glycolic acid) nanoparticles decorated with multifunctional brush polymers. Particle & Particle Systems Characterization. 39(5).

Clarke, B. 2008. Normal bone anatomy and physiology. Clinical Journal of the American Society of Nephrology: CJASN 3 Suppl. 3: S131–S139.

Crowley, C., J. Min-Leong Wong, D.M. Fisher and W.S. Khan. 2013. A systematic review on preclinical and clinical studies on the use of scaffolds for bone repair in skeletal defects. Current Stem Cell Research and Therapy. 8(3): 243–252.

Cui, H., W. Zhu, M. Nowicki, X. Zhou, A. Khademhosseini and L.G. Zhang. 2016. Hierarchical fabrication of engineered vascularized bone biphasic constructs via dual 3D bioprinting: integrating regional bioactive factors into architectural design. Advanced Healthcare Materials. 5(17): 2174–2181.

Cunniffe, G.M., T. Gonzalez-Fernandez, A. Daly, B.N. Sathy, O. Jeon, E. Alsberg, et al., 2017. Three-dimensional bioprinting of polycaprolactone reinforced gene activated bioinks for bone tissue engineering. Tissue Engineering Part A. 23(17–18): 891–900.

Currey, J.D. 2002. Modeling and reconstruction. pp. 337–379 *In*: Currey, J.D. (ed.). Bones: Structure and Mechanics, 1st Ed. Princeton University Press. http://www.jstor.org/stable/j.ctt4cg9wv

Daly, A.C., G.M. Cunniffe, B.N. Sathy, O. Jeon, E. Alsberg and D.J. Kelly. 2016. 3D bioprinting of developmentally inspired templates for whole bone organ engineering. Advanced Healthcare Materials. 5(18): 2353–2362.

Daly, A.C. and D.J. Kelly. 2019. Biofabrication of spatially organised tissues by directing the growth of cellular spheroids within 3D printed polymeric microchambers. Biomaterials 197: 194–206.

Dapunt, U., T. Giese, F. Lasitschka, J. Reinders, B. Lehner, J. Kretzer, et al., 2014. On the inflammatory response in metal-on-metal implants. Journal of Translational Medicine. 12(1): 74.

Dare, A.J and G. Hu. 2017. China's evolving fracture burden. The Lancet Global Health. 5(8): e736–e737.

Das, S., F. Pati, Y.J. Choi, G. Rijal, J.H. Shim, S.W. Kim, et al., 2015. Bioprintable, cell-laden silk fibroin-gelatin hydrogel supporting multilineage differentiation of stem cells for fabrication of three-dimensional tissue constructs. Acta Biomaterialia. 11(1): 233–246.

Delmas, P.D. and L. Malaval. 1993. The proteins of bone. pp. 673–724. *In*: G.R. Mundy and T.J. Martin (eds). Physiology and Pharmacology of Bone. Berlin: Springer.

Demirtaş, T.T., G. Irmak and M. Gümüşderelioğlu. 2017. A bioprintable form of chitosan hydrogel for bone tissue engineering. Biofabrication. 9(3): 35003.

Dong, L., S.J. Wang, X.R. Zhao, Y.F. Zhu and J.K. Yu. 2017. 3D-printed poly (ε-caprolactone) scaffold integrated with cell-laden chitosan hydrogels for bone tissue engineering. Scientific Reports. 7(1): 4–12.

Dorozhkin, S.V. 2009. Calcium orthophosphate cements and concretes. Materials. 2(1): 221–291.

Dorozhkin, S.V. 2016. Calcium phosphates. pp. 91–118. *In*: I.V. Antoniac (ed.). Handbook of Bioceramics and Biocomposites. Cham: Springer International Publishing.

Du, M., B. Chen, Q. Meng, S. Liu, X. Zheng, C. Zhang, et al., 2015. 3D bioprinting of BMSC-laden methacrylamide gelatin scaffolds with CBD-BMP2-collagen microfibers. Biofabrication. 7(4): 044104.

Duarte C., D. Filipa, A. Blaeser, K. Buellesbach, K.S. Sen, W. Xun, et al., 2016. Bioprinting organotypic hydrogels with improved mesenchymal stem cell remodeling and mineralization properties for bone tissue engineering. Advanced Healthcare Materials. 5(11): 1336–1345.

Einhorn, T.A. 1998. The cell and molecular biology of fracture healing. Clinical Orthopaedics and Related Research. 355(Suppl.): S7–S21.

Einhorn, T.A. and L.C. Gerstenfeld. 2015. Fracture healing: mechanisms and interventions. Nature Reviews Rheumatology. 11(1): 45–54.

Eliaz, N. and N. Metoki. 2017. Calcium phosphate bioceramics: a review of their history, structure, properties, coating technologies and biomedical applications. Materials. 10(4): 334.

Eshraghi, S. and S. Das. 2010. Mechanical and microstructural properties of polycaprolactone scaffolds with one-dimensional, two-dimensional, and three-dimensional orthogonally oriented porous architectures produced by selective laser sintering. Acta Biomaterialia. 6(7): 2467–2476.

Eugen, G., M. Claus, S. Anna-Maria, D. Niklas, S. Philipp, E. Andrea, et al., 2023. Degradation of 3D-printed magnesium phosphate ceramics *in vitro* and a prognosis on their bone regeneration potential. Bioactive Materials. 19: 376–391.

Fang, Z., J. Chen, J. Pan, G. Liu and C. Zhao. 2021. The development tendency of 3D-printed bioceramic scaffolds for applications ranging from bone tissue regeneration to bone tumor therapy. Frontiers in Bioengineering and Biotechnology. 9 (December).

Farr, J.N., L.J. Melton, S.J. Achenbach, E.J. Atkinson, S. Khosla and S. Amin. 2017. Fracture incidence and characteristics in young adults aged 18 to 49 years: a population-based study. Journal of Bone and Mineral Research. 32(12): 2347–2354.

Feng, X. 2009. Chemical and biochemical basis of cell-bone matrix interaction in health and disease. Current Chemical Biology. 3(2): 189–196.

Florencio-Silva, R., G.R.S. Sasso, E. Sasso-Cerri, M.J. Simões and P.S. Cerri. 2015. Biology of bone tissue: structure, function, and factors that influence bone cells. BioMed Research International. 2015.

Fontcuberta-Rigo, M., M. Nakamura and P. Puigbò. 2023. Phylobone: a comprehensive database of bone extracellular matrix proteins in human and model organisms. Bone Research. 11(1): 44.

Frade, B.B., R.B. Dias, S.G. Piperni and D.C. Bonfim. 2023. The role of macrophages in fracture healing: a narrative review of the recent updates and therapeutic perspectives. Stem Cell Investigation. 10: 4–4.

Francis Suh, J.K. and H.W.T. Matthew. 2000. Application of chitosan-based polysaccharide biomaterials in cartilage tissue engineering: a review. Biomaterials. Vol. 21.

Gao, G., A.F. Schilling, T. Yonezawa, J. Wang, G. Dai and X. Cui. 2014. Bioactive nanoparticles stimulate bone tissue formation in bioprinted three-dimensional scaffold and human mesenchymal stem cells. Biotechnology Journal. 9(10): 1304–1311.

Gao, G., T. Yonezawa, K. Hubbell, G. Dai and X. Cui. 2015. Inkjet-bioprinted acrylated peptides and peg hydrogel with human mesenchymal stem cells promote robust bone and cartilage formation with minimal printhead clogging. Biotechnology Journal. 10(10): 1568–1577.

Gautam, S., C.F. Chou, A.K. Dinda, P.D. Potdar and N.C. Mishra. 2014. Fabrication and characterization of PCL/gelatin/chitosan ternary nanofibrous composite scaffold for tissue engineering applications. Journal of Materials Science. 49(3): 1076–1089.

Gentile, P., V. Chiono, I. Carmagnola and P. Hatton. 2014. An overview of poly(lactic-co-glycolic) acid (PLGA)-based biomaterials for bone tissue engineering. International Journal of Molecular Sciences. 15(3): 3640–3659.

Gharibshahian, M., M. Salehi, N. Beheshtizadeh, M. Kamalabadi-Farahani, A. Atashi, M.-S. Nourbakhsh, et al., 2023. Recent advances on 3d-printed pcl-based composite scaffolds for bone tissue engineering. Frontiers in Bioengineering and Biotechnology. 11: 1168504.

Giuseppe, M.D., N. Law, B. Webb, R.A. Macrae, L.J. Liew, T.B. Sercombe, et al., 2018. Mechanical Behaviour of Alginate-Gelatin Hydrogels for 3D Bioprinting. Journal of the Mechanical Behavior of Biomedical Materials. 79: 150–157.

Golafshan, N., E. Vorndran, S. Zaharievski, H. Brommer, F.B. Kadumudi, A. Dolatshahi-Pirouz, et al., 2020. Tough magnesium phosphate-based 3D-printed implants induce bone regeneration in an equine defect model. Biomaterials. 261: 120302.

Gong, C., S. Fang, K. Xia, J. Chen, L. Guo and W. Guo. 2021. Enhancing the mechanical properties and cytocompatibility of magnesium potassium phosphate cement by incorporating oxygen-carboxymethyl chitosan. Regenerative Biomaterials. 8(1).

Götz, L.-M., K. Holeczek, J. Groll, T. Jüngst, and U. Gbureck. 2021. Extrusion-based 3D printing of calcium magnesium phosphate cement pastes for degradable bone implants. Materials. 14(18): 5197.

Groll, J., J.A. Burdick, D.W. Cho, B. Derby, M. Gelinsky, S.C. Heilshorn, et al. 2019. A definition of bioinks and their distinction from biomaterial inks. Biofabrication. 11(1): 13001.

Gu, Y., A. Forget and V.P. Shastri. 2022. Biobridge: an outlook on translational bioinks for 3D bioprinting. Advanced Science. 9(3): 2103469.

Guo, T., T.R. Holzberg, C.G. Lim, F. Gao, A. Gargava, J.E. Trachtenberg, et al., 2017. 3D printing PLGA: a quantitative examination of the effects of polymer composition and printing parameters on print resolution. Biofabrication. 9(2): 024101.

Gupta, A.A., S.K. Mishra, S. Uikey and D. Maravi. 2019. A study—incidence and pattern of musculoskeletal injuries among patients attending the emergency of tertiary health care center in central India. Journal of Orthopaedic Diseases and Traumatology. 2(1): 11.

Gupta, T., S.B. Ghosh, S. Bandyopadhyay-Ghosh and M. Sain. 2023. Is it possible to 3D bioprint load-bearing bone implants? A critical review. Biofabrication. 15(4).

Heid, S. and A.R. Boccaccini. 2020. Advancing bioinks for 3D bioprinting using reactive fillers: a review. Acta Biomaterialia. 113: 1–22.

Heinemann, S., S. Rössler, M. Lemm, M. Ruhnow and B. Nies. 2013. Properties of injectable ready-to-use calcium phosphate cement based on water-immiscible liquid. Acta Biomaterialia. 9(4): 6199–6207.

Hellwinkel, J.E., T. Miclau, M.T. Provencher, C.S. Bahney and Z.M. Working. 2020. The life of a fracture: biologic progression, healing gone awry, and evaluation of union. JBJS Reviews. 8(8): e1900221.

Holzapfel, B.M., M. Rudert and D.W. Hutmacher. 2017. Gerüstträgerbasiertes knochen-tissue-engineering. Orthopade. 46(8): 701–710.

Huang, J., H. Fu, Z. Wang, Q. Meng, S. Liu, H. Wang, et al., 2016. BMSCs-laden gelatin/sodium alginate/carboxymethyl chitosan hydrogel for 3D bioprinting. RSC Advances. Vol. 6.

Huang, Y., X. Wang, D. Zhou, W. Zhou, F. Dai and H. Lin. 2021a. Macrophages in heterotopic ossification: from mechanisms to therapy. NPJ Regenerative Medicine. 6(1): 70.

Huang, Y., X. Wang, D. Zhou, W. Zhou, F. Dai and H. Lin. 2021b. Macrophages in heterotopic ossification: from mechanisms to therapy. NPJ Regenerative Medicine. 6(1): 70.

Huang, P., J. Xu, L. Xie, G. Gao, S. Chen, Z. Gong, et al. 2023. Improving hard metal implant and soft tissue integration by modulating the 'inflammatory-fibrous complex' response. Bioactive Materials. 20: 42–52.

Huh, J.T., J.U. Lee, W.J. Kim, M. Yeo and G.H. Kim. 2018. Preparation and characterization of gelatin/α-TCP/SF biocomposite scaffold for bone tissue regeneration. International Journal of Biological Macromolecules. 110: 488–496.

Izadifar, Z., T. Chang, W. Kulyk, X. Chen and B.F. Eames. 2016. Analyzing biological performance of 3D-printed, cell-impregnated hybrid constructs for cartilage tissue engineering. Tissue Engineering—Part C: Methods. 22(3): 173–188.

Jodati, H., B. Yılmaz and Z. Evis. 2020. A review of bioceramic porous scaffolds for hard tissue applications: effects of structural features. Ceramics International. 46(10): 15725–15739.

Jorgensen, A.M., J.J. Yoo and A. Atala. 2020. Solid organ bioprinting: strategies to achieve organ function. Chemical Reviews. 120(19): 11093–11127.

Kang, S.-W., S.-J. Lee, J.-S. Kim, E.-H. Choi, B.-H. Cha, J.-H. Shim, et al., 2011. Effect of a scaffold fabricated thermally from acetylated PLGA on the formation of engineered cartilage. Macromolecular Bioscience. 11(2): 267–274.

Kang, H.W., S.J. Lee, I.K. Ko, C. Kengla, J.J. Yoo and A. Atala. 2016. A 3D bioprinting system to produce human-scale tissue constructs with structural integrity. Nature Biotechnology. 34(3): 312–319.

Kanter, B., M. Geffers, A. Ignatius and U. Gbureck. 2014. Control of *in vivo* mineral bone cement degradation. Acta Biomaterialia. 10(7): 3279–3287.

Kanter, B., A. Vikman, T. Brückner, M. Schamel, U. Gbureck and A. Ignatius. 2018. Bone regeneration capacity of magnesium phosphate cements in a large animal model. Acta Biomaterialia. 69: 352–361.

Karageorgiou, V. and D. Kaplan. 2005. Porosity of 3D biomaterial scaffolds and osteogenesis. Biomaterials. 26(27): 5474–5491.

Ke, P., X.N. Jiao, X.H. Ge, W.M. Xiao and B. Yu. 2014. From macro to micro: structural biomimetic materials by electrospinning. RSC Advances. 4(75): 39704–39724.

Kenkre, J.S. and J.H.D. Bassett. 2018. The bone remodelling cycle. Annals of Clinical Biochemistry. 55(3): 308–327.

Keramaris, N.C., G.M. Calori, V.S. Nikolaou, E.H. Schemitsch and P.V. Giannoudis. 2008. Fracture vascularity and bone healing: a systematic review of the role of VEGF. Injury. 39 (Suppl. 2): S45–S57.

Keriquel, V., H. Oliveira, M. Rémy, S. Ziane, S. Delmond, B. Rousseau, et al. 2017. *In situ* printing of mesenchymal stromal cells, by laser-assisted bioprinting, for *in vivo* bone regeneration applications. Scientific Reports. Vol. 7.

Kim, J., I.S. Kim, T.H. Cho, K.B. Lee, S.J. Hwang, G. Tae, et al., 2007. Bone regeneration using hyaluronic acid-based hydrogel with bone morphogenic protein-2 and human mesenchymal stem cells. Biomaterials. 28(10): 1830–1837.

Kim, Y.B. and G.H. Kim. 2015. PCL/alginate composite scaffolds for hard tissue engineering: fabrication, characterization, and cellular activities. ACS Combinatorial Science. 17(2): 87–99.

Kim, B.S., J. Jang, S. Chae, G. Gao, J.S. Kong, M. Ahn, et al., 2016a. Three-dimensional bioprinting of cell-laden constructs with polycaprolactone protective layers for using various thermoplastic polymers. Biofabrication. 8(3): 1–14.

Kim, Y.B., H. Lee, G.H. Yang, C.H. Choi, D.W. Lee, H. Hwang, et al., 2016b. Mechanically reinforced cell-laden scaffolds formed using alginate-based bioink printed onto the surface of a PCL/alginate mesh structure for regeneration of hard tissue. Journal of Colloid and Interface Science. 461: 359–368.

Kim, T., C.W. See, X. Li and D. Zhu. 2020. Orthopedic implants and devices for bone fractures and defects: past, present and perspective. Engineered Regeneration. 1: 6–18.

Koester, K.J., J.W. Ager and R.O. Ritchie. 2008. The true toughness of human cortical bone measured with realistically short cracks. Nature Materials. 7(8): 672–677.

Kolesky, D.B., K.A. Homan, M.A. Skylar-Scott and J.A. Lewis. 2016. three-dimensional bioprinting of thick vascularized tissues. Proceeding of the National Academy of Sciences of the United States of America. 113(12): 3179–3184.

Krujatz, F., S. Dani, J. Windisch, J. Emmermacher, F. Hahn, M. Mosshammer, et al., 2022. Think outside the box: 3D bioprinting concepts for biotechnological applications—recent developments and future perspectives. Biotechnology Advances. 58: 107930.

Kumawat, V.S., A. Vyas, S. Bandyopadhyay-Ghosh and S.B. Ghosh. 2020. Selectively modified nanostructured fluorcanasite glass-ceramic with enhanced micromechanical properties. Journal of Non-Crystalline Solids. 547: 120303.

Kundu, J., J.-H. Shim, J. Jang, S.-W. Kim and D.-W. Cho. 2015. An additive manufacturing-based PCL-alginate-chondrocyte bioprinted scaffold for cartilage tissue engineering. Journal of Tissue Engineering and Regenerative Medicine. 9(11): 1286–1297.

Lavorato, A., S. Raimondo, M. Boido, L. Muratori, G. Durante, F. Cofano, et al., 2021. Mesenchymal stem cell treatment perspectives in peripheral nerve regeneration: systematic review. International Journal of Molecular Sciences. 22(2): 1–22.

Law, N., B. Doney, H. Glover, Y. Qin, Z.M. Aman, T.B. Sercombe, et al., 2018. Characterisation of hyaluronic acid methylcellulose hydrogels for 3D bioprinting. Journal of the Mechanical Behavior of Biomedical Materials. 77: 389–399.

Lee, H., S. Ahn, L.J. Bonassar and G. Kim. 2013. Cell(MC3T3-E1)-printed poly (ε-caprolactone)/alginate hybrid scaffolds for tissue regeneration. Macromolecular Rapid Communications. 34(2): 142–149.

Lee, J., M.M. Farag, E.K. Park, J. Lim and H.-S. Yun. 2014. A simultaneous process of 3D magnesium phosphate scaffold fabrication and bioactive substance loading for hard tissue regeneration. Materials Science and Engineering C. 36(1): 252–260.

Levato, R., O. Dudaryeva, C.E. Garciamendez-Mijares, B.E. Kirkpatrick, R. Rizzo, J. Schimelman, et al., 2023. Light-based vat-polymerization bioprinting. Nature Reviews Methods Primers. 3(1): 47.

Li, X., B. Liu, B. Pei, J. Chen, D. Zhou, J. Peng, et al., 2020. Inkjet bioprinting of biomaterials. Chemical Reviews. 120(19): 10793–10833.

Li, N., R. Guo and Z.J. Zhang. 2021. Bioink formulations for bone tissue regeneration. Frontiers in Bioengineering and Biotechnology. 9: 630488.

Li, M., D. Sun, J. Zhang, Y.Y. Wang, Q. Wei and Y.Y. Wang. 2022. Application and development of 3d bioprinting in cartilage tissue engineering. Biomaterials Science. 10(19): 5430–5458.

Lin, K., R. Sheikh, S. Romanazzo and I. Roohani. 2019. 3D printing of bioceramic scaffolds-barriers to the clinical translation: from promise to reality, and future perspectives. Materials. 12(7): 2660.

Linh, N.T.B., C.D.G. Abueva, D.W. Jang and B.T. Lee. 2020. Collagen and bone morphogenetic protein-2 functionalized hydroxyapatite scaffolds induce osteogenic differentiation in human adipose-derived stem cells. Journal of Biomedical Materials Research—Part B Applied Biomaterials. 108(4): 1363–1371.

Löffler, J., A. Noom, A. Ellinghaus, A. Dienelt, S. Kempa and G.N. Duda. 2023. A comprehensive molecular profiling approach reveals metabolic alterations that steer bone tissue regeneration. Communications Biology. 6(1): 327.

Longoni, A., L. Knežević, K. Schepers, H. Weinans, A.J.W.P. Rosenberg and D. Gawlitta. 2018. The impact of immune response on endochondral bone regeneration. NPJ Regenerative Medicine. 3(1): 22.

Loozen, L.D., F. Wegman, F.C. Öner, W.J.A. Dhert and J. Alblas. 2013. Porous bioprinted constructs in bmp-2 non-viral gene therapy for bone tissue engineering. Journal of Materials Chemistry B. 1(48): 6619.

Lorenzo, J. 2017. The many ways of osteoclast activation. Journal of Clinical Investigation. 127(7): 2530–2532.

Luo, Y., A. Lode, A.R. Akkineni and M. Gelinsky. 2015. Concentrated gelatin/alginate composites for fabrication of predesigned scaffolds with a favorable cell response by 3D plotting. RSC Advances. 5: 43480–43488.

Luo, Y., G. Luo, M. Gelinsky, P. Huang and C. Ruan. 2017. 3D bioprinting scaffold using alginate/polyvinyl alcohol bioinks. Materials Letters. 189: 295–298.

Luo, Y., Y. Li, X. Qin and Q. Wa. 2018. 3D printing of concentrated alginate/gelatin scaffolds with homogeneous nano apatite coating for bone tissue engineering. Materials and Design. 146: 12–19.

Mandrycky, C., Z. Wang, K. Kim and D.H. Kim. 2016. 3D bioprinting for engineering complex tissues. Biotechnology Advances. 34(4): 422–434.

Manoukian, O.S., N. Sardashti, T. Stedman, K. Gailiunas, A. Ojha, A. Penalosa, et al., 2019. Biomaterials for tissue engineering and regenerative medicine. pp. 462–482. *In*: R. Narayan (ed.). Encyclopedia of Biomedical Engineering. Oxford: Elsevier.

Marenzana, M. and T.R. Arnett. 2013. The key role of the blood supply to bone. Bone Research. 1(3): 203–215.

Martínez-Vázquez, F.J., M.V. Cabañas, J.L. Paris, D. Lozano and M. Vallet-Regí. 2015. Fabrication of novel si-doped hydroxyapatite/gelatine scaffolds by rapid prototyping for drug delivery and bone regeneration. Acta Biomaterialia. 15: 200–209.

Maruyama, M., C. Rhee, T. Utsunomiya, N. Zhang, M. Ueno, Z. Yao, et al., 2020. Modulation of the inflammatory response and bone healing. Frontiers in Endocrinology. 11: 386.

Mass, E., F. Nimmerjahn, K. Kierdorf and A. Schlitzer. 2023. Tissue-specific macrophages: how they develop and choreograph tissue biology. Nature Reviews Immunology. 23(9): 563–579.

McBeth, C., J. Lauer, M. Ottersbach, J. Campbell, A. Sharon and A.F. Sauer-Budge. 2017. 3D bioprinting of GelMA scaffolds triggers mineral deposition by primary human osteoblasts. Biofabrication. 9(1): 15009.

Meng, X., B. Wielockx, M. Rauner and A. Bozec. 2021. Hypoxia-inducible factors regulate osteoclasts in health and disease. Frontiers in Cell and Developmental Biology. 9: 658893.

Mestres, G. and M.-P. Ginebra. 2011. Novel magnesium phosphate cements with high early strength and antibacterial properties. Acta Biomaterialia. 7(4): 1853–1861.

Milazzo, M., N.C. Negrini, S. Scialla, B. Marelli, S. Farè, S. Danti, et al., 2019. Additive manufacturing approaches for hydroxyapatite-reinforced composites. Advanced Functional Materials. 29: 190305.

Mohseni, M., A. Jahandideh, G. Abedi, A. Akbarzadeh and S. Hesaraki. 2018. Assessment of tricalcium phosphate/collagen (TCP/collagene) nanocomposite scaffold compared with hydroxyapatite (HA) on healing of segmental femur bone defect in rabbits. Artificial Cells, Nanomedicine and Biotechnology. 46: 242–249.

Moy, P.K. and T. Aghaloo. 2019. Risk factors in bone augmentation procedures. Periodontology 2000. 81(1): 76–90.

Murphy, C., K. Kolan, W. Li, J. Semon, D. Day and M. Leu. 2017. 3D bioprinting of stem cells and polymer/bioactive glass composite scaffolds for bone tissue engineering. International Journal of Bioprinting. 3(1): 005.

Murphy, S.V., P.D. Coppi and A. Atala. 2020. Opportunities and challenges of translational 3D bioprinting. Nature Biomedical Engineering. 4(4): 370–380.

Muskan, D.G. and N.P. Negi. 2022. 3D bioprinting: printing the future and recent advances. Bioprinting. 27: e00211.

Muzzarelli, R.A.A., F. Greco, A. Busilacchi, V. Sollazzo and A. Gigante. 2012. Chitosan, hyaluronan and chondroitin sulfate in tissue engineering for cartilage regeneration: a review. Carbohydrate Polymers. 89(3): 723–739.

Naghieh, S. and X. Chen. 2021. Printability–a key issue in extrusion-based bioprinting. Journal of Pharmaceutical Analysis. 11(5): 564–579.

Narayanan, L.K., P. Huebner, M.B. Fisher, J.T. Spang, B. Starly and R.A. Shirwaiker. 2016. 3D-bioprinting of polylactic acid (PLA) nanofiber-alginate hydrogel bioink containing human adipose-derived stem cells. ACS Biomaterials Science and Engineering. Vol. 2.

Nazzal, M.K., A.J. Morris, R.S. Parker, F.A. White, R.M. Natoli, M.A. Kacena, et al., 2024. Do not lose your nerve, be callus: insights into neural regulation of fracture healing. Current Osteoporosis Reports. 22(1): 182–192.

Neufurth, M., X. Wang, H.C. Schröder, Q. Feng, B. Diehl-Seifert, T. Ziebart, et al., 2014. Engineering a morphogenetically active hydrogel for bioprinting of bioartificial tissue derived from human osteoblast-like SaOS-2 cells. Biomaterials. 35: 8810–8819.

Neufurth, M., X. Wang, S. Wang, R. Steffen, M. Ackermann, N.D. Haep, et al., 2017. 3D printing of hybrid biomaterials for bone tissue engineering: calcium-polyphosphate microparticles encapsulated by polycaprolactone. Acta Biomaterialia. 64): 377–388.

Nudelman, F., A.J. Lausch, N.A.J.M. Sommerdijk and E.D. Sone. 2013. *In vitro* models of collagen biomineralization. Journal of Structural Biology. 183(2): 258–269.

Ong, C.S., P. Yesantharao, C.Y. Huang, G. Mattson, J. Boktor, T. Fukunishi, et al., 2018. 3D bioprinting using stem cells. Pediatric Research. 83(1–2): 223–231.

Osidak, E.O., V.I. Kozhukhov, M.S. Osidak and S.P. Domogatsky. 2020. Collagen as bioink for bioprinting: a comprehensive review. International Journal of Bioprinting. 6(3): 270.

Ostrowski, N., A. Roy and P.N. Kumta. 2016. Magnesium phosphate cement systems for hard tissue applications: a review. ACS Biomaterials Science and Engineering. 2(7): 1067–1083.

Ozbolat, I.T. 2015. Bioprinting scale-up tissue and organ constructs for transplantation. Trends in Biotechnology. 33(7): 395–400.

Panzavolta, S., B. Bracci, K. Rubini and A. Bigi. 2011. Optimization of a biomimetic bone cement: role of DCPD. Journal of Inorganic Biochemistry. 105(8): 1060–1065.

Park, J.Y., J.C. Choi, J.H. Shim, J.S. Lee, H. Park, S.W. Kim, et al., 2014. A comparative study on collagen type I and hyaluronic acid dependent cell behavior for osteochondral tissue bioprinting. Biofabrication. 6(3): 35004.

Park, J.Y., J.H. Shim, S.A. Choi, J. Jang, M. Kim, S.H. Lee, et al., 2015. 3D printing technology to control BMP-2 and VEGF delivery spatially and temporally to promote large-volume bone regeneration. Journal of Materials Chemistry B. 3: 5415–5425.

Park, J., S.J. Lee, S. Chung, J.H. Lee, W.D. Kim, J.Y. Lee, et al., 2017. Cell-laden 3D bioprinting hydrogel matrix depending on different compositions for soft tissue engineering: characterization and evaluation. Materials Science and Engineering C. 71: 678–684.

Park, J., V.K. Kaliannagounder, S.R. Jang, D. Yoon, A.I. Rezk, D.P. Bhattarai, et al., 2021. Electroconductive polythiophene nanocomposite fibrous scaffolds for enhanced osteogenic differentiation via electrical stimulation. ACS Biomaterials Science and Engineering. 8: 1975–1986.

Penolazzi, L., E. Tavanti, R. Vecchiatini, E. Lambertini, F. Vesce, R. Gambari, et al., 2010. Encapsulation of mesenchymal stem cells from wharton's jelly in alginate microbeads. Tissue Engineering Part C: Methods. 16(1): 141–155.

Pfeiffenberger, M., A. Damerau, A. Lang, F. Buttgereit, P. Hoff and T. Gaber. 2021. Fracture Healing Research—Shift towards *in vitro* Modeling? Biomedicines. 9(7): 748.

Phillippi, J.A., E. Miller, L. Weiss, J. Huard, A. Waggoner and P. Campbell. 2008. Microenvironments engineered by inkjet bioprinting spatially direct adult stem cells toward muscle- and bone-like subpopulations. Stem Cells. 26(1): 127–134.

Plochocki, J.H., J.P. Rivera, C. Zhang and S.A. Ebba. 2008. Bone modeling response to voluntary exercise in the hindlimb of mice. Journal of Morphology. 269(3): 313–318.

Poldervaart, M.T., H. Gremmels, K.V. Deventer, J.O. Fledderus, F.C. Öner, M.C. Verhaar, et al., 2014. Prolonged presence of VEGF promotes vascularization in 3D bioprinted scaffolds with defined architecture. Journal of Controlled Release. 184(1): 58–66.

Porter, J.R., T.T. Ruckh and K.C. Popat. 2009. Bone tissue engineering: a review in bone biomimetics and drug delivery strategies. Biotechnology Progress. 25(6): 1539–1560.

Qin, C. and C. Wu. 2022. Inorganic biomaterials-based bioinks for three-dimensional bioprinting of regenerative scaffolds. VIEW. 3(4): 20210018.

Raffa, M.L., V.-H. Nguyen, P. Hernigou, C.-H. Flouzat-Lachaniette and G. Haiat. 2021. Stress shielding at the bone-implant interface: influence of surface roughness and of the bone-implant contact ratio. Journal of Orthopaedic Research. 39(6): 1174–1183.

Raja, N. and H.S. Yun. 2016. A simultaneous 3D printing process for the fabrication of bioceramic and cell-laden hydrogel core/shell scaffolds with potential application in bone tissue regeneration. Journal of Materials Chemistry B. 4: 4707–4716.

Ravanbakhsh, H., V. Karamzadeh, G. Bao, L. Mongeau, D. Juncker and Y.S. Zhang. 2021. Emerging technologies in multi-material bioprinting. Advanced Materials. 33: 2104730.

Reznikov, N., M. Bilton, L. Lari, M.M. Stevens and R. Kröger. 2018. Fractal-like hierarchical organization of bone begins at the nanoscale. Science. 360(6388): eaao2189.

Rezwan, K., Q.Z. Chen, J.J. Blaker and A.R. Boccaccini. 2006. Biodegradable and bioactive porous polymer/inorganic composite scaffolds for bone tissue engineering. Biomaterials. 27(18): 3413–3431.

Rhee, S., J.L. Puetzer, B.N. Mason, C.A. Reinhart-King and L.J. Bonassar. 2016. 3D bioprinting of spatially heterogeneous collagen constructs for cartilage tissue engineering. ACS Biomaterials Science and Engineering. 2: 1800–1805.

Rodriguez-Palomo, A., Østergaard, M. and Birkedal, H. 2024. Bone hierarchical structure: heterogeneity and uniformity. Advanced Functional Materials. 34: 2307026. https://doi.org/10.1002/adfm.202307026

Romanazzo, S., T.G. Molley, S. Nemec, K. Lin, R. Sheikh, J.J. Gooding, et al., 2021. Synthetic bone-like structures through omnidirectional ceramic bioprinting in cell suspensions. Advanced Functional Materials. 31(13): 2008216.

Rouzic, M.L., T. Chaussadent, G. Platret and L. Stefan. 2017. Mechanisms of K-struvite formation in magnesium phosphate cements. Cement and Concrete Research. 91: 117–122.

Salgado, A.J., O.P. Coutinho and R.L. Reis. 2004. Bone tissue engineering: state of the art and future trends. Macromolecular Bioscience. 4(8): 743–765.

Sanz-Sánchez, I., I. Sanz-Martín, A. Ortiz-Vigón, A. Molina and M. Sanz. 2022. Complications in bone-grafting procedures: classification and management. Periodontology 2000. 88(1): 86–102.

Sarker, M.D., S. Naghieh, A.D. McInnes, L. Ning, D.J. Schreyer and X. Chen. 2019. Biofabrication of peptide-modified alginate scaffolds: printability, mechanical stability and neurite outgrowth assessments. Bioprinting. 14: e00045.

Sawkins, M.J., P. Mistry, B.N. Brown, K.M. Shakesheff, L.J. Bonassar and J. Yang. 2015. Cell and protein compatible 3D bioprinting of mechanically strong constructs for bone repair. Biofabrication. 7(3): 35004.

Schlundt, C., T.E. Khassawna, A. Serra, A. Dienelt, S. Wendler, H. Schell, et al. 2018. Macrophages in bone fracture healing: their essential role in endochondral ossification. Bone. 106: 78–89.

Schwab, A., R. Levato, M. D'Este, S. Piluso, D. Eglin and J. Malda. 2020. Printability and shape fidelity of bioinks in 3D bioprinting. Chemical Reviews. 120(19): 11028–11055.

Semba, J.A., A.A. Mieloch and J.D. Rybka. 2020. Introduction to the state-of-the-art 3D bioprinting methods, design, and applications in orthopedics. Bioprinting. 18: e00070.

Shanjani, Y., C.C. Pan, L. Elomaa and Y. Yang. 2015. A novel bioprinting method and system for forming hybrid tissue engineering constructs. Biofabrication. 7(4): 45008.

Sheen, J.R., A. Mabrouk and W. Garla. 2023. Fracture healing overview. [Updated 2023 Apr 8]. *In*: StatPearls [Internet]. Treasure Island (FL): StatPearls Publishing; 2025. Available from: https://www.ncbi.nlm.nih.gov/books/NBK551678/

Sheikh, Z., M.N. Abdallah, A.A. Hanafi, S. Misbahuddin, H. Rashid and M. Glogauer. 2015. Mechanisms of *in vivo* degradation and resorption of calcium phosphate based biomaterials. Materials. 8(11): 7913–7925.

Shim, J.H., K.M. Jang, S.K. Hahn, J.Y. Park, H. Jung, K. Oh, et al., 2016. Three-dimensional bioprinting of multilayered constructs containing human mesenchymal stromal cells for osteochondral tissue regeneration in the rabbit knee joint. Biofabrication. 8(1): 14102.

Shoichet, M.S., R.H. Li, M.L. White and S.R. Winn. 1996. Stability of hydrogels used in cell encapsulation: an *in vitro* comparison of alginate and agarose. Biotechnology and Bioengineering. 50(4): 374–381.

Simunovic, F. and G. Finkenzeller. 2021. Vascularization strategies in bone tissue engineering. Cells. 10(7): 1749.

Sun, X., Y. Liu, Y. Wei and Y. Wang. 2022. Chirality-induced bionic scaffolds in bone defects repair—a review. Macromolecular Bioscience. 22(5): 202100502.

Szwed-Georgiou, A., P. Płociński, B. Kupikowska-Stobba, M.M. Urbaniak, P. Rusek-Wala, K. Szustakiewicz, et al. 2023. Bioactive materials for bone regeneration: biomolecules and delivery systems. ACS Biomaterials Science and Engineering. 9(9): 5222–5254.

Tao, R., B. Mi, Y. Hu, S. Lin, Y. Xiong, X. Lu, et al., 2023. Hallmarks of peripheral nerve function in bone regeneration. Bone Research. 11(1): 6.

Tavafoghi, M., M.A. Darabi, M. Mahmoodi, R. Tutar, C. Xu, A. Mirjafari, et al., 2021. Multimaterial bioprinting and combination of processing techniques towards the fabrication of biomimetic tissues and organs. Biofabrication. 13(4): 042002.

Teixeira, M.C., K.K. Singh, B.A.G. de Melo, P. Severino, J.C. Cardoso and E.B. Souto. 2022. 3D bioprinting: an innovative technique for biofabrication applied to regenerative medicine and tissue engineering. pp. 195–232. *In*: M.H. Santana, E.B. Souto and R. Shegokar (eds). Nanotechnology and Regenerative Medicine: History, Techniques, Frontiers, and Applications. Elsevier.

Thamma, U., T.J. Kowal, M.M. Falk and H. Jain. 2021. Nanostructure of bioactive glass affects bone cell attachment via protein restructuring upon adsorption. Scientific Reports. 11(1): 5763.

Toosi, S. and J. Behravan. 2020. Osteogenesis and bone remodeling: a focus on growth factors and bioactive peptides. BioFactors. 46(3): 326–340.

Turnbull, G., J. Clarke, F. Picard, P. Riches, L. Jia, F. Han, et al., 2018. 3D bioactive composite scaffolds for bone tissue engineering. Bioactive Materials. 3(3): 278–314.

Vandrovcová, M., T. Douglas, D. Hauk, B. Grössner-Schreiber, J. Wiltfang, L. Bačáková, et al. 2011. Influence of collagen and chondroitin sulfate (CS) coatings on poly-(lactide-co-glycolide) (PLGA) on MG 63 osteoblast-like cells. Physiological Research. 60(5): 797–813.

Verma, R., S.R. Mishra, V. Gadore and Md. Ahmaruzzaman. 2023. Hydroxyapatite-based composites: excellent materials for environmental remediation and biomedical applications. Advances in Colloid and Interface Science. 315: 102890.

Vorndran, E., A. Ewald, F.A. Müller, K. Zorn, A. Kufner and U. Gbureck. 2011. Formation and properties of magnesium–ammonium–phosphate hexahydrate biocements in the Ca–Mg–PO_4 system. Journal of Materials Science: Materials in Medicine. 22(3): 429–436.

Vyas, A., S.B. Ghosh, S. Bandyopadhyay-Ghosh, A.K. Agrawal, D. Khare and A.K. Dubey. 2022. Digital light processing mediated 3D printing of biocomposite bone scaffolds: physico-chemical interactions and *in-vitro* biocompatibility. Polymer Composites. 43(5): 3175–3188.

Vyas, A., S. Mondal, V.S. Kumawat, S.B. Ghosh, D. Mishra, J. Sen, et al., 2024. Biomineralized fluorocanasite-reinforced biocomposite scaffolds demonstrate expedited osteointegration of critical-sized bone defects. Journal of Biomedical Materials Research—Part B Applied Biomaterials. 112(1).

Woodruff, M.A. and D.W. Hutmacher. 2010. The return of a forgotten polymer—polycaprolactone in the 21st century. Progress in Polymer Science. 35(10): 1217–1256.

Wu, W., A. Deconinck and J.A. Lewis. 2011. Omnidirectional printing of 3D microvascular networks. Advanced Materials. 23: H178–H183.

Wu, S., X. Liu, K.W.K. Yeung, C. Liu and X. Yang. 2014. Biomimetic porous scaffolds for bone tissue engineering. Materials Science and Engineering R: Reports. 80(1): 1–36.

Wu, A.-M., C. Bisignano, S.L. James, G.G. Abady, A. Abedi, E. Abu-Gharbieh, et al. 2021. Global, regional, and national burden of bone fractures in 204 countries and territories, 1990–2019: a systematic analysis from the Global Burden of Disease Study 2019. The Lacet Healthy Longevity. 2(9): e580–e592.

Wu, Y., Y. Zhang, M. Yan, G. Hu, Z. Li, W. He, et al., 2024. Research progress on the application of inkjet printing technology combined with hydrogels. Applied Materials Today. 36: 102036.

Xu, H.H., P. Wang, L. Wang, C. Bao, Q. Chen, M.D. Weir, et al., 2017. Calcium phosphate cements for bone engineering and their biological properties. Bone Research. 5(1): 17056.

Xu, Y., F. Zhang, W. Zhai, S. Cheng, J. Li and Y. Wang. 2022. Unraveling of advances in 3d-printed polymer-based bone scaffolds. Polymers. 14(3): 566.

Yang, S.S., W.H. Choi, B.R. Song, H. Jin, S.J. Lee, S.H. Lee, et al., 2015. Fabrication of an osteochondral graft with using a solid freeform fabrication system. Tissue Engineering and Regenerative Medicine. 12(4): 239–248.

Yang, X., Y. Wang, Y. Zhou, J. Chen and Q. Wan. 2021. The application of polycaprolactone in three-dimensional printing scaffolds for bone tissue engineering. Polymers. 13(16): 2754.

Yao, Q., B. Wei, N. Liu, C. Li, Y. Guo, A.N. Shamie, et al., 2015. Chondrogenic regeneration using bone marrow clots and a porous polycaprolactone-hydroxyapatite scaffold by three-dimensional printing. Tissue Engineering—Part A. 21(7–8): 1388–1397.

Yazdanpanah, Z., M.E. Bahrololoom and B. Hashemi. 2015. Evaluating morphology and mechanical properties of glass-reinforced natural hydroxyapatite composites. Journal of the Mechanical Behavior of Biomedical Materials. 41: 36–42.

Yin, X., C. Zhou, J. Li, R. Liu, B. Shi, Q. Yuan, et al., 2019. Autophagy in bone homeostasis and the onset of osteoporosis. Bone Research. 7(1): 28.

You, F., X. Wu, N. Zhu, M. Lei, B.F. Eames and X. Chen. 2016. 3D printing of porous cell-laden hydrogel constructs for potential applications in cartilage tissue engineering. ACS Biomaterials Science and Engineering. 2: 1200–1210.

Yu, C., X. Ma, W. Zhu, P. Wang, K.L. Miller, J. Stupin, et al., 2019. Scanningless and continuous 3D bioprinting of human tissues with decellularized extracellular matrix. Biomaterials. 194(12): 1–13.

Zeng, D., L. Xia, W. Zhang, H. Huang, B. Wei, Q. Huang, et al., 2012. Maxillary sinus floor elevation using a tissue-engineered bone with calcium-magnesium phosphate cement and bone marrow stromal cells in rabbits. Tissue Engineering—Part A. 18 (7–8): 870–881.

Zeng, J.-H., S.-W. Liu, L. Xiong, P. Qiu, L.-H. Ding, S.-L. Xiong, et al., 2018. Scaffolds for the repair of bone defects in clinical studies: a systematic review. Journal of Orthopaedic Surgery and Research. 13(1): 33.

Zhai, P., X. Peng, B. Li, Y. Liu, H. Sun and X. Li. 2020. The application of hyaluronic acid in bone regeneration. International Journal of Biological Macromolecules. 151: 1224–1239.

Zhang, J.T., F. Tancret and J.M. Bouler. 2011. Mechanical properties of calcium phosphate cements (CPC) for bone substitution: influence of fabrication and microstructure. Key Engineering Materials. 493–494r): 409–414.

Zhang, Y.S., K. Yue, J. Aleman, K. Mollazadeh-Moghaddam, S.M. Bakht, J. Yang, et al., 2017. 3D bioprinting for tissue and organ fabrication. Annals of Biomedical Engineering. 45(1): 148–163.

Zhang, Y.S., G. Haghiashtiani, T. Hübscher, D.J. Kelly, J.M. Lee, M. Lutolf, et al., 2021. 3D extrusion bioprinting. Nature Reviews Methods Primers. 1(1): 75.

Zhang, Y., T. Shu, S. Wang, Z. Liu, Y. Cheng, A. Li, et al., 2022. The osteoinductivity of calcium phosphate-based biomaterials: a tight interaction with bone healing. Frontiers in Bioengineering and Biotechnology. 10: 1–9.

Zhou, H., A.K. Agarwal, V.K. Goel and S.B. Bhaduri. 2013. Microwave assisted preparation of magnesium phosphate cement (MPC) for orthopedic applications: a novel solution to the exothermicity problem. Materials Science and Engineering: C. 33(7): 4288–4294.

Zhou, H., B. Liang, H. Jiang, Z. Deng and K. Yu. 2021. Magnesium-based biomaterials as emerging agents for bone repair and regeneration: from mechanism to application. Journal of Magnesium and Alloys. 9(3): 779–804.

Zhu, N., M.G. Li, D. Cooper and X.B. Chen. 2011. Development of novel hybrid poly(l-lactide)/chitosan scaffolds using the rapid freeze prototyping technique. Biofabrication. 3(3): 34105.

Zhu, Z., Y.-M. Wang, J. Yang and X.-S. Luo. 2017. Hyaluronic acid: a versatile biomaterial in tissue engineering. Plastic and Aesthetic Research. 4: 219–227.

Zioupos, P. and J.D. Currey. 1998. Changes in the stiffness, strength, and toughness of human cortical bone with age. Bone. 22(1): 57–66.

Chapter 5

Biofabrication of Poly(Hydroxybutyrate-*co*-Hydroxyvalerate)/ Poly(Lactic-*co*-Glycolic Acid) Electrospun Nanofibers as the Mediation of Strontium-Hydroxyapatite for Bone Tissue Engineering

Mohd Izzat Hassan[1], Naznin Sultana[2], Kugambikai Vangetaraman[1], Dida Faadihilah Khrisna[1], Rathosivan Gopal[1] and Syafiqah Saidin*[1,3]

[1]Dept. of Biomedical Engineering & Health Sciences, Faculty of Electrical Engineering, Universiti Teknologi Malaysia, 81310 UTM Johor Bahru, Johor, Malaysia

[2]Medical Academy, Prairie View A&M University, TX77446, Texas, USA

[3]IJN-UTM Cardiovascular Engineering Centre, Institute of Human Centered Engineering, Universiti Teknologi Malaysia, 81310 UTM Johor Bahru, Johor, Malaysia

*For Correspondence: syafiqahs@utm.my

5.1 INTRODUCTION

Electrospinning is a favorable technique for the biofabrication of tissue engineered scaffolds which are regarded to facilitate cellular expression, growth, and differentiation. This technique allows the combination of biomaterials to achieve a significant progress in the production of nanofibrous scaffolds potential for tissue regeneration (Behere and Ingavle 2022). The key advantage of electrospinning is the ability to form nanofibrous structures with high surface area to volume ratio that possess high porosity, suitable to be employed as tissue engineered-scaffolds (Chinnappan et al. 2022).

5.2 ELECTROSPINNING TECHNIQUE FOR TISSUE-ENGINEERED NANOFIBROUS SCAFFOLD FORMATION

Formhals laid the foundation for electrospinning process which had a resurgence of popularity in the 1990s (Anjum et al. 2022). Electrospinning is an effective technique for mass production of fibers, ranging from nanometers to micrometers. This technique can be used to develop a two-dimensional (2D) fiber mat. In producing fiber's structure, electrospinning applies an effortless and efficient technique that utilizes high electrostatic forces to draw a continuous filament from a polymeric solution (Jun et al. 2018).

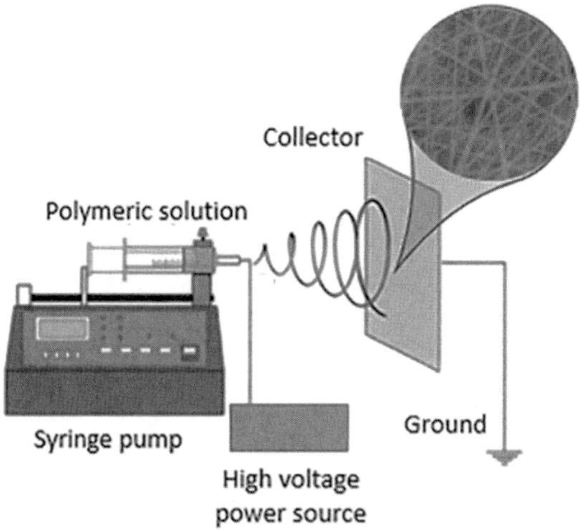

Figure 5.1 Basic electrospinning set up consists of syringe pump, high voltage power source, syringe loaded with polymeric solution and metal collector.

5.2.1 Working Principle of Electrospinning

The fundamental of electrospinning setup includes syringe, syringe pump, high voltage power source, and collector with tunable parameters to produce desired fiber morphology as shown in Figure 5.1. In general, the operating principle of electrospinning is based on the potential difference to create a liquid jet, entailed by mechanical stretching, that cause fiber deposition due to the jet elongation (Xie et al. 2020). By supplying high voltage between the spinneret and the grounded collector, liquid jet will be formed at the tip of the syringe needle, forming a Taylor cone. The drawn out of the liquid jet from the tip of the syringe needle is influenced by the greater electrostatic forces that emanate above the surface tension of the polymeric solution. The jet then elongates and upon solvent evaporation, fibers are deposited on the collector (Vass et al. 2020).

5.2.2 Electrospinning Parameters

Numerous precise electrospinning setting can be set up as a framework to acquire required features of nanofibers in facilitating cellular activities (Behere and Ingavle 2022). These parameters are vital in tuning the fiber morphology and biomechanical performance (Parham et al. 2020). Electrospinning parameters can be majorly classified into solution parameters, processing parameters, and ambient parameters. Their effects toward fiber morphology are described in Table 5.1.

5.3 BIOMATERIALS FOR ELECTROSPINNING BIOFABRICATION

The European Society for Biomaterials (ESB) defines "biomaterial" as "a material intended to interface with biological systems to evaluate, treat, augment or replace any tissue, organ or function of the body" (Socci et al. 2023). Biomaterials, either natural or synthetic, are essential for inducing healing and restoring the initial biological activity (Chelu and Musuc 2023). Biomaterials such as metals, polymers, and ceramics had been widely implemented in the biomedical field as tissue-engineered scaffolds (Qu et al. 2019).

Polymers are organic large macromolecules composed of several chains of atom bonded covalently, with the existence of repeating units found in abundance known as 'mers' (Burke et al. 2019). Polymers are well established biomaterials implemented to fabricate scaffolds for regeneration of both soft and hard tissues (Gautam et al. 2022). Polymeric materials can be processed into a variety of forms, including fibers, films, bars, and viscous liquids. They can be categorized into non-biodegradable and biodegradable polymers (Al-Shalawi et al. 2023). Non-biodegradable polymers are typically inert such as ultra-high molecular weight polyethylene (UHMWPE), which serve its use as sliding surfaces of artificial joint, catheters, and heart valves (Festas et al. 2019). In contrast, biodegradable polymers are currently of great interest due to its occurrence to eliminate the issue of implant removal once the implant's function is fulfilled (Festas et al. 2019). Examples of biodegradable polymers which are widely exploited in the biomedical field include poly(lactic-*co*-glycolic acid) (PLGA), poly(3-hydroxybutyrate-cohydroxyvalerate) (PHBV), poly(caprolactone) (PCL), etc.

Table 5.1 Effects on fiber morphology based on its respective electrospinning parameters

Parameters	Effects on Fiber Morphology	References
Solution parameter		
Concentration	Lesser chain entanglement between molecules due to very low concentration causes the formation of beaded fibers. Progressive increment of the concentration of the polymeric solution is associated with the concomitant appearance of beaded fibers.	(Liu et al. 2020, Ibrahim and Klingner 2020)
Molecular weight	Increasing the molecular weight of electrospinning solution reduces the number of beads. However, a further rise in molecular weight could result in the formation of coarse and flat fibers.	(Liu et al. 2020, Inamuddin et al. 2020)
Viscosity	Less viscous solution increases the mobility of polymer chains to be in aligned direction during electrospinning, to form narrowed fibers with high degree of orientation, crystallinity, and strength. Too viscous solution tends to clog the needle.	(Ibrahim and Klingner 2020, Subrahmanya et al. 2021)
Surface tension	Lower surface tension favors the creation of nanofibers with smooth surfaces without beads while higher surface tension forms liquid droplets, effectuating beaded fibers.	(Liu et al. 2020, Sandri et al. 2020)
Conductivity	Higher conductivity forms long protruded fibers while poor conductivity solution impedes jet elongation to produce defect free fibers due to low electrostatic forces.	(Sandri et al. 2020, Vasconcelos and Bölgen 2022)
Process parameter		
Applied voltage	Higher applied voltages result in stronger electrostatic repulsive forces on the ejected jet which reduces fiber diameter and enables the solvent to evaporate quickly, thus preventing beads formation. Very high applied voltage may break the fibers due to excessive electrostatic repulsive forces.	(Ibrahim and Klingner 2020, Long et al. 2019)
Distance between needle and collector	Larger diameter and defective fibers are formed at shorter distance, while increasing distances lead to reduced fiber diameter due to more time for the jet to stretch. However, increasing the distance can also form thicker fibers as the electric field strength causes less jet acceleration.	(Liu et al. 2020, Sandri et al. 2020)
Flow rate	Higher flow rate may break the fibers due to the periodic drop of solution and further increment in flow rate forms the wet fibers. Reduced flow rate produces narrowed and uniform fiber as there is sufficient time for the jet to stretch and the solvent to evaporate thoroughly.	(Liu et al. 2020, Ibrahim and Klingner 2020)
Ambient parameter		
Humidity	Less humid environment expedites solvent evaporation and fiber solidification, resulting in increased fiber diameter, while high humid facilitates the discharge of the jet.	(Long et al. 2019, Purabgola and Balasubramaniam 2020)
Temperature	Higher temperatures increase the molecule's energy in the solution, hence increasing its electrical con-ductivity while decreasing its viscosity and surface tension, thus resulting in reduction in fiber diameter.	(Williams et al. 2018, Vasconcelos et al. 2022)

Ceramics constitute a diverse class of nonmetallic, inorganic crystalline materials (George et al. 2021). Ceramics are used as engineering materials in electronic industry and optical field, while in biomedical field, ceramics have been used in dental sciences and orthopedics (Gautam et al. 2022). Ceramic scaffolds have emerged as a prominent choice for bone regeneration applications due to their exceptional biocompatibility, nonimmunogenic nature, and osteoinductive properties, which facilitate cell recruitment and promote osteogenic differentiation (Dolcimascolo et al. 2019). While ceramic scaffolds offer several advantages for tissue engineering applications, their inherent brittleness, and slow degradation pose limitations to their widespread usage (Dolcimascolo et al. 2019). Ceramics can be classified into bioinert ceramics such as alumina and zirconia which do not interact extensively with the surrounding living tissues but primarily provide physical support. Another classification of bioceramics is bioactive ceramics such as bioglass and hydroxyapatite (HA) which contribute in stimulating strong osteoinductive response and boosting bone tissue formation (Sadia et al. 2023).

Polymeric and ceramic materials are playing an increasingly important role in bone tissue regeneration to improve bone remodeling while mitigating the risk of osteolysis (Ielo et al. 2022). Polymers are advantageous to be deployed in bone tissue engineering applications, owing to their biodegradability and mechanical properties. Meanwhile, ceramics emulate bone apatite which mimick bone-like structure to support genuine tissue regeneration. Hence, the adoption of electrospinning technique to fabricate electrospun nanofibrous scaffolds, consisting of both types of materials for synergistic functional properties, is valuable to develop tissue-engineered scaffolds specifically for bone development. In this chapter, the use of polymeric and ceramic materials are highlighted in fabricating poly(hydroxybutyrate-*co*-hydroxyvalerate)/poly(lactic-*co*-glycolic acid) (PHBV/PLGA) electrospun nanofibers incorporated with strontium-hydroxyapatite (Sr-HA) for bone tissue regeneration.

5.3.1 Poly(lactic-*co*-glycolic acid) (PLGA)

Poly(lactic-*co*-glycolic acid) has been approved by Food and Drug Administration (FDA) to be implemented in the human physiological environment (Jin et al. 2021). This polymer has attracted significant attention from researchers due to its tolerable biodegradability. Its main composition of glycolic acid and lactic acid provide the tunable degradation by adjusting the ratio of lactic acid to glycolic acid (Zhao et al. 2021). Its biodegradability primarily occurs through the processes of hydrolysis and ester bond autocatalytic degradation (Jin et al. 2021). The degradation rate of PLGA decreases with the increase in lactide:glycolide (LA:GA) ratio (Sun et al. 2022, Wang et al. 2019). Besides, PLGA exhibits admirable biocompatibility and minimal cytotoxicity, supporting its use in biomedical applications (Wang et al. 2019, Ma et al. 2023). A study conducted by Ajmal et al. using nanofiber PLGA-gelatin showed excellent cytocompatibility with the observation of full re-epithelialization, projecting its potential use for wound healing applications (Ajmal et al. 2023). However, PLGA is also known for its poor bioactivity, hydrophobicity, and substandard mechanical properties for load-bearing applications (Ma et al. 2023).

5.3.2 Poly(3-hydroxybutyrate-*co*-hydroxyvalerate) (PHBV)

Poly(3-hydroxybutyrate-*co*-3-hydroxyvalerate) is a semicrystalline linear polyester consisting of hydroxybutyrate copolymer of random segments of hydroxyvalerate (Mollaqasem et al. 2020). It is a biodegradable biopolymer, produced by microorganisms through a fermentation phase. Poly(3-hydroxybutyrate-*co*-3-hydroxyvalerate) has garnered significant attention in the biomedical field due to its remarkable properties, including low immunogenicity, excellent biocompatibility, high biodegradability, and superior mechanical properties (Nahanmoghadam et al. 2021). This polymer demonstrates nontoxicity as its degradation products—water, carbon dioxide, and D-3-hydroxybutyric acid—belong to natural constituents of the human blood (Pryadko et al. 2021). Another extraordinary property of PHBV is the piezoelectricity which is comparable to bone tissues that play an essential role in bone regeneration (Tahmasebi et al. 2020). A study by Karbowniczek et al. revealed that the electrospun PHBV blended with HA particles has fostered an environment that promoted cell integration as well as enhanced cell anchoring, projecting great potential in bone regeneration (Karbowniczek et al. 2021).

5.3.3 Hydroxyapatite (HA)

Hydroxyapatite with the chemical formula of $Ca_{10}(PO_4)_6(OH)_2$ is one type of calcium phosphate (Paterlini et al. 2020). Hydroxyapatite can be produced from variety of sources such as bovine bones, fish bones, fish scales, shells, plants, and mineral sources (Hussin et al. 2022). Hydroxyapatite closely resembles the hard tissue of bone and teeth, and also has favorable properties such as biocompatibility, nonimmunogenic, slow resorption, bioactivity, and desirable osteoconductivity, making it to be widely used in orthopedic and dental applications (Tapsir et al. 2018). Hydroxyapatite possesses similar X-ray diffraction patterns and chemical composition to those of natural bone (Hussin et al. 2022). Its crystallinity provides flexibility, strength, and toughness (Púa et al. 2023). However, despite these unique properties, the weak mechanical behavior of HA including intrinsic brittleness, poor tensile strength, low fracture toughness, and low wear resistance limit its range of applications which can be solved by the incorporation of polymer or composites (Nosrati et al. 2021). A study conducted by Lv et al. reported that the presence of HA in poly-L-lactic acid (PLLA) electrospun nanofibers promoted osteogenic induction in bone mesenchymal cells (BMSC) of rats, exhibiting its potential to repair torn rotator cuff (Lv et al. 2022).

5.3.4 Strontium-hydroxyapatite (Sr-HA)

Strontium (Sr) is one of the trace elements in the human bones with 99% of strontium (36–140 mg/kg) being deposited in the femur, lumbar spine, and iliac bone, while the remaining are precipitated in the extracellular fluid (You et al. 2022). Strontium is involved in the control of bone metabolism by increasing osteoblast formation and suppressing osteoclast activity (Zhang et al. 2020). It can be incorporated into the HA lattice and plays a vital role in bone mineralization

(Surmenev et al. 2019). The similar radii of Sr^{2+} (1.02 Å) and Ca^{2+} (1.00 Å) allows Sr^{2+} to spontaneously substitute Ca^{2+} in the HA lattice (Liu et al. 2022). The substitution of Sr into HA, known as strontium-hydroxyapatite((Sr-HA), exhibits remarkable bioactivity due to the ability of both materials to promote osteoblast activity and suppress osteoclast activity, resulting in reduced bone resorption and ultimately increasing bone formation (Melnik et al. 2019).

5.4 BIOFABRICATION OF SR-HA INCORPORATED PHBV/PLGA ELECTROSPUN NANOFIBERS

Polymer-ceramic composites are frequently utilized to improve bone integration between the graft material and the native host tissues (Basaran et al. 2022). In addressing bone regeneration, tissue-engineered composite scaffolds composed of biodegradable polymers of PHBV and PLGA, incorporated with bioactive and osteoconductive Sr-HA, were developed using an electrospinning technique. The fabrication of this bone scaffold begins by the synthesis of nano-emulsified Sr-HA, entailed by the preparation of electrospinning solution through the blending of PHBV, PLGA, and Sr-HA. The polymer-ceramic electrospinning composite was then electrospinned to produce PHBV/PLGA/Sr-HA nanofibrous scaffolds.

Materials for the preparation of Sr-HA, ammonium hydrogen phosphate ($(NH_4)_2HPO_4$), ammonium hydrogen carbonate (NH_4HCO_3), sodium hydroxide (NaOH), strontium nitrate ($Sr(NO_3)_2$), calcium nitrate tetrahydrate ($Ca(NO_3)_2.4H_2O$), acetone, and ammonium hydroxide ($NH_4(OH)$) were obtained from Sigma-Aldrich, USA. The materials used to fabricate the nanofibrous scaffolds included PHBV (PHV content 12 mol%), PLGA (lactide:glycolide = 85:15 with a molecular weight of 50,000–75,000). Analytical grade trifluoroethanol solvent was supplied by Sigma Aldrich, USA.

5.4.1 Synthesis of Nano-emulsified Sr-HA

Aqueous and oil solutions were prepared by dissolving $(NH_4)_2HPO_4$ (4.0375 g) and NH_4HCO_3 (1.1860 g) in deionized water (700 mL) while $Sr(NO_3)_2$ (1.1927 g) and $Ca(NO_3)_2.4H_2O$ (10.7343 g) in acetone (100 ml), respectively. The aqueous solution was maintained at pH 11 by adding NaOH with the final adjustment volume to 1 L. The two solutions were mixed for 0.5 min at room temperature, filtered and washed three times with deionized water to remove the organic solvent. The formed precipitates were then frozen and freeze-dried subsequently for 3 days to obtain Sr-HA powders.

5.4.2 Fabrication of Electrospun Nanofibers

Both 26% w/v solution of PHBV and PLGA were prepared in trifluoroethanol at 50:50 ratio. Nano-emulsified Sr-HA powders were then added to the solution at a concentration of 10% w/w. The resulting solution was loaded into a 5 mL

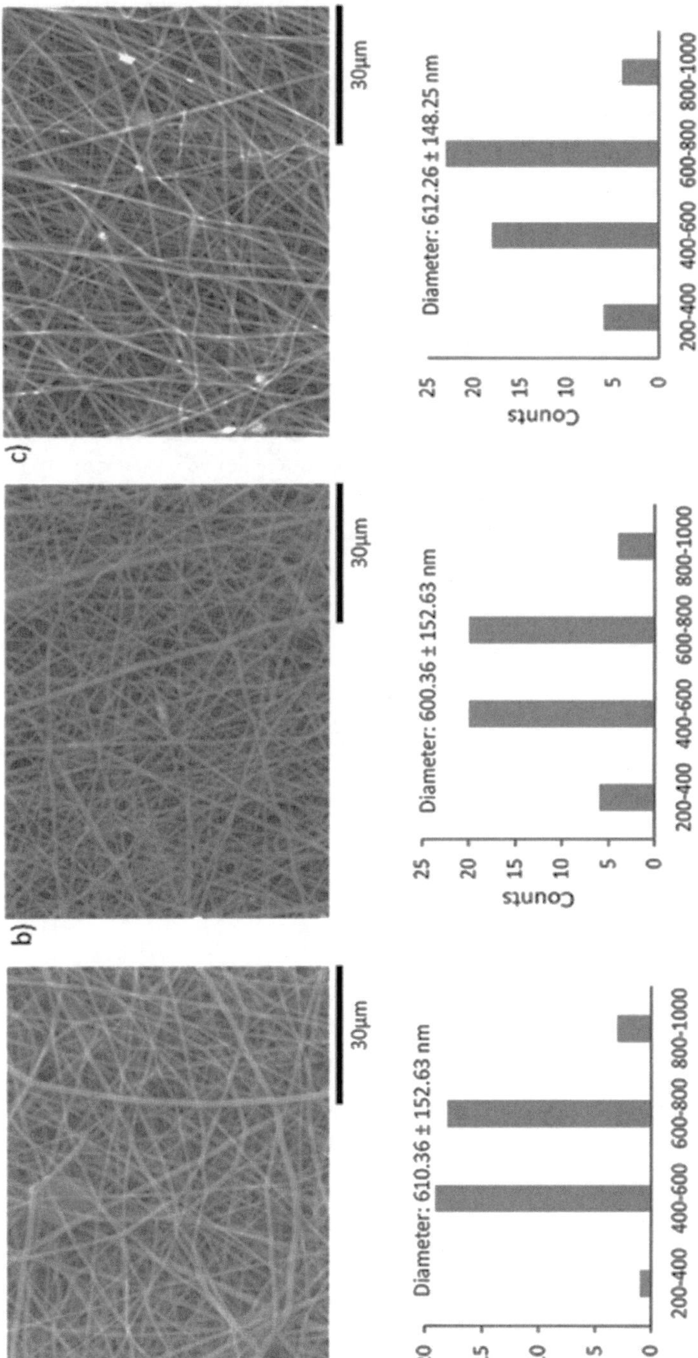

Figure 5.2 Morphology of nanofibrous scaffolds and fiber diameter distribution observed under SEM: (a) PHBV/PLGA, (b) PHBV/PLGA/HA, and (c) PHBV/PLGA/Sr-HA.

syringe and electrospun at a constant applied voltage of 20 kV, a flow rate of 3 mL/h, and a needle tip to collector distance of 10 cm.

5.5 PHYSICOCHEMICAL PROPERTIES

5.5.1 Nanofiber Morphology

Based on the previous studies in Table 5.1, the morphology of electrospun nanofibers is influenced by various polymer solution properties and processing parameters. In this study, the PHBV/PLGA solution was fixed at 26% w/v and 50:50 ratio for the formation of smooth and homogenous elongated nanofibers with random orientation, visualized under scanning electron microscopy (SEM) analysis. Figure 5.2 presents the morphology of PHBV/PLGA, PHBV/PLGA/HA, and PHBV/PLGA/Sr-HA along with their respective fiber diameter distribution and mean fiber diameters. There were no apparent changes in terms of mean fiber diameter observed between each electrospun nanofibrous scaffold. The incorporation of low concentration of 10% (w/w)Sr-HA nanoparticles produced uniformly well mixed homogenous electrospinning solution, eventually hindering nanoparticle agglomeration and fiber breaking (Zhang et al. 2022).

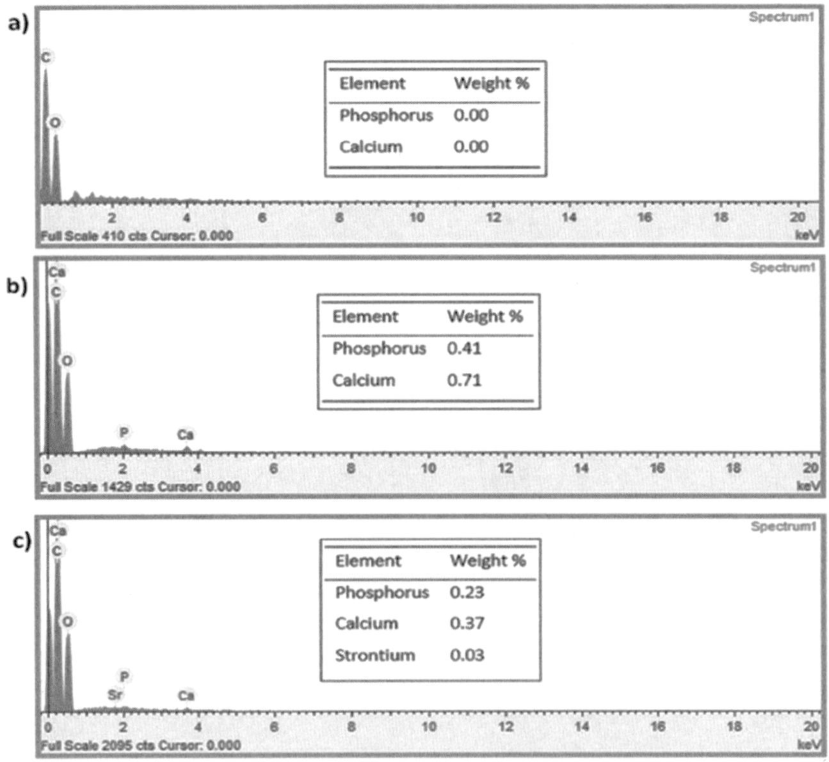

Figure 5.3 Records of elemental composition through EDX analysis: (a) PHBV/PLGA, (b) PHBV/PLGA/HA and (c) PHBV/PLGA/Sr-HA.

5.5.2 Nanofiber Composition

The successful incorporation of Sr-HA nanoparticles within the composite PHBV/PLGA nanofibrous scaffolds was verified through energy dispersive X-ray (EDX) analysis as shown in Figure 5.3. There was no appearance of calcium (Ca) and phosphorus (P) elements in the PHBV/PLGA nanofibrous scaffolds. The incorporation of HA into the polymer matrix caused noticeable Ca and P elements with the Ca/P ratio of 1.73, slightly above the theoretical stoichiometric of HA which is 1.67. The EDX analysis on the PHBV/PLGA/Sr-HA nanofibrous scaffolds showed presentable Ca and P elements with the Ca/P ratio of 1.61 and the Sr record of 0.03%.

5.5.3 Chemical Functionalities of Nanofibers

Attenuated total reflectance-Fourier transform spectroscopy (ATR-FTIR) analysis is widely used to assess the chemical bonding and functionalities of materials. The presences of PHBV and PLGA in the nanofibrous scaffolds were annotated at ATR-FTIR wavenumber of 1700 – 1800 cm^{-1} and 1040 – 1300 cm^{-1}, attributed to C=O stretching vibration and C–O stretching bands, respectively (Figure 5.4). Narrow wavenumber between 2900 – 3000 cm^{-1} corresponded to C–H stretching bands while 1300 – 1500 cm^{-1} corresponded to C–H bending vibrations, which were associated to the chemical structures of PHBV and PLGA (Hassan and Sultana 2019). Slightly higher intensity of PO_4^{3-} group at 1127 cm^{-1}, 1129 cm^{-1}, and 1053 cm^{-1} correlated to the incorporation of HA and Sr-HA in the composite nanofibrous scaffolds (Diez et al. 2021).

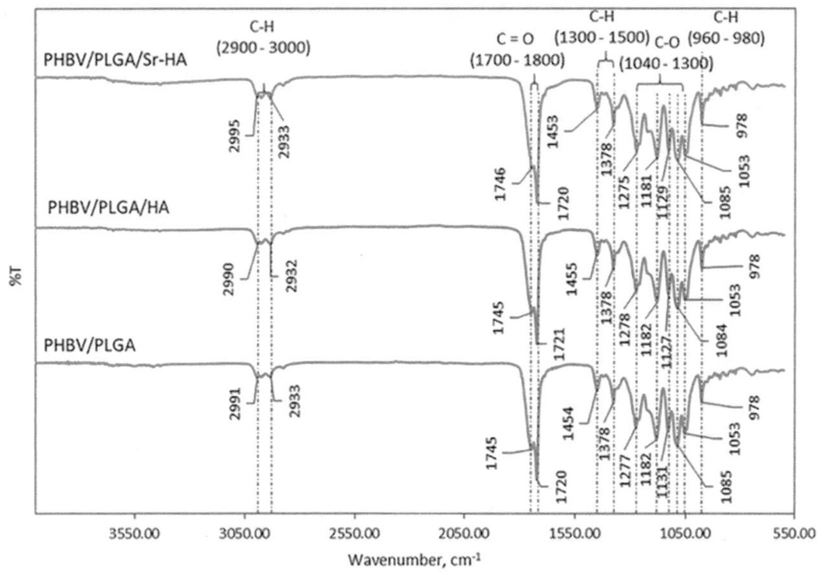

Figure 5.4 Annotation of chemical functionalities of PHBV/PLGA, PHBV/PLGA/HA, and PHBV/PLGA/Sr-HA on ATR-FTIR spectra.

5.6 BONE REGENERATION CAPABILITY OF Sr-HA INCORPORATED PHBV/PLGA NANOFIBERS

5.6.1 Bioactivity Formation of Flake-like Crystal Apatite

Bioactivity analysis can be used to predict the *in vivo* bone-binding ability which is essential for scaffold integration with the bone (Baino and Yamaguchi 2020). Combining PHBV/PLGA with bioactive materials such as HA and Sr-HA has induced the formation of bone-like apatite on the electrospun nanofiber surfaces. Figure 5.5 visualizes the morphologies of PHBV/PLGA, PHBV/PLGA/HA, and PHBV/PLGA/Sr-HA nanofibrous scaffolds after the immersion in simulated body fluid (SBF) solution for 7 days. The PHBV/PLGA nanofibrous scaffolds can be considered inert (Riau et al. 2020) as evidenced by the absence of apatite formation on the nanofiber surfaces. Meanwhile, the PHBV/PLGA/HA and PHBV/PLGA/Sr-HA nanofibrous scaffolds revealed excellent ability of apatite formation on the nanofiber surfaces through the display of flake-like crystal structure. The binding of PO_4^{3-} and CO_3^{2-} ions with Ca^{2+} ions in the SBF induced crystal growth, and led to the formation of apatite HA crystals, which is the main mineral composition in the human bone (Riau et al. 2020).

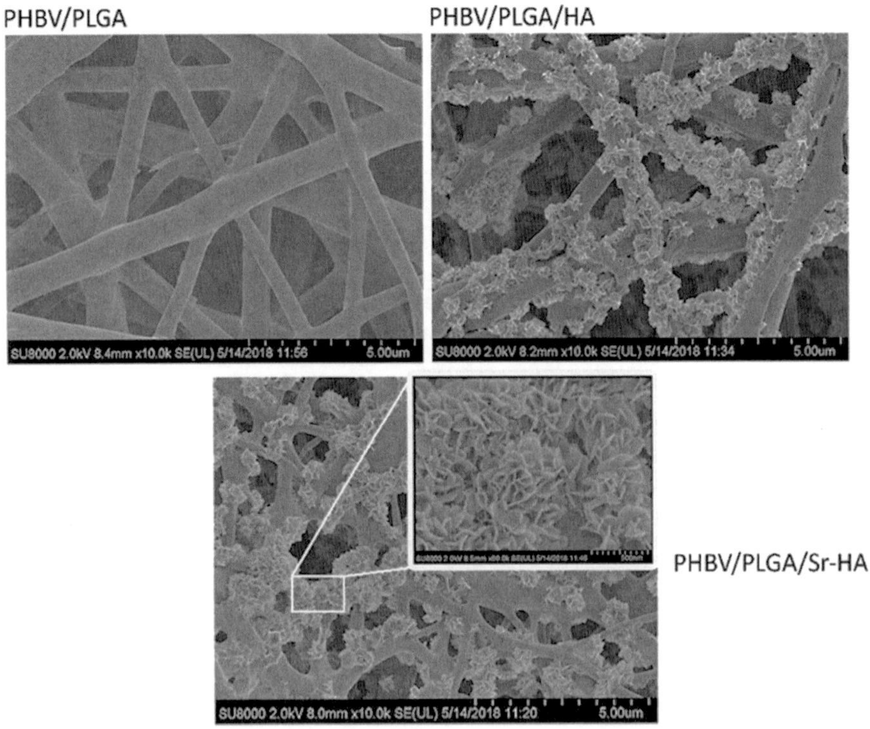

Figure 5.5 Formation of flake-like crystal apatite on PHBV/PLGA/HA and PHBV/PLGA/Sr-HA nanofiber surfaces, indicating active bone integration through mineral deposition.

5.6.2 Human Fetal Osteoblast (hFOB) Regeneration Structure

Osteoblast cells are also known as bone-forming cells that are responsible for osteogenesis and ossification. Human fetal osteoblast (hFOB) cells were observed adhered on/within the surfaces of all nanofibrous scaffolds during early incubation period. The hFOB cells exhibited flattened morphology, covering more integration area on the PHBV/PLGA/Sr-HA and PHBV/PLGA/HA nanofibrous scaffolds compared to the PHBV/PLGA as shown in Figure 5.6. The presence of bioceramics such as HA and Sr-HA have promoted bioactivity and accelerated the hFOB growth and integration (Kaur et al. 2020). Additionally, most cells were in multipolar polygonal morphologies with protruded filopodia and lamellipodia. The nano-topography of the electrospun scaffolds has also contributed in cell adherence by providing a large contact area for the cells to attach (Ding et al. 2019).

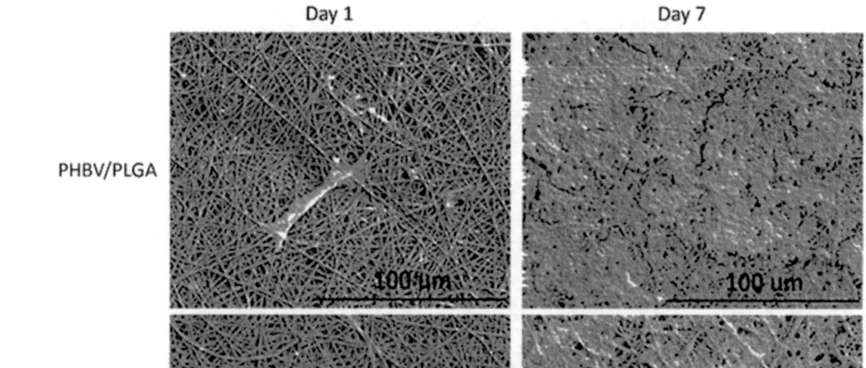

Figure 5.6 Transformation of multipolar polygonal hFOB into confluent monolayer integration on PHBV/PLGA, PHBV/PLGA/HA, and PHBV/PLGA/Sr-HA nanofibrous scaffolds after longer incubation period.

Longer incubation period, up to 7 days, led to the confluent monolayers of hFOB cells on all nanofibrous scaffolds with the cells randomly distributed over the entire surface. A good sign for cell-scaffold interactions was noticed, indicating preferential environment for the proliferation and growth of hFOB cells due to the presence of bioactive compounds such as HA and Sr-HA (Tsai et al. 2022).

The cytoskeleton of hFOB viewed under fluorescence microscope in Figure 5.7 extended its filopodia and lamellipodia on all scaffolds, supporting cell attachment and proliferation. Better cell spread morphology was observed on the nanofiber composite scaffolds of PHBV/PLGA/HA and PHBV/PLGA/Sr-HA due to the presence of HA and Sr-HA which enhanced proliferation of osteoblastic cells (Sarian et al. 2022). After 7 days of culture, the organization of actin stress fibers became more prominent, indicating higher proliferation of cells.

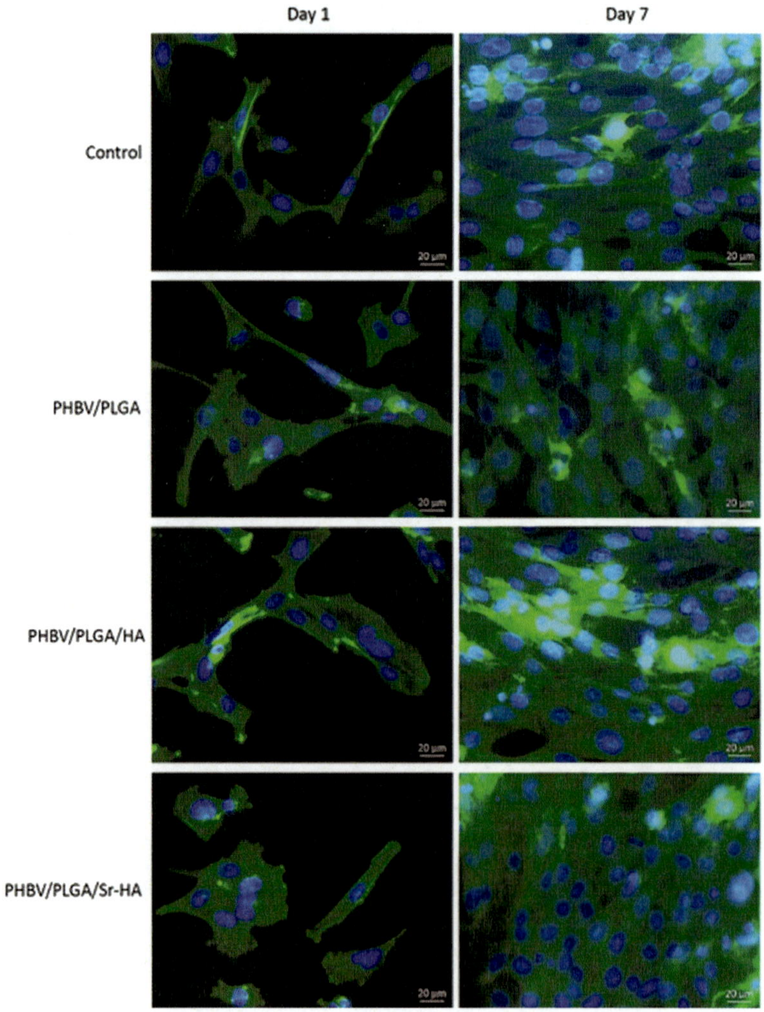

Figure 5.7 Fluorescence images of hFOB showing cytoskeletal actin filament (green) and nuclei (blue) on cultured control (without nanofibrous scaffold), PHBV/PLGA, PHBV/PLGA/HA, and PHBV/PLGA/Sr-HA.

The nanofiber's topography provided a high surface area-to-volume ratio that allowed cell attachment and proliferation of osteoblast, eventually promoting bone tissue regeneration (Barnthip et al. 2022). The incorporation of HA and Sr-HA contributed to bone mineralization where the rough electrospun nanofiber surfaces provided anchorage platform for cell adhesion (Ebrahimi et al. 2022, Tsai et al. 2022). Additionally, the role of individual material greatly supported osteoblastic cellular activities by influencing cell's capacity to proliferate within the nanofibrous scaffolds.

5.6.3 Human Fetal Osteoblast (hFOB) Regeneration Capability

2,5-diphenyl-2H-tetrazolium bromide (MTT) assay gauges the metabolic activity of cells which corresponds to the cell viability (Kamiloglu et al. 2020). The proliferation of hFOB cells on electrospun nanofibrous scaffolds was compared with tricalcium phosphate (TCP) which is known for its capability to regenerate hFOB (Figure 5.8). At day 1, the absorbance of cells on the PHBV/PLGA was significantly lower ($p < 0.05$) compared to the cells grown on the TCP (control), PHBV/PLGA/HA, and PHBV/PLGA/Sr-HA. The surface hydrophobicity and lack of bioactivity led to poor cell proliferation on the PHBV/PLGA nanofibrous scaffolds (Hassan and Sultana 2019). The hFOB cells cultured on the PHBV/PLGA/Sr-HA nanofibrous scaffolds demonstrated significantly enhanced proliferation ($p < 0.05$) in contrast to the PHBV/PLGA and PHBV/PLGA/HA. The PHBV/PLGA/Sr-HA and PHBV/PLGA/HA possessed higher cell affinity due to the presence of bioactive HA and Sr-HA, respectively, that greatly nourished bioactivity of the nanofibrous scaffolds by providing suitable environment to support cell adhesion, spreading, and proliferation (Bulina et al. 2022).

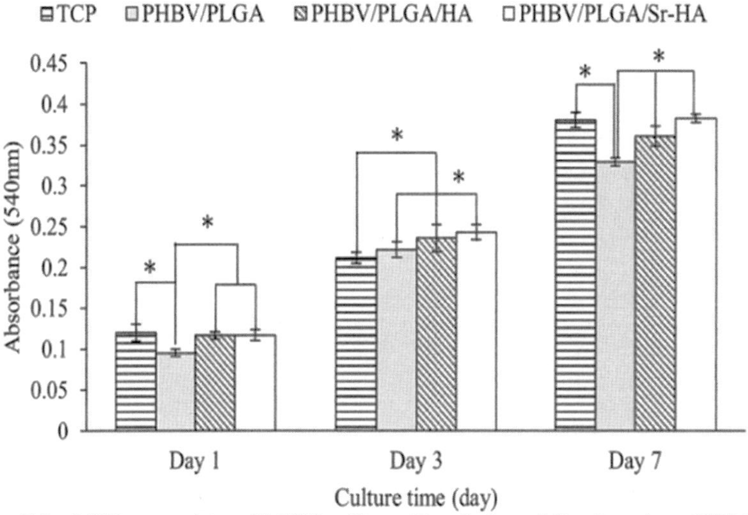

Figure 5.8 MTT assay data of hFOB cells on Day 1, 3, and 7 cultured on TCP, PHBV/PLGA, PHBV/PLGA/HA, and PHBV/PLGA/Sr-HA electrospun nanofibrous scaffolds.

The osteogenic differentiation of hFOB cells on the scaffolds of TCP, PHBV/PLGA, PHBV/PLGA/HA, and PHBV/PLGA/Sr-HA were evaluated by measuring alkaline phosphatase activity (ALP) activities at day 7, 14, and 21 as presented in Figure 5.9. ALP aids in the formation of calcium apatite found in the bone matrix and can be used as an early marker of osteoblast activity which converts organic pyrophosphate to inorganic phosphate (Ebrahimi et al. 2022). The ALP activity functions to determine osteogenesis as it is an important protein during the initiation of osteoblasts differentiation (Su et al. 2020).

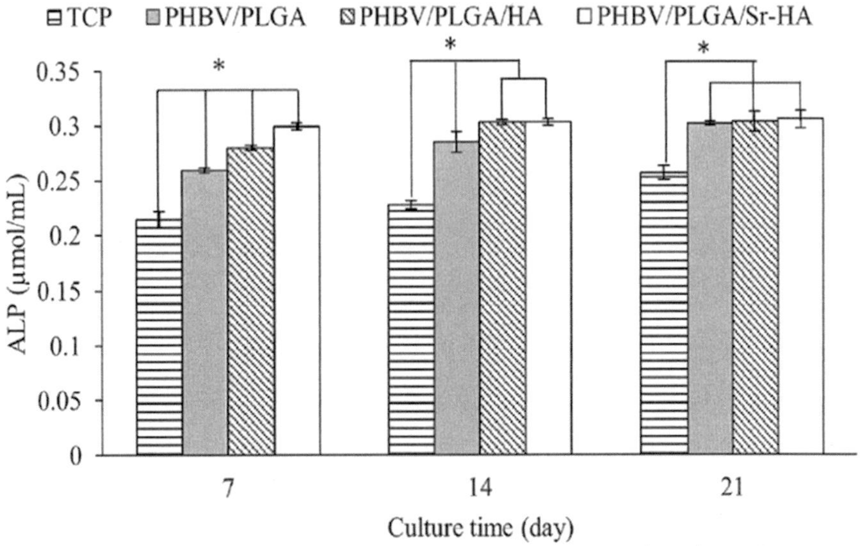

Figure 5.9 ALP activities of hFOB cells on day 7, 14, and 21, cultured on TCP, PHBV/PLGA, PHBV/PLGA/HA, and PHBV/PLGA/Sr-HA electrospun nanofibrous scaffolds.

The PHBV/PLGA/Sr-HA nanofibrous scaffolds have the highest levels of ALP activity ($p < 0.05$) compared to the other scaffolds. By day 14, the ALP activity has increased in all scaffolds but no apparent increase to be seen for the PHBV/PLGA/Sr-HA nanofibrous scaffolds. There was no significant variation in the ALP activity across the PHBV/PLGA, PHBV/PLGA/HA and PHBV/PLGA/Sr-HA scaffolds after 21 days of incubation ($p > 0.05$). The release of Sr ions into the local environment has triggered the activation of calcium-sensing receptors which induced hFOB differentiation, eventually promoting bone tissue regeneration (Cheshmedzhieva et al. 2021).

5.7 CONCLUSION

Biofabrication of nanofibers using electrospinning technique has been investigated extensively for various biomedical applications such as for the construction of tissue-engineered scaffolds, owing to its ability to support cell growth by mimicking the extracellular matrix. This chapter discusses the fabrication of electrospun nanofibrous bone scaffolds composed of PHBV/PLGA polymeric matrix incorporated with nano-emulsified Sr-HA bioceramic for bone tissue regeneration. The smooth and homogenous fiber distribution of PHBV/PLGA/Sr-HA has been observed with identical Sr-HA nanoparticles. The nanofibrous morphology of electrospun PHBV/PLGA/Sr-HA similar to the physiological tissue structure has greatly aided cellular adhesion with excellent osteoblasts regeneration capability. High affinity towards ALP has been achieved through the incorporation of Sr-HA within electrospun polymeric fibers of PHBV/PLGA, thus inducing tissue regeneration and projecting its potential use in bone tissue engineering application.

ACKNOWLEDGEMENT

The content was supported by Hi-Tech F4 Program: Project III from Department of Deputy Vice Chancellor (Research & Innovation), Universiti Teknologi Malaysia, through grant number Q.J130000.4609.00Q11.

REFERENCES

Ajmal G., G.V. Bonde, P. Mittal, V.K Pandey, N.Yadav and B. Mishra. 2023. PLGA/Gelatin-based electrospun nanofiber scaffold encapsulating antibacterial and antioxidant molecules for accelerated tissue regeneration. Materials Today Communications. 35: 105633.

Al-Shalawi, F.D., A.H.M. Ariff, D.W. Jung, M.K.A.M. Ariffin, C.L.S. Kim, D. Brabazon, et al., 2023. Biomaterials as implants in the orthopedic field for regenerative medicine: Metal versus synthetic polymers. Polymers (Basel). 15: 2601.

Anjum, S., F. Rahman, P. Pandey, D.K Arya, M. Alam, P. Rajinikanth, et al., 2022. Electrospun biomimetic nanofibrous scaffolds: A promising prospect for bone tissue engineering and regenerative medicine. International Journal of Molecular Sciences. 23: 9206.

Baino, F. and S. Yamaguchi. 2020. The use of simulated body fluid (SBF) for assessing materials bioactivity in the context of tissue engineering: review and challenges. Biomimetics (Basel). 5: 57.

Barnthip, N., J. Teeka, P. Kantha, S. Teepoo and W. Damjuti. 2022. Fabrication and characterization of polycaprolactone/cellulose acetate blended nanofiber mats containing sericin and fibroin for biomedical application. Scientific Reports 12: 22370.

Basaran, S.H., A. Bayrak, G. Tanriverdi, B. Tanriverdi and M.C. Avkan. 2022. Partial load-bearing rabbit ulnar segmental defects are regenerated with biocompatible grafts with or without bone marrow-derived mesenchymal stem cells. Turkish Journal of Trauma and Emergency Surgery. 28. 1066–1072.

Behere, I. and G. Ingavle. 2022. *In vitro* and *in vivo* advancement of multifunctional electrospun nanofiber scaffolds in wound healing applications: Innovative nanofiber designs, stem cell approaches, and future perspectives. Journal of Biomedical Materials Research A. 110: 443–461.

Bulina, N.V., O.B. Vinokurova, I.Y. Prosanov, A.M. Vorobyev, K.B. Gerasimov, I. Borodulina, et al., 2022. Mechanochemical synthesis of strontium- and magnesium-substituted and cosubstituted hydroxyapatite powders for a variety of biomedical applications. Ceramics International. 48: 35217–35226.

Burke, G., E. Kenny, M. Dalton, D.M. Devine, E. Hoctor, I. Major, et al., 2019. Biodegradation and biodegradable polymers. pp. 1–17. *In*: D. Declan (ed.). Bioresorbable Polymers: Biomedical Applications. Walter de Gruyter GmbH, Berlin/Boston.

Chelu, M. and A.M. Musuc. 2023. Advanced biomedical applications of multifunctional natural and synthetic biomaterials. Processes. 11: 2696.

Cheshmedzhieva, D., S. Ilieva, E.A. Permyakov, S.E. Permyakov and T. Dudev. 2021. Ca^{2+}/Sr^{2+} Selectivity in calcium-sensing receptor (CaSR): implications for strontium's anti-osteoporosis effect. Biomolecules. 11: 1576.

Chinnappan, B.A., M. Krishnaswamy, H. Xu and M.E. Hoque. 2022. Electrospinning of biomedical nanofibers/nanomembranes: effects of process parameters. Polymers (Basel). 14: 3719.

Diez, C., M.A. Rojo, J. Martín-Gil, P. Martín-Ramos, M. Garrosa and D. Córdoba-Diaz. 2021. Infrared spectroscopic analysis of the inorganic components from teeth exposed to psychotherapeutic drugs. Minerals. 12: 28.

Ding, J., J. Zhang, J. Li, D. Li, C. Xiao, H. Xiao, et al., 2019. Electrospun polymer biomaterials. Progress in Polymer Science. 90: 1–34.

Dolcimascolo, A., G. Calabrese, S. Conoci and R. Parenti. 2019. Innovative Biomaterials for Tissue Engineering. pp. 1–18. *In:* M. Barbeck, O. Jung, R. Smeets and T. Koržinskas (eds). Innovative Biomaterials for Tissue Engineering. IntechOpen.

Ebrahimi, Z., S. Irani, A. Ardeshirylajimi and E. Seyedjafari. 2022. Enhanced osteogenic differentiation of stem cells by 3D printed PCL scaffolds coated with collagen and hydroxyapatite. Scientific Reports. 12: 12359.

Festas, A.J., A. Ramos and J.P. Davim. 2019. Medical devices biomaterials—a review. Proceedings of the Institution of Mechanical Engineers L: Journal of Materials Design. 234: 218–228.

Gaurav, G., S. Kumar and K. Kumar. 2022. Processing of biomaterials for bone tissue engineering: state of the art. Materials Today Proceedings. 50: 2206–2217.

George, G., T. Senthil, Z. Luo and S. Anandhan. 2021. Sol-gel electrospinning of diverse ceramic nanofibers and their potential applications. pp. 689–764. *In*: Y. Dong, A. Baji and S. Ramakrishna (eds). Electrospun Polymers and Composites. Woodhead Publishing.

Haider, A., S. Haider and I. Kang. 2018. A comprehensive review summarizing the effect of electrospinning parameters and potential applications of nanofibers in biomedical and biotechnology. Arabian Journal of Chemistry. 11: 1165–1188.

Hassan, M.I. and N. Sultana. 2019. *In vitro* cell viability of PHBV/PLGA nanofibrous membrane for tissue engineering. Malasian Journal of Fundamental and Applied Sciences 15: 522–527.

Hussin, M.S.F., H.Z. Abdullah, M.I. Idris and M.A.A. Wahap. 2022. Extraction of natural hydroxyapatite for biomedical applications—a review. Heliyon. 8: e10356.

Ibrahim, H.M. and A. Klingner. 2020. A review on electrospun polymeric nanofibers: Production parameters and potential applications. Polymer Testing. 90: 106647.

Ielo, I., G. Calabrese, G. De Luca and S. Conoci. 2022. Recent advances in hydroxyapatite-based biocomposites for bone tissue regeneration in orthopedics. International Journal of Molecular Sciences. 23: 9721.

Jin, S., X. Xia, J. Huang, C. Yuan, Y. Zuo, Y. Li, et al., 2021. Recent advances in PLGA-based biomaterials for bone tissue regeneration. Acta Biomater. 127: 56–79.

Jun, I., H.S. Han, J.R. Edwards and H. Jeon. 2018. Electrospun fibrous scaffolds for tissue engineering: viewpoints on architecture and fabrication. International Journal of Molecular Sciences. 19: 745.

Kamiloglu, S., G. Sari, T. Ozdal and E. Capanoglu. 2020. Guidelines for cell viability assays. Food Frontiers. 1: 332–349.

Karbowniczek, J.E., L. Kaniuk, K. Berniak, A. Gruszczynski and U. Stachewicz. 2021. Enhanced cells anchoring to electrospun hybrid scaffolds with PHBV and HA particles for bone tissue regeneration. Frontiers in Bioengineering and Biotechnology. 9: 632029.

Kaur, P., K.J. Singh, A.K. Yadav, S. Kaur, R. Kaur and S. Kaur. 2020. Growth of bone like hydroxyapatite and cell viability studies on CeO_2 doped $CaO–P_2O_5–MgO–SiO_2$ bioceramics. Materials Chemistry and Physics. 243: 122352.

Liu, Y., M. Hao, Z. Chen, L. Liu, Y. Liu, W. Yang, et al., 2020. A review on recent advances in application of electrospun nanofiber materials as biosensors. Current Opinion in Biomedical Engineering. 13: 174–189.

Liu, Y., B. Zhang, F. Liu, Y. Qiu, W. Mu, L. Chen, et al., 2022. Strontium doped electrospinning fiber membrane with antibacterial and osteogenic properties prepared by pulse electrochemical method. Engineering and Regenerative. 3: 339–351.

Long, Y., X. Yan, X. Wang, J. Zhang and M. Yu. 2019. Electrospinning: The Setup and Procedure. pp. 21–52. *In*: B. Ding, X. Wang, J. Yu (eds). Electrospinning: Nanofabrication and Applications. William Andrew Publishing.

Lv, Y., X. Sang, Z. Tian, S. Jiang, C. Li, Q. Guo, et al., 2022. Electrospun hydroxyapatite loaded L-polylactic acid aligned nanofibrous membrane patch for rotator cuff repair. International Journal of Biological Macromolecules. 217: 180–187.

Ma, R., Y. Su, R. Cao, K. Wang and P. Yang. 2023. Enhanced osteogenic activity and bone repair ability of PLGA/MBG scaffolds doped with zif-8 nanoparticles loaded with BMP-2. International Journal of Nanomedicine. 18: 5055–5072.

Melnik, E.V., S.N. Shkarina, S.I. Ivlev, V. Weinhardt, T. Baumbach, et al., 2019. *In vitro* degradation behaviour of hybrid electrospun scaffolds of polycaprolactone and strontium-containing hydroxyapatite microparticles. Polymer Degradation and Stability. 167: 21–32.

Nahanmoghadam, A., M. Asemani, V. Goodarzi and S. Ebrahimi-Barough. 2021. *In vivo* investigation of PCL/PHBV/Hydroxyapatite nanocomposite scaffold in regeneration of critical-sized bone defects. Fibers Polymer. 22: 2507–2516.

Nosrati, H., R. Sarraf-Mamoory, M.C. Perez, D.Q.S. Le, R.Z. Emameh and C.E. Bünger. 2021. Characteristics of hydroxyapatite-reduced graphene oxide composite powders synthesized via hydrothermal method in the absence and presence of diethylene glycol. Open Ceramics. 5: 100067.

Parham, S., A.Z. Kharazi, H.R. Bakhsheshi-Rad, H. Ghayour, A.F. Ismail, H. Nur, et al., 2020. Electrospun nano-fibers for biomedical and tissue engineering applications: A comprehensive review. Materials (Basel). 13: 2153.

Paterlini, V., M. Bettinelli, R. Rizzi, A.E. Khouri, M. Rossi, G.D. Ventura, F. Capitelli, et al., 2020. Characterization and luminescence of Eu^{3+}- and Gd^{3+} - doped hydroxyapatite $Ca_{10}(PO_4)_6(OH)_2$. Crystals. 10: 806.

Pryadko, A., M.A. Surmeneva and R.A. Surmenev. 2021. Review of hybrid materials based on polyhydroxyalkanoates for tissue engineering applications. Polymers (Basel). 13: 1738.

Púa, G., L.D. Carmen, J.C.R. Montenegro, A.M.F Reyes, H.Z. Rodríguez and V.N.P Méndez. 2023. Biomaterials for orthopedic applications and techniques to improve corrosion resistance and mechanical properties for magnesium alloy: a review. Journal of Materials Science. 58: 3879–3908.

Purabgola, A. and K. Balasubramaniam. 2020. Physical Characterization of Electrospun Fibers. pp. 71–112. *In*: Inamuddin, R. Boddula, M.I. Ahamed and A.M. Asiri (eds). Electrospun Materials and Their Allied Applications. Scrivener Publishing. Beverly.

Qu, H., H. Fu, Z. Han and Y. Sun. 2019. Biomaterials for bone tissue engineering scaffolds: a review. RSC Advances. 9: 26252–26262.

Riau, A.K., S.S. Venkatraman and J.S. Mehta. 2020. Biomimetic vs. direct approach to deposit hydroxyapatite on the surface of low melting point polymers for tissue engineering. Nanomaterials (Basel). 10: 2162.

Richbourg, N.R., N.A. Peppas and V.I. Sikavitsas. 2019. Tuning the biomimetic behavior of scaffolds for regenerative medicine through surface modifications. Journal of Tissue Engineering and Regenerative Medicine. 13: 1275–1293.

Sadia, M., M.A.M. Zaki, S.K. Jaganathan, M.F.M. Shakhih, A.S Kamarozaman, N. Ab'lah, et al., 2023. Blending of *Moringa oleifera* into biodegradable polycaprolactone/silver electrospun membrane for hemocompatibility improvement. Arabian Journal for Science and Engineering. 48: 7323–7336.

Sandri, G., S. Rossi, M.C. Bonferoni, C. Caramella and F. Ferrari. 2020. Electrospinning Technologies in Wound Dressing Applications. pp. 315–226. *In*: J. Boateng (ed.). Therapeutic Dressings and Wound Healing Applications. John Wiley & Sons Ltd.

Sarian, M.N., N. Iqbal, P. Sotoudehbagha, M. Razavi, Q.U. Ahmed, C. Suktojo, et al., 2022. Potential bioactive coating system for high-performance absorbable magnesium bone implants. Bioactive Materials. 12: 42–63.

Silva, M., F.N. Ferreira, N.M. Alves and M.C. Paiva. 2020. Biodegradable polymer nanocomposites for ligament/tendon tissue engineering. Journal of Nanobiotechnology. 18: 23.

Socci, M.C., G. Rodriguez, E. Oliva, S. Fushimi, K. Takabatake, H. Nagatsuka, et al., 2023. Polymeric materials, advances and applications in tissue engineering: a review. Bioengineering (Basel). 10: 218.

Su, T., T.H. Huang, C.T. Kao, H.Y. Ng, Y.C. Chiu and T.T. Hsu. 2020. The calcium channel affect osteogenic differentiation of mesenchymal stem cells on strontium-substituted calcium silicate/poly-ε-caprolactone scaffold. Processes. 8: 198.

Subrahmanya, T.M., A.B. Arshad, P.T Lin, J. Widakdo, H.K. Makari, H.F.M. Austria, et al., 2021. A review of recent progress in polymeric electrospun nanofiber membranes in addressing safe water global issues. RSC Advances. 11: 9638–9663.

Sun, F., X. Sun, H. Wang, C. Li, Y. Zhao, J. Tian, et al., 2022. Application of 3D-Printed, PLGA-based scaffolds in bone tissue engineering. International Journal of Molecular Sciences. 23: 5831.

Surmenev, R.A., S. Shkarina, D.S. Syromotina, E.V. Melnik, R. Shakrin, I.I. Selezneva, et al., 2019. Characterization of biomimetic silicate- and strontium-containing hydroxyapatite microparticles embedded in biodegradable electrospun polycaprolactone scaffolds for bone regeneration. European Polymer Journal. 113: 67–77.

Tahmasebi, A., A.S. Moghadam, S.E. Enderami, M. Islami, M. Kaabi, E. Saburi, et al., 2020. Aloe vera-derived gel-blended PHBV nanofibrous scaffold for bone tissue engineering. ASAIO Journal. 66: 966–973.

Tapsir, Z., F.H. Jamaludin, B. Pingguan-Murphy and S. Saidin. 2018. Immobilisation of hydroxyapatite-collagen on polydopamine grafted stainless steel 316L: Coating adhesion and *in vitro* cells evaluation. Journal of Biomaterials Applications. 32: 987–995.

Tsai, S.W., Y.W. Hsu, W.L. Pan, A. Vadivelmurugan, P.A. Hwang and F.Y. Hsu. 2022. Influence of the components and orientation of hydroxyapatite fibrous substrates on osteoblast behavior. Journal of Functional Biomaterials. 13: 168.

Tunthasen, R., P. Pripatnanont and J. Meesane. 2022. *In vitro* biocompatibility of a novel semi-rigid shell barrier system: as a new application for guided bone regeneration. Polymers (Basel). 14: 2451.

Vahid, K.M., A. Asefnejad, M.R. Nourani, V. Goodarzi and M.R Kalaee. 2020. Incorporation of graphene oxide and calcium phosphate in the PCL/PHBV core-shell nanofibers as bone tissue scaffold. Journal of Applied Polymer Science. 138: 49797.

Vasconcelos, F., R.L. Reis, A. Martins and N.M. Neves. 2022. Biomedical applications of fibers produced by electrospinning, microfluidic spinning and combinations of both. pp. 251–295. *In:* A.Vaseashta and Nimet Bölgen (eds). Electrospun Nanofibers. Springer, Switzerland.

Vass, P., E. Szabo, A. Domokos, E. Hirsch, D. Galata, B. Farkaz, et al., 2020. Scale-up of electrospinning technology: applications in the pharmaceutical industry. Wiley Interdiscip. Rev. Nanomedicine and Nanobiotechnology. 12: e1611.

Wang, Z., R. Liang, X. Jiang, J. Xie, P. Cai, H. Chen, et al., 2019. Electrospun PLGA/PCL/OCP nanofiber membranes promote osteogenic differentiation of mesenchymal stem cells (MSCs). Material Science and Engineering C. Materials Biology and Applications 104: 109796.

Williams, G.R., B.T. Raimi-Abraham and C.J. Luo. 2018. Nanofibres in Drug Delivery: UCL Press, London.

Xie, X., Y. Chen, X. Wang, X. Xu, Y. Shen, A.U.R. Khan, et al., 2020. Electrospinning nanofiber scaffolds for soft and hard tissue regeneration. Journal of Material Science and Technology. 59: 243–261.

You, J., Y. Zhang and Y. Zhou. 2022. Strontium functionalized in biomaterials for bone tissue engineering: a prominent role in osteoimmunomodulation. Front, Bioengineering and Biotechnology. 10: 928799.

Zhang, Q., Y. Ji, W. Zheng, M. Yan, D. Wang, M. Li, et al., 2020. Electrospun nanofibers containing strontium for bone tissue engineering. Journal of Nanomaterials. 2020: 1–14.

Zhang, M., W. Song, Y. Tang, X. Xu, Y. Huang and D. Yu. 2022. Polymer-based nanofiber-nanoparticle hybrids and their medical applications. Polymers (Basel). 14: 351.

Zhao, D., T. Zhu, J. Li, L.Cui, Z. Zhang, X. Zhuang, et al., 2021. Poly(lactic-co-glycolic acid)-based composite bone-substitute materials. Bioactive Materials. 6: 346–360.

Chapter **6**

The Development and Application of Extrusion-based 3D Bioprinting

Sheril Amira Othman[1], Chin Fhong Soon*[1,2], Norhidayah Che Ani[1], Naznin Sultana[3], Farshid Sefat[4] and Kian Sek Tee[2]

[1]Microelectronics and Nanotechnology-Shamsuddin Research Center (MiNT-SRC), Universiti Tun Hussein Onn Malaysia, 86400 Parit Raja, Johor, Malaysia

[2] Faculty of Electrical and Electronic Engineering, Universiti Tun Hussein Onn Malaysia, 86400 Parit Raja, Johor, Malaysia

[3]Medical Academy, Prairie View A&M, TX77446, USA

[4]School of Engineering, University of Bradford, BD7 1DP Bradford, United Kingdom

6.1 INTRODUCTION

Conventional two-dimensional (2D) cell models have been widely used in biological studies to understand cell formation, functions, and pathology (Breslin and O'Driscoll 2013, Loessner et al. 2010). 2D cell culture involves growing cells on a flat surface, typically in culture dishes or plates, allowing researchers to observe and manipulate them under a microscope. This technique has been employed for many years and has contributed significantly to our understanding of cellular behavior and various biological processes.

2D cell models offer several advantages: simplicity, cost-effectiveness, and ease of handling. They allow for precise control of experimental conditions and

*For Correspondence: soon@uthm.edu.my

enable high-throughput screening of compounds or drugs. Additionally, 2D cell cultures can be used for various assays, including cell proliferation, migration, differentiation, and gene expression studies. However, it is essential to note that 2D cell models have limitations (Hoarau-Véchot et al. 2018). Cells grown in 2D lack the natural (3D) architecture found in living tissues, which can influence cell behavior and function (Kapałczyńska et al. 2018). They also lack the cellular interactions and microenvironmental cues present *in vivo*, which can affect the accuracy and relevance of experimental results. (Anton et al. 2015).

To address these limitations, researchers have increasingly used 3D cell models, such as organoids and tissue-engineered constructs, which better mimic the complexity of *in vivo* tissues. These 3D models provide a more physiologically relevant environment and allow for the study of cell-cell interactions, tissue organization, and response to stimuli (Anton et al. 2015). These native-like 3D cells are becoming an essential scientific tool in cell research. Therefore, researchers are encouraged to develop new techniques that can produce 3D cell models, such as 3D bioprinting technology.

3D bioprinting involves the 3D printing of biological tissues and organs through the process of layering the living cells, which begins by creating an architectural design of the fundamental composition of the target tissue or organ with the help of computer-aided design (CAD) (Shafiee and Atala 2016). The 3D printer that applies in bioprinting is commonly known as a 3D bioprinter. Based on the architectural design, the bioprinter will deposit a thin layer of cells using a printer head that moves according to the required configuration. The bioprinter uses the bioink to build the 3D structure of the organic materials. The bioink can be made of non-synthetic biomaterials mostly made up of dissolvable hydrogel components such as collagen, chitosan, pluronic, and hyaluronic acid (HA) (Gungor-Ozkerim et al. 2018). Hydrogel is a hydrophilic network of polymer chains that is highly absorbent and possesses flexibility similar to the natural tissues (Choudhary et al. 2018a). This hydrogel is considered to have great potential for medical purposes due to its ability to absorb large quantities of water; thus, in bioprinting, it is often used to support and protect the cells as well as act as a filler to fill empty spaces within the tissues (Chai et al. 2017).

The application of bioprinting technology in the biomedical field was inspired by the early development of the 3D printing system back in the 1900s (Jose et al. 2016). It has been rapidly commercialized due to the significant interests gained from biomaterials' development, medicine, and pharmaceutics industries. The current market offers bioprinters such as Organovo ExVive™, Aether I, FABION, and BioBot™, which work similarly but differ in the printing mechanism and bioink used (Choudhury et al. 2018b, Pereira et al. 2018). This rising technology has been adopted in tissue engineering and regenerative medicine to create functional living tissues and organs. It has great potential to keep pace with the growing demands for suitable organs and acts as an alternative to organ transplantation (Li et al. 2013). This new approach can create a structural and molecular environment that mimics significantly the properties of the native organ to support the recipient's cells. The 3D bioprinting allows the accurate evaluation of cell microenvironment that accelerates the development of cancer therapeutics in oncology studies (Charbe et al. 2017). The 3D tissue model produced by bioprinting techniques can also be used in medicine and pharmacology research for drug treatments and the study of diseases (Ventola 2014).

There are four types of bioprinting developed to produce 3D cells, namely stereolithography bioprinting (Morris et al. 2017, Raman and Bashir 2015), drop-on-demand or inkjet bioprinting (Ferris et al. 2013, Ng et al. 2017), extrusion bioprinting (Lee et al. 2016, Liu et al. 2017) and laser-assisted bioprinting (Ali et al. 2014, Keriquel et al. 2017). Stereolithography bioprinting works with photopolymers, which are unstable over time and might harm the cell. For inkjet-based bioprinting, the droplets tend to spread inconsistently and are unequal in size (Pantani and Turng 2015). Previous research on laser-assisted bioprinting showed that this method did not experience any nozzle clogging due to the nozzle printing setup (Sundaramurthi et al. 2016). However, it is an expensive method to use.

Extrusion-based bioprinting is indeed a simple and effective method for printing a wide range of biocompatible materials with different viscosities, and it offers shorter fabrication times than other bioprinting techniques. However, previous work mainly focused on printing scattered single cells in the bioink, which limited the ability of the cells to communicate due to the long distances between them (Gillentine et al. 2017).

To address this limitation, this research project proposed the inclusion of microtissues that simulate tumors embedded deep within tissues, with a thickness of 1 mm. The main objective was to develop a microextrusion-based bioprinter capable of extruding and printing 3D microtissue capsules coated with sodium alginate/gelatin hydrogels as the bioink. In this project, microspheres of HeLa or cervical cancer cells are first encapsulated with calcium alginate coating using a flicking machine to produce microtissue spheroids. These spheroids are mixed into the sodium alginate/gelatin bioink and printed to create 3D tubular structures or tubules. Calcium chloride ($CaCl_2$) is a crosslinker that strengthens the printed hydrogel structure. The system and methodology developed in this project can potentially improve the prediction of drug diffusion through multilayered tissue structures. Incorporating microtissues and using larger diameter nozzles to reduce shear stress can improve cell viability and functionality of printed structures. This research aims to advance extrusion-based bioprinting technology by incorporating microtissues and improving printing to fabricate more complex and functional 3D tissue models.

6.2 MATERIALS AND METHODS

6.2.1 Development of Extruding-based 3D Bioprinting System

Figure 6.1 shows a conceptual model of an extrusion-based 3D bioprinting system. The model was designed using Google SketchUp 2017 software (Trimble Navigation Limited, version 17.2.2555). The system consists of a controller board, a syringe infusion pump, or a customized extruder machine with a 5 ml commercial syringe (Terumo, Philippines) and a commercial 3D printer (Prusa i3, Czech Republic). The commercial 3D printer uses the Fused Deposition Model (FDM) printing technique, which allows the printer to move in the X-axis, Y-axis, and Z-axis to print a new slice. In this design, the syringe needle was connected to a silicon Tygon tube (diameter = 2.5 mm) ending with a conical pipette tip with a nozzle size of ~1 mm. The nozzle is the orifice for the diffusion of the bioink. A three-dimensional (3D) pipette tip holder was created and attached to the print head of the 3D printer. The bioink was loaded into the syringe and flowed through the silicone tube. The customized extruder comprises a motorized infusion pump with a syringe holder and a 3D plunger holder. To facilitate the deposition of the bioink, a glass substrate is placed on the 3D printer stage.

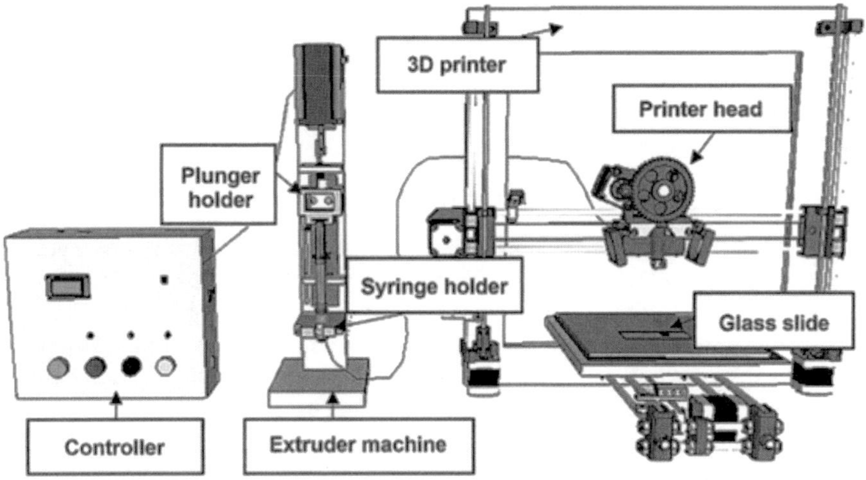

Figure 6.1 Conceptual model of the system.

The controller box consists of buttons, power supply switches, 16 × 2 LCD, LEDs, and ports for the Arduino Mega 2560 microcontroller, Universal Serial Bus (USB), and motor driver connectors. The four different buttons are the forward and reverse buttons, the select button for flow rate, and the reset button. The power supply switches are located at the side of the controller board. The flow rate of the bioprinting ink can be seen on the 16 × 2 LCD. The three LEDs indicate the functionality of the select, forward, and reverse buttons. The LED at the upper corner of the board shows the power indicator.

The plunger holder was designed to fit a 15 mm syringe plunger. The two holes fix the plunger holder's position on the linear slider.

In addition, a pipette tip holder was also designed to affix the pipette tip. This holder was attached firmly to the printer head. The holder was designed with a length of 40 mm so that the overall structure of the tip can stand properly. A 30 mm linear structure was built at the side of the holder as the attachment site of the printer head.

6.2.2 Synthesis of Sodium Alginate/Gelatin Bioink

The experimental procedure started with weighing 1 g of sodium alginate (Sigma-Aldrich, Dorset, United Kingdom) and 1 g of fish gelatin (Sigma-Aldrich, Dorset, United Kingdom). The sodium alginate was dissolved in 10 ml of Dulbecco's Modified Eagle Media (DMEM) to make 10% w/v alginate concentration. Then, 2 g, 3 g, 4 g and 5 g of fish gelatin solids were mixed into 10 ml of distilled water to make different concentrations of 10% w/v, 20% w/v, 30% w/v, 40% w/v and 50% w/v. The gelatin must be submerged for three to five minutes to ensure the solids absorb the water. This process is called the blooming of gelatin. It is one of the integral steps to ensure the smooth texture of the gelatin. Therefore, when it is heated, the gelatin will dissolve evenly. Then, the fish gelatin mixture was heated at 65°C. Subsequently, the sodium alginate and gelatin solutions were mixed and loaded into a 5 ml syringe. The needle of the syringe was connected to a Tygon silicon tube. It is essential to ensure no bubbles when the bioink is loaded into the syringe. The presence of bubbles would destroy the surface of the printed constructs, and the printing would not be smooth.

6.2.3 Evaluation of Printability

An extrusion-based 3D printing system was evaluated for sodium alginate/gelatin bioink printability. The extrusion was programmed to allow adjustment of flow rate values from 1 to 5 ml/min under diffusion and infusion modes. The 3D printer was programmed by G-code, a numerical control programming language, to generate hexagonal print patterns. The G code instructs the printer head to create a hexagonal shape using a series of G, X, Y, and Z commands. The 3D bioprinter moves at a printing speed set between 10 to 50 mm/s based on the G-codes designed for the desired structures. Table 6.1 shows the experimental parameters of the sodium alginate/gelatin bioink printability test.

Table 6.1 The 3D bioprinting parameters.

No.	Parameter	Value
1.	Flow rate	1–5 ml/min
2.	Printing speed	10–50 mm/s
3.	Nozzle tip diameter	1 mm

6.2.4 Viscosity Measurement

Five alginate/gelatin bioink samples were prepared with different concentration ratios (10:10, 10:20, 10:30, 10:40, and 10:50% v/v). The gelatin solutions were first prepared by mixing 2, 4, 6, 8, and 10 g of fish gelatin solids in 20 ml of distilled water until they bloomed. Then, the gelatin solution was stirred at 65°C until the solids were dissolved in the distilled water. Subsequently, 10% w/v of alginate solution was added to the gelatin solution and stirred to create a homogenous bioink solution. SV-10 Vibro Viscometer (A&D Company, Tokyo, Japan) was used to measure the viscosity of the mixed solution in pascal (Pa) unit for 60 minutes. The readings were taken every 5 minutes for 60 minutes. The alginate/gelatin bioink mixture was filled in a 45 ml sample case. The tuning-fork vibration method was applied using two sensor plates attached to the viscometer for viscosity measurement of the mixed bioink solutions. The two sensor plates resonated at a frequency of 30 Hz with an amplitude of less than 1 mm. Similar experiments were repeated for all mixed solutions.

6.2.5 Contact Angle Measurement

Five sheets of alginate/gelatin with different concentration ratios were polymerized in a calcium chloride bath for 30 minutes. Then, the samples were polymerized in a calcium chloride bath for 30 minutes. The samples were then measured using the Video Contact Angle system (VCA Optima, AST Products., USA). 1 μL deionized water was deposited via a 100 μL motorized syringe. The angle measurements were performed right after the water droplet was deposited and were repeated three times for each sample.

6.2.6 Fourier Transform Infrared Spectroscopy

Two samples with five different alginate/gelatin concentration ratios were prepared. One set of the samples was polymerized with calcium chloride ($CaCl_2$), while another was the alginate/gelatin hydrogels without the crosslinking of $CaCl_2$. After that, the hydrogels were placed on the flat plane of the Fourier Transform Infrared spectrometer (FTIR) (Spectrum 100, PerkinElmer, USA) to determine the differences before and after the crosslinking of alginate/gelatin hydrogel. The transmission property of the hydrogels was recorded with the range of wavenumbers from 4000 to 600 cm^{-1} and with a resolution of 8 cm^{-1}.

6.2.7 Dynamic Mechanical Analysis

A dynamic mechanical analyzer with a compression fixture (DMA Q800, TA Instruments, USA) was used to determine the mechanical properties of the crosslinked alginate/gelatin hydrogel. The viscoelasticity of the five different concentrations of alginate/gelatin was tested by applying the frequency sweep test. The test was fixed at an isothermal temperature of 37°C to study the changes in viscoelasticity behavior towards the deformation under various frequencies ranging from 1 to 10 Hz. An applied force of 0.01 N and an initial displacement amplitude of 10 μm were set in this experiment. The crosslinked samples' elastic response, or the storage modulus, G', and the viscous response, or the loss modulus, G'', were recorded.

6.2.8 Degradation test

Hexagonal multilayer structures made up of alginate/gelatin bioink were incubated at 37°C to observe the degradation rate of the composite hydrogel. After printing the hexagonal structures, the samples were weighed and polymerized with $CaCl_2$ for 30 minutes. Then, the samples were immersed in DMEM, which was changed daily for 14 days. The samples were washed with distilled water, dried, and weighed at a predetermined time. The samples were repeatedly weighed three times every 24 hours. The degradation rate of calcium alginate/gelatin bioink was calculated by using Equation (1):

$$D\% = \frac{w_i - w_f}{w_i} \times 100\% \tag{6.1}$$

where W_i is the initial weight of the sample, and W_f is the weight of the sample after a while.

6.2.9 Henrietta Lacks Cell Preparation

Henrietta Lacks (HeLa), or cervical cell lines, were obtained from ATCC. It was maintained in a 25 cm^2 tissue culture flask which contains Dulbecco's Modified Eagle's Medium (DMEM) (Corning®, USA) supplemented with fetal bovine serum (10%) (Promocell, UK), 10 ml of L-glutamine (2 mm) (Sigma Aldrich, UK), 1 ml of Amphotericin B (2.5 mg/l) (Sigma Aldrich, UK) and 5 ml of Penicillin Streptomycin (100 mg/ml) (Sigma Aldrich, UK). When the cell reached 80% confluency, the old medium needed to be discarded from the culture flask. The cell subculture process occurred in an ECSO Class II biological safety cabinet. The cell inside the flask was washed three times using Hank's Balanced Salt Solution (HBSS) (Sigma Aldrich, UK). 1 ml of EDTA-trypsin (0.25%) (Biowest, France) was used to dissociate the cell from the surface of the flask. The cell was incubated in a carbon dioxide (CO_2) incubator for five minutes. After five minutes, the cells detached during the trypsinization process. 5 ml of DMEM was added into the flask to stop the trypsinization process. The mixture was put inside a centrifuge tube and centrifuged for five minutes at 1200 rpm. A pellet of cells can be seen at the bottom of the centrifuge tube, and the cells were resuspended in 5 ml of medium.

6.2.10 Microencapsulation of HeLa Cells

HeLa cells can be encapsulated after obtaining the cell pellet from the cell subculture. First, 2% sodium alginate solution and 1% calcium chloride were prepared. The sodium alginate solution was mixed with the cell pellet inside a centrifuge tube. The calcium chloride was put inside a petri dish to polymerize alginate microbeads. U-100 0.5 ml insulin syringe (Becton, Dickinson and Company, USA) was used to infuse 0.35 ml of alginate solution with cell suspension. Then, the syringe was fixed inside the flicking machine. The slider speed was set to medium and 80 rpm for the flicking speed to produce microbeads of cells ranging from 270 to 430 μm. The flicker disperses the alginate with cells into the calcium chloride solution and polymerizes it to become microtissues. The vibrator motor was activated to vibrate the ionic solution bath during the experiment to avoid the microbeads from clumping together.

6.2.11 Live and Dead Cells Staining

Cell viability in calcium alginate microtissues embedded in calcium alginate/gelatin bioink was investigated using the live/dead® viability kit for mammalian cells (Invitrogen, Paisley, UK). 20 µl of green-fluorescent Calcein AM and 5 µl of red-fluorescent ethidium homodimer-1 (EthD-1) reagents were diluted in 10 ml HBSS. Calcein AM will react to the intracellular esterase activity, while EthD-1 indicates the loss of plasma membrane integrity. The microtissues doped with alginate/gelatin bioink were cleaned thrice using HBSS before the staining process. A filter paper was used to absorb any excess liquid. The microtissues were incubated in 2 µm calcein AM and 4 µm EthD-1 stain solutions for 20 minutes and placed in a dark condition. Subsequently, the stained microtissues were observed and captured by a BX53 fluorescence microscope (Olympus, Tokyo, Japan) mounted with a DP73 CCD digital camera.

6.3 RESULTS AND DISCUSSION

6.3.1 Extrusion-based 3D Bioprinter System

Figure 6.2 shows the complete assembly of an extrusion-based 3D bioprinting system consisting of three major parts: the controller board, the extruder, and the 3D bioprinter. The controller board that controls the extruder machine's single-axis motorized linear slider movement is as shown in Figure 6.3. It features a reset button, infusion and diffusion buttons, a select button, LEDs, an LCD

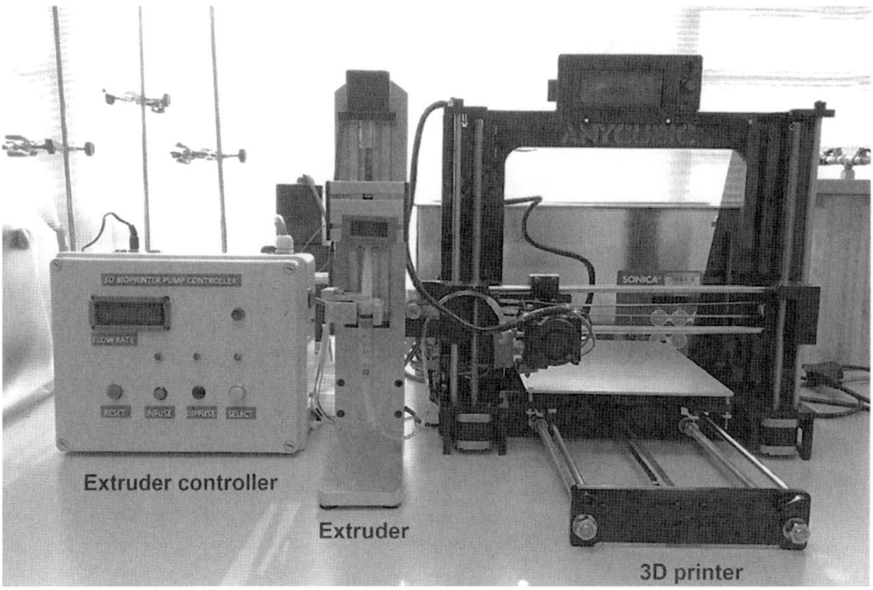

Figure 6.2 The prototype of a developed 3D bioprinting system.

screen, and a power indicator. A reset button was created for an immediate return to the initial zero position of the extruder. Infusion and diffusion buttons were created to control the downward and upward movement of the extruder. A select button was made to select a bioink flow or extrusion rate between 1 to 5 ml/min. LEDs were created as indicators for each button. The LCD was used to display the system flow rates for selections. A power indicator was designed to indicate whether the on/off switch powers the extruder machine.

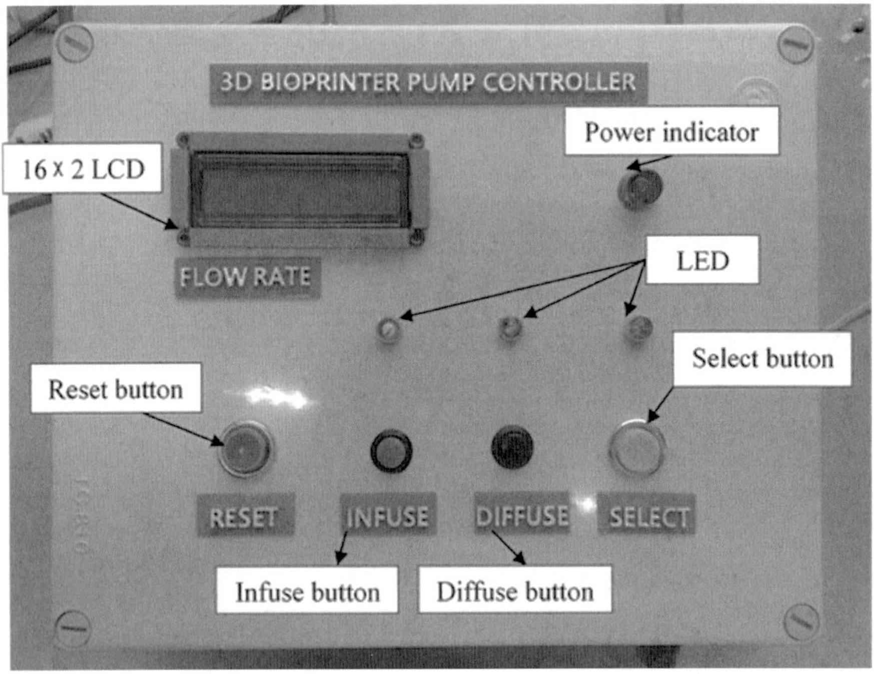

Figure 6.3 The bioink extruder controller board.

Figure 6.4(a) shows a custom extruder system attached to a 3D-printed carriage or plunger holder. The extruder system is made based on the infusion pump mechanism. Four control signal wires, namely A+, A-, B+, and B-, connect the stepper motor of the bioink extruder to the controller board. The plunger holder was attached to the linear slider carriage, as shown in Figure 6.4(b). The slide will move along with the carriage in either the infusion or diffusion direction.

Figure 6.5(a) shows how the bioink extruder was integrated into the existing 3D printer. The syringe nozzle was connected to a silicone tube, and the other end was attached to a pipette, where the bioprinting ink would be extruded. The pipette tip holder was designed as shown in Figure 6.5(b), with a linear shape used as the attachment site between the pipette tip holder and the printing head beside the tip.

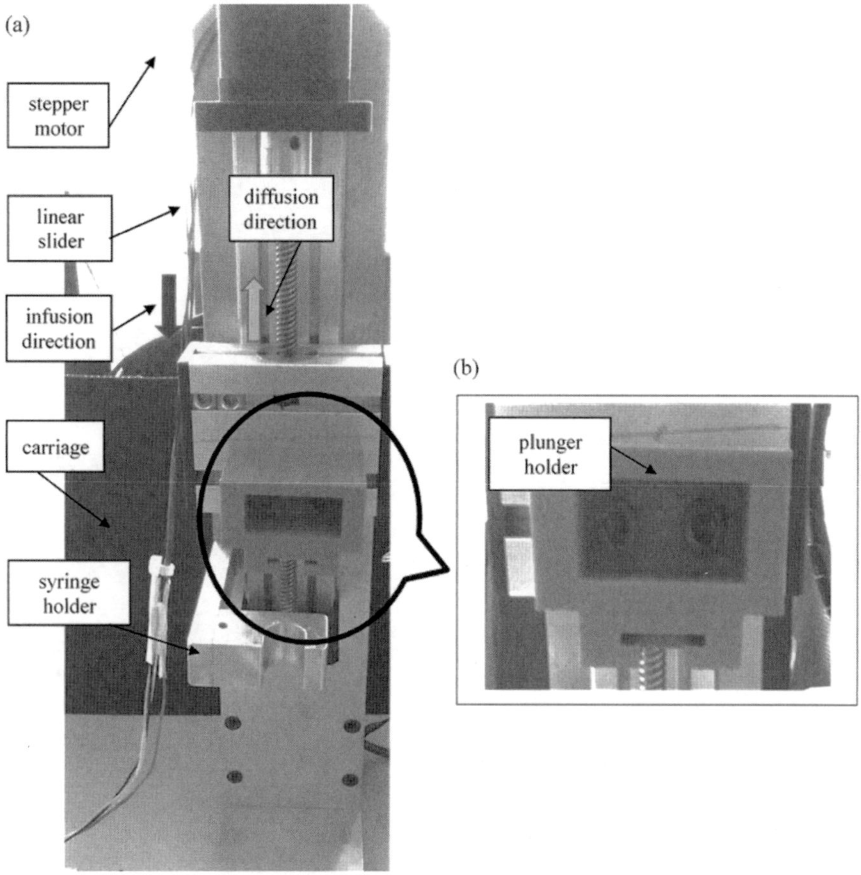

Figure 6.4 (a) The bioink extruder machine and (b) Plunger holder.

6.3.2 The Effects of Printing Speed and Flow Rates on the Width of the Printed Filament

Figure 6.6 shows the relationship between the printing speed of a 3D printer, the flow rates of the extruder, and the width of the printed sodium alginate/gelatin bioink filament. The filament width of the printed bioink is directly proportional to the flow rate and associated with the printing speed. When the printing speed is constant, a higher flow rate results in a thicker filament width, and vice versa. Similarly, higher printing speeds are associated with smaller filament widths within a similar flow rate, and vice versa.

The flow rates ranging from 1 to 5 ml/min yield filament widths between 1 and 6 mm at printing speeds ranging from 50 to 10 mm/s, respectively. As observed in Figure 6.7, specific filament widths are more significant than the nozzle diameter (nozzle size = 1.0 mm). This phenomenon can be attributed to the dead swelling or Barus effect, a common phenomenon when dealing with non-Newtonian materials such as sodium alginate/gelatin bioink. During extrusion, the bioink is subjected to shear stress, which leads to the stretching of its molecular chains at the nozzle. This stretching phenomenon contributes to the swelling of the bioink, resulting in a filament width that exceeds the diameter of the nozzle.

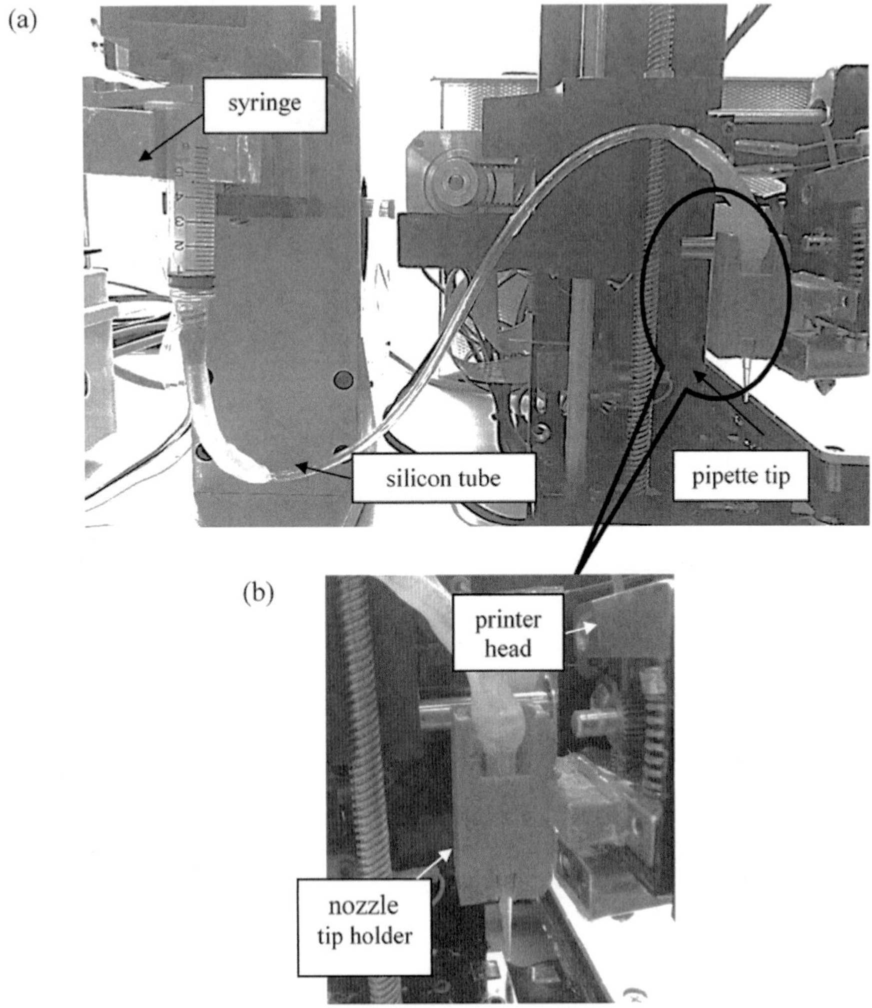

Figure 6.5 (a) Integration with 3D printer and (b) Nozzle tip holder attached to the 3D printer head.

The lowest flow rate of 1 ml/min enables printing of the smallest filament widths ranging from 0.8 to 1 mm due to the low shear stress applied to the bioink. The increased pressure enhances the die swell effect at a higher flow rate of 5 ml/min, leading to a rapid flow of the viscoelastic bioink. Consequently, high flow rates produce bulging shapes on the printed structures. When the flow rate exceeds 5 ml/min, broken and thick filaments are obtained, making it challenging to produce printable biofilaments.

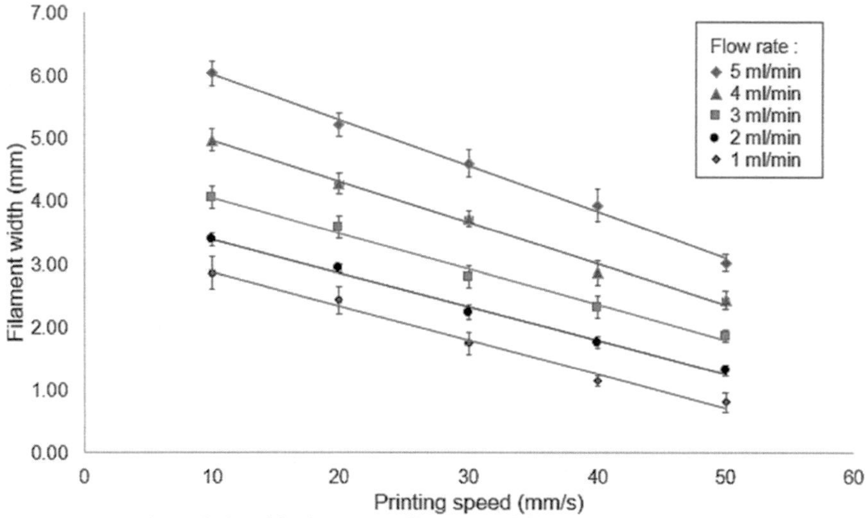

Figure 6.6 The relationship between filament width and printing speed of the 3D bioprinting concerning different flow rate. (Othman et al. 2021). (*Reproduced with permission of Springer Nature*)

Figure 6.7 shows the filaments printed at different print speeds, and the flow rate remains constant at 1 ml/min. The figure indicates that thin filaments can be produced at 40 and 50 mm/s printing speeds. However, thicker filaments are printed at lower printing speeds of 10, 20, and 30 mm/s. For instance, at a printing speed of 50 mm/s and a flow rate of 1 ml/min, a filament as thin as 0.8 mm can be printed. On the other hand, a filament as thick as 3 mm can be produced when utilizing a print speed of 10 mm/s and a flow rate of 1 ml/min.

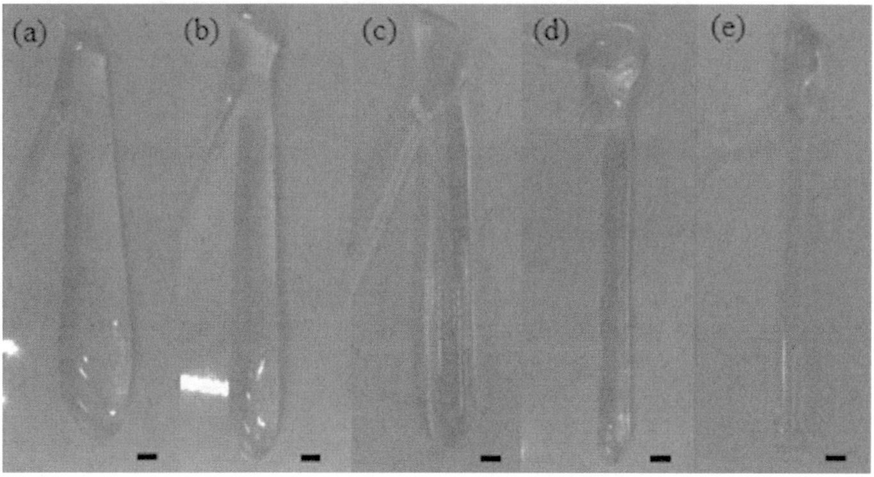

Figure 6.7 Filament width printed at a speed of (a) 10 mm/s, (b) 20 mm/s, (c) 30 mm/s, (d) 40 mm/s, and (e) 50 mm/s for 1 ml/min of flow rate (Scale bar: 1 mm) (Othman et al. 2021). (*Reproduced with permission of Springer Nature*).

6.3.3 The Influence of Nozzle Distance on the Printability of Bioink

Figure 6.8 shows the printed filament at various nozzle distances of 1, 2, 3, 4, and 5 mm. It was observed that nozzle distances of 1 and 2 mm produced filaments printed with high resolution, maintaining the intended hexagonal shape. However, as the nozzle distance increases to 3, 4, and 5 mm, the printability of the filament becomes less precise. The structure failed to retain its hexagonal shape and lost its edges at a nozzle distance of 5 mm. Increasing the nozzle distance resulted in the formation of a significant strand of bioink that was poorly adhered to the surface. These observations indicate that the nozzle distance can influence the direction of bioink dispensing, thus affecting the printability of the filament. Therefore, this study used a nozzle distance of 1 mm to ensure satisfactory printability of sodium alginate/gelatin filaments.

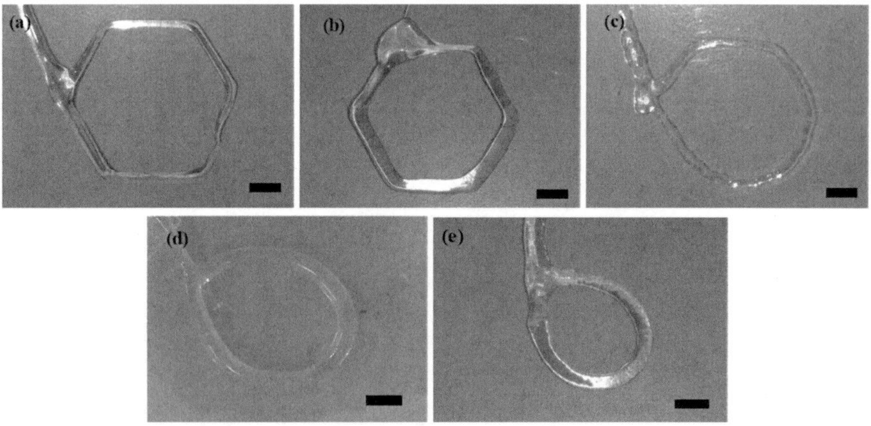

Figure 6.8 The printed filament at various nozzle distances (a) 1, (b) 2, (c) 3, (d) 4, and (e) 5 mm (Scale bar: 5 mm).

6.3.4 Printability of Multilayer Structure

The printability of the multilayer structure was observed by printing the layers until the structure collapsed. Different numbers of layers (2, 4, 6, 8, and 10) in hexagon shape at 1 mm per layer were printed to observe the sturdiness in building a three-dimensional structure (Figure 6.9). The hexagon structure is printed at a diameter of 15 mm. The printing process was based on the fused deposition modeling method (FDM). It is virtually printed layer-by-layer on the X, Y, and Z axes according to the programmed G-code. In this research, the printability of multilayer structures was possible due to the programmed G-code. Each layer's X and Y axes were repeated to ensure the layers were printed at the exact coordinates. The Z displacements were changed upon printing each of the layers. After a layer is completed, the platform will move according to the Z displacement, and another layer will be constructed. In this manner, the entire 3D hexagonal model was generated layer by layer. The height increased uniformly for each layer during printing, proving the bioprinter's ability to print multilayers. The bioink structures stand sturdy until the tenth layer. The structure started to collapse when it reached the 10th layer of sodium alginate/gelatin hydrogel, as in Figure 6.9(e). This is a common challenge when printing gel-like structures, as it can be difficult to prevent sagging or collapse before the hydrogel solidifies. The weak physical crosslinking bonds in the NaAlg/gelatin hydrogel contribute to the loss of structural integrity.

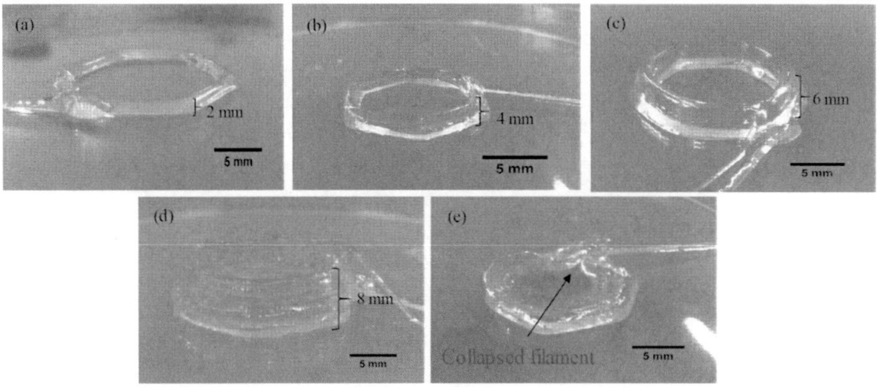

Figure 6.9 Hexagon structure printed at (a) 2, (b) 4, (c) 6, (d) 8, and (e) 10 layers of sodium alginate/gelatine filaments. (Othman et al. 2021). (*Reproduced with permission of Springer Nature*).

6.3.5 The Viscosity of Sodium Alginate/Gelatin Bioink

Figure 6.10 shows the viscosity-time graph of NaAlg/gelatin composites with different concentration ratios. The viscosity of the NaAlg/gelatin bioink is influenced by the concentration of NaAlg: gelatin, temperature, and time. The solutions become more viscous as the tuning fork rotates over one hour. Higher concentrations of bioink compositions are usually associated with higher viscosity, which can be attributed to the relative increase in molecular weight and polymer chains within the bioink. The NaAlg used in this study has a molecular weight of 396 kDa, while the gelatin bioink has a molecular weight of 60 kDa (Sigma-Aldrich, CAS no. 9005-38-3, W201502 and CAS no. 9000-70-8, G7041, respectively).

For the 10:10 v and 10:20% v/v NaAlg/gelatin bioink compositions, the changes in viscosity were less significant over 60 minutes. The viscosity gradually increased for the 10: 30% NaAlg/gelatin bioink. A tremendous increase can be observed after 5 minutes for the 10: 40% v/v NaAlg/gelatin bioink. After one hour, the viscosity of the 10: 50% v/v NaAlg/gelatin bioink reached 12 Pa.s, indicating the highest viscosity among the four bioink compositions. Based on the viscosity results in Figure 6.10, it is recommended to complete the printing of a structure within 30 minutes to avoid bioink hardening and nozzle clogging.

In this study, the 10: 50% v/v sodium alginate/gelatin bioink might seem complicated to extrude, but this bioink concentration was proven to be printed with a well-defined structure. This is because the bioink becomes shear thinning in response to the extrusion's shear force and can visco-elastically recover after the extrusion. This flow behavior showed that the NaAlg/gelatin bioink is highly thixotropic.

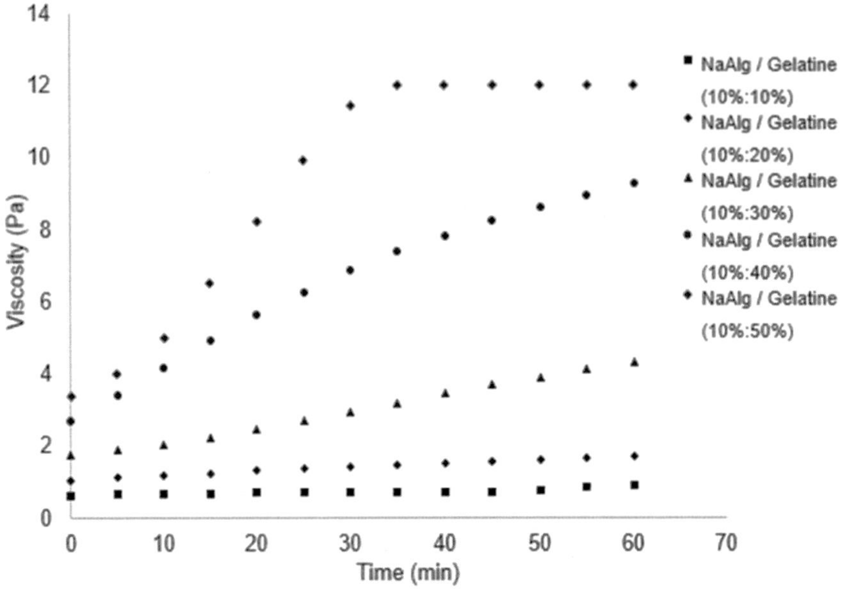

Figure 6.10 Viscosity measurement of a bioink composite of NaAlg/Gelatine. (Othman et al. 2021). (Reproduced with permission of Springer Nature)

6.3.6 Surface Behavior of NaAlg/Gel Bioink after Polymerization in $CaCl_2$

Figure 6.11 presents the results of dynamic contact angle measurements for NaAlg/Gelatin at different concentrations after polymerization in a $CaCl_2$ bath for 30 minutes. The contact angles measured were 28.10° for 10:10% v/v sodium alginate/gelatin hydrogel and 40.80° for 10:20% v/v sodium alginate/gelatin hydrogel, as shown in Figure 6.11. Both angles are within 90°, indicating that the surfaces exhibit high wettable behavior. The wetting tendency increases, causing the contact angle to become smaller. This is attributed to the low surface tension of NaAlg/Gel.

The contact angle of 10:30% v/v NaAlg/gelatin hydrogel increased to 59.50°. At a value ratio of 10:40% NaAlg /gelatin hydrogel, the contact angle formed was 65.80°, indicating a decrease in surface wettability with an increase in NaAlg/Gel concentration. The contact angle for 10:50% v/v NaAlg /gelatin hydrogel was the highest at 72.30°. The wettability of the NaAlg /gelatin hydrogel allows for protein absorption and nutrient delivery to the surface.

Blending gelatin with other polymers, such as NaAlg, can increase the printed structure's hydrophobic behavior while maintaining hydrophilicity suitable for functioning as a bioink. Hydrophilic hydrogels promote cell growth on the surface, while hydrophobic surfaces offer better cell adhesion. Therefore, the 10:50% NaAlg /gelatin hydrogel exhibits an appropriate balance between hydrophilicity and hydrophobicity surface behavior. This balance is crucial as excessive hydrophilic hydrogel may hinder cell-to-cell interactions, while highly hydrophobic surfaces can reduce biocompatibility and printability.

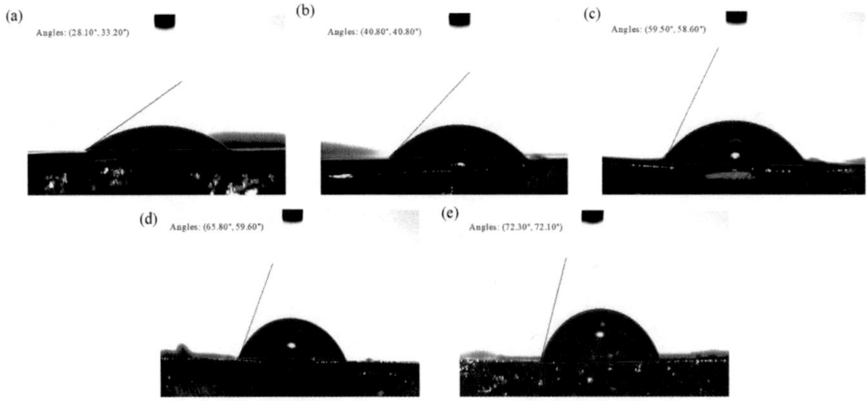

Figure 6.11 The contact angle for NaAlg/Gelatin at different concentrations after polymerization in a $CaCl_2$ bath for 30 minutes: (a) 10:10 (b) 10:20 (c) 10:30 (d) 10:40 and (e) 10:50% v/v. (Othman et al. 2021). (Reproduced with permission of Springer Nature)

6.3.7 Functional Groups of Sodium Alginate/Gelatine Hydrogel

Figure 6.12 shows the FTIR spectra of pure gelatin, sodium alginate, sodium alginate/gelatin composites, and calcium alginate/gelatin hydrogels. The FTIR spectrum of gelatin peaks at 3273 cm^{-1}, indicating the presence of the –OH group. The gelatin spectrum also displays absorbance bands at 1635 cm^{-1} and 1543 cm^{-1}, corresponding to the N–H stretching vibration peaks of amide I and II, respectively. The peak at 1449 cm^{-1} represents the asymmetrical stretching of COO–. A broad absorbance band is observed at 3322 cm^{-1} for sodium alginate, corresponding to the functional group of –OH. The peak intensity at 1631 cm^{-1} indicates the symmetrical stretching of COO–, while the peak at 1032 cm^{-1} is associated with the stretching of the C–O group. In the FTIR spectrum of sodium alginate/gelatin hydrogels, a peak at 3300 cm^{-1} (O–H vibration) was observed due to the –OH group. The peak intensity at 1640 cm^{-1} indicates the presence of the $-CONH_2$ characteristic, where the opposing group in alginate interacts with the positive charges in gelatin. The absorption band at 1547 cm^{-1} represents the characteristic peak of amide II in the gelatin protein structure. Additionally, the absorption band at 1039 cm^{-1} demonstrates the C–H group in the guluronic units of the alginate polysaccharide structure, with C–O–C stretching vibration. The peak at 1457 cm^{-1} represents the asymmetrical stretching of the COO– bond linked to the hydrogen of alginate. The peak signal at 1039 cm^{-1} is attributed to the vibrations of the C–O group in gelatin. Only minor differences exist between the sodium alginate/gelatin hydrogel peaks and the crosslinked calcium alginate/gelatin hydrogels. Based on Figure 6.12(d), the peak at 1633 cm^{-1} is attributed to the stretching of C–O and C–N in the amide I group. The peak at 1039 cm^{-1} in Figure 6.12 (c) is replaced by a peak at 1029 cm^{-1} in Figure 6.12(d). This slight reduction most likely illustrates the weakening of the C–O bond due to stretching to accommodate the Ca^{2+} ion.

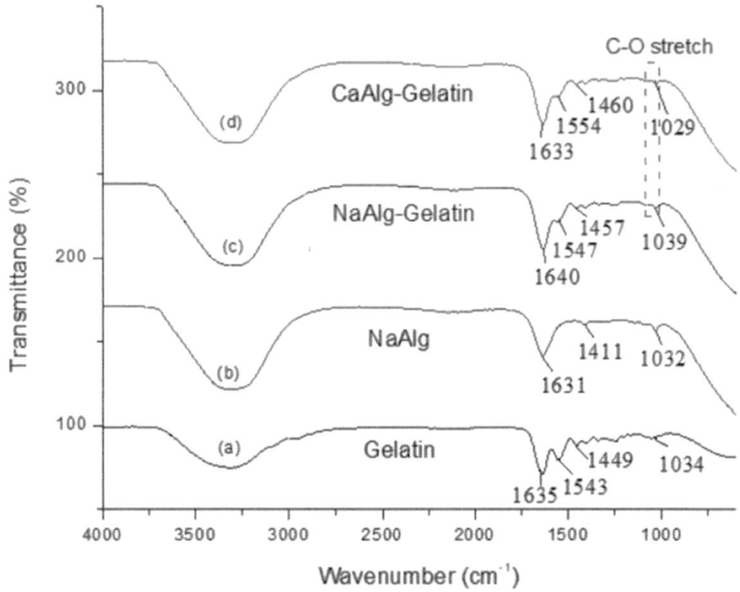

Figure 6.12 FTIR spectra of (a) pure gelatine, (b) sodium alginate, (c) mixture of sodium alginate/gelatine hydrogel, (d) crosslinked calcium alginate/gelatine hydrogel. (Othman et al. 2021). (Reproduced with permission of Springer Nature)

Figure 6.13 shows the FTIR spectra of five different sodium alginate/gelatine concentrations. No significant difference in peaks can be seen for all the concentrations. However, more peaks are visible at 1543 cm^{-1}, 1545 cm^{-1}, and 1547 cm^{-1}, respectively, when sodium alginate/gelatine is added. This might be due to the presence of amide I in the gelatine protein as the gelatine concentration increases.

6.3.8 Dynamic Mechanical Analysis of Crosslinked Calcium Alginate/Gelatine Hydrogel

Figure 6.14 presents the frequency sweep test results of calcium alginate/gelatin hydrogel at different concentrations. As shown in Figure 6.14, the effect of adding a higher concentration of gelatine is the increase of storage modulus (G′) than the loss modulus (G″) of the composite hydrogen over a compressive frequency range of 1–10 Hz at an isothermal temperature of 37°C. At 50% of gelatine concentration, G′ and G″ reached to 0.074 and 0.028 MPa, respectively, at 10 Hz. In contrast, 10% of gelatine concentration yielded the lowest G′ and G″ moduli at 0.031 MPa and 0.014 MPa, respectively, at 10 Hz. This is because the looser chain of the alginate network enables the gelatine to develop helices and network alongside the calcium chloride ($CaCl_2$). At higher moduli, this may suggest that higher concentration sodium alginate/gelatine bioink could provide higher viscoelasticity after the extrusion. This is important in bioprinting as it allows the bioink to be in the liquefied state for extrusion and then rapidly solidified during bioink deposition to form stable 3D structures.

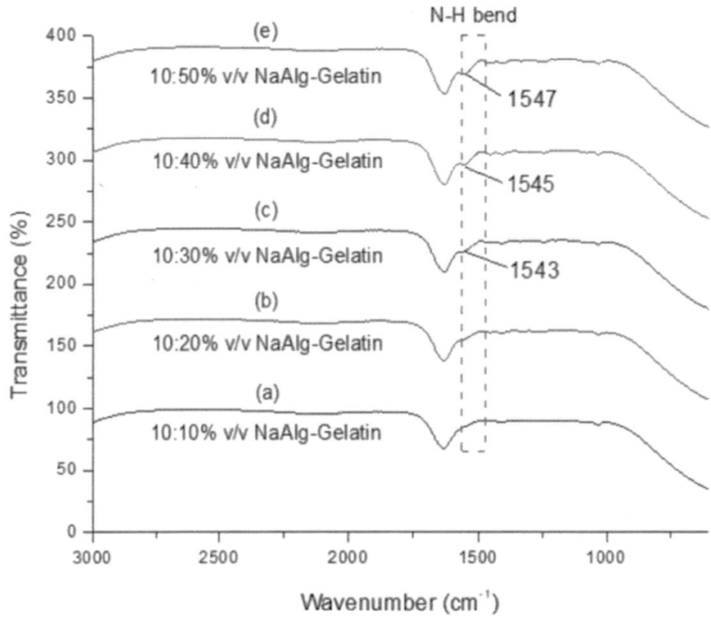

Figure 6.13 FTIR spectra of sodium alginate/gelatine hydrogel at different concentrations: (a) 10:10% v/v (b) 10:20% v/v (c) 10:30% v/v (d) 10:40% v/v and (e) 10:50% v/v. (Othman et al. 2021). (Reproduced with permission of Springer Nature)

Table 6.2 shows the comparison of storage modulus (G′) and loss modulus (G″) of different calcium alginate/gelatine concentrations at various frequencies. In the table, the storage modulus is higher than the loss modulus in all bioink concentrations. Thus, it can be concluded that the bioink exhibits more elastic properties. However, overall, the loss modulus (G″) can increase slowly, suggesting that the composite bioink became stiffer due to the crosslinking driven by Ca^{2+}.

This work showed that the increase in gelatine concentration increased storage modulus while exhibiting a low loss modulus. This is due to the nature of the network formed by gelatine, which is more viscoelastic. The sodium alginate solution in the bioink formed a liquid phase gel, and adding the gelatine solution can impact the viscoelastic properties. Because of that, gelatine bioink is classified as stiff solid bioink when exposed to room temperature for too long, and the combination with sodium alginate can tune the gelatine's stiffness property.

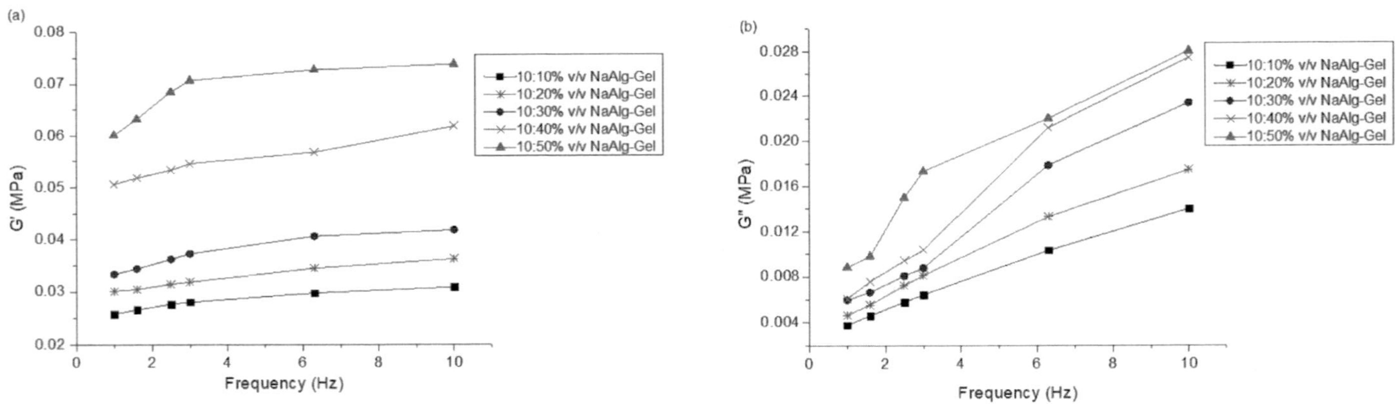

Figure 6.14 Frequency sweep test results with (a) storage modulus (G′) and (b) loss modulus (G″) for different concentrations of crosslinked calcium alginate/gelatine hydrogel. (Othman et al. 2021). (*Reproduced with permission of Springer Nature*).

Table 6.2 The comparison of storage (G′) and loss modulus (G″) of each concentration at different frequency (Hz)

Na Alg/Gel (v/v)	1.0 Hz		1.6 Hz		2.5 Hz		3.0 Hz		6.3 Hz		10 Hz	
	G′ MPa	G″ MPa	G′ MPa	G″ MPa	G′ MPa	G″ MPa	G′ MPa	G″ MPa	G′ MPa	G″ MPa	G′ MPa	G″ MPa
10: 10%	0.026	0.004	0.027	0.005	0.028	0.006	0.028	0.006	0.029	0.010	0.031	0.014
	G′ > G″		G′ > G″		G′ > G″		G′ > G″		G′ > G″		G′ > G″	
10: 20%	0.030	0.005	0.030	0.006	0.031	0.007	0.032	0.008	0.034	0.013	0.036	0.017
	G′ > G″		G′ > G″		G′ > G″		G′ > G″		G′ > G″		G′ > G″	
10: 30%	0.033	0.006	0.034	0.007	0.036	0.008	0.037	0.009	0.041	0.018	0.042	0.023
	G′ > G″		G′ > G″		G′ > G″		G′ > G″		G′ > G″		G′ > G″	
10: 40%	0.051	0.006	0.052	0.008	0.053	0.009	0.055	0.010	0.057	0.021	0.062	0.027
	G′ > G″		G′ > G″		G′ > G″		G′ > G″		G′ > G″		G′ > G″	
10: 50%	0.060	0.009	0.063	0.010	0.068	0.015	0.071	0.017	0.073	0.022	0.074	0.028
	G′ > G″		G′ > G″		G′ > G″		G′ > G″		G′ > G″		G′ > G″	

6.3.9 The Degradation of Calcium Alginate/Gelatine Hydrogel

The degradability of calcium alginate/gelatine hydrogel is an important test to evaluate the mechanical stability of biomaterial *in vitro*. Figure 6.15 shows the degradation percentage of calcium alginate/gelatine hydrogel in 14 days before undergoing the complete degradation process. In Figure 6.15, the degradation rate increased over days associated with weight loss, and the crosslinked calcium alginate constructs were wholly degraded after 14 days. There was more than a 40% increase in degradation over 7 days and over 91% in degradation rate on Day 14.

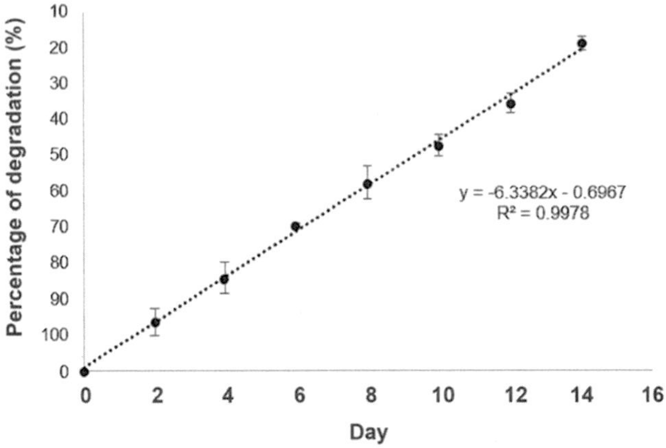

Figure 6.15 The percentage of degradation of printed calcium alginate/gelatine structure (Othman et al. 2021). (*Reproduced with permission of Springer Nature*).

Figure 6.16 shows the degradation progress of a printed hexagonal calcium alginate/gelatine structure from Day 0 to Day 14. It is observed that the structure started to dissociate and lose its shape from Day 8 onwards. The structure seemed more soluble over days due to the weakened network in the sodium alginate and gelatine chain by the infusion of water molecules. Hydrogels like sodium alginate and gelatine contain large pores capable of accommodating living cells and eventually will degrade while releasing the growth factors for cells.

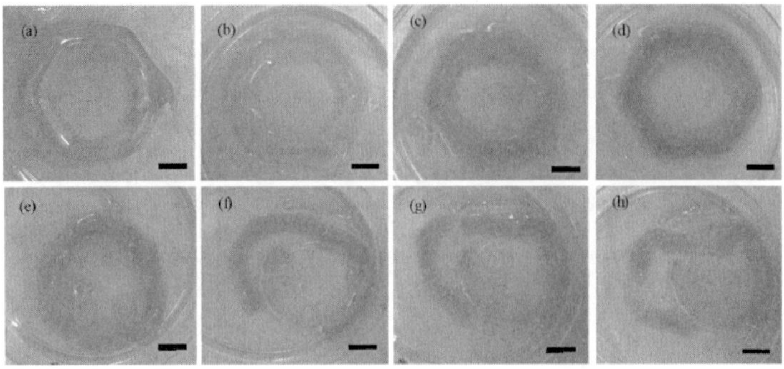

Figure 6.16 Degradation of bioprinted structures of calcium alginate/gelatine on (a) Day 0, (b) 2, (c) 4, (d) 6, (e) 8, (f) 10, (g) 12, and (h) Day 14 (Scale bar:5 mm).

6.3.10 HeLa Microtissues Printed in Sodium Alginate/Gelatine Bioink

The printing of microtissues with sodium alginate/gelatine was performed after all the printing process parameters were optimized. Figure 6.17(a) shows the printed structure of calcium alginate/gelatine with the inclusion of HeLa microtissues while Figure 6.17(b) is the image of the construct after crosslinking with ionic solution, $CaCl_2$. More than 30 microtissues were used to ensure all the tissues were printed and could be seen in the printed structure.

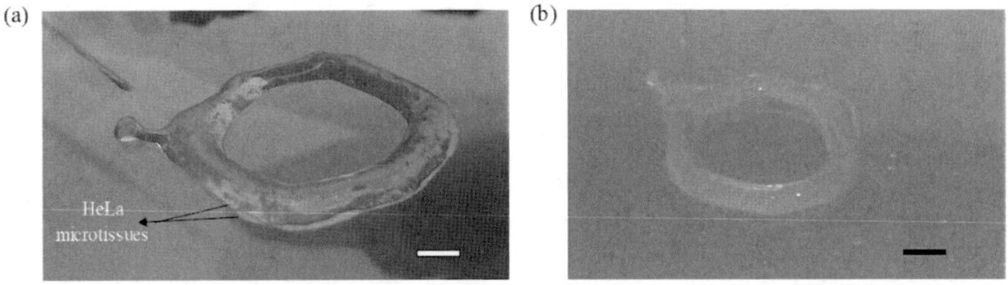

Figure 6.17 (a) 3D printed sodium alginate/gelatin structure with HeLa microtissue. (b) 3D printed calcium alginate/gelatin structure with HeLa microtissues (Scale bar: 5 mm). (Othman et al. 2021). (Reproduced with permission of Springer Nature)

The microscopic image of microtissues inside the sodium alginate/gelatine bioink is shown in Figure 6.18. It was found that the microtissues happened to be clustered together and randomly positioned inside the bioink. Some parts of the bioink were void without microtissues. However, the printing techniques can be enhanced to ensure the microtissues are more closely positioned. The microtissues of HeLa cells stimulate tumor growth *in vivo*, and the position must be close enough to communicate with the surrounding cells.

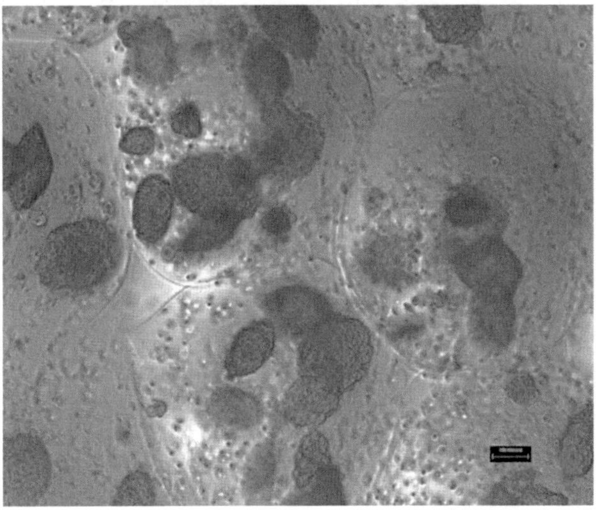

Figure 6.18 The phase contrast microscopy image of HeLa microtissues printed in calcium alginate/gelatine structure (Scale bar: 100 μm).

6.3.11 Live and Dead Cell Assay

After printing, the two-layer printed structure laden with the microtissues was polymerized in a calcium chloride solution for 30 minutes. Then, it was placed in a petri dish containing high glucose DMEM culture media and incubated for 7 days. The live and dead assay was performed after 7 days. Figure 6.19 shows the image of the HeLa microtissue after staining.

The live and dead cells are stained (see Figure 6.19) which reveals that most of the microtissue cells in the control (see Figure 6.19(a)) and calcium alginate/gelatine bioink were stained green and remained alive. It can be seen that the majority of the cells in the microtissues remain viable after growing in the biofilaments (Figure 6.19(b)). Very few cells were found non-viable and stained red. The 2 mm thick sodium alginate/gelatine structure allows the diffusion of media, interchanging of gases, and excretion of catabolites. This is supported by the degradation results, in which the media exchange is continuous despite the thick layer of filament laden with microtissues. Hence, this work has successfully demonstrated a new model to mimic the embedding of cancer tumors deep-separated from direct exposure to culture media. The printed biofilament will also contribute to a better prediction of drug diffusions and responses for cancer treatment.

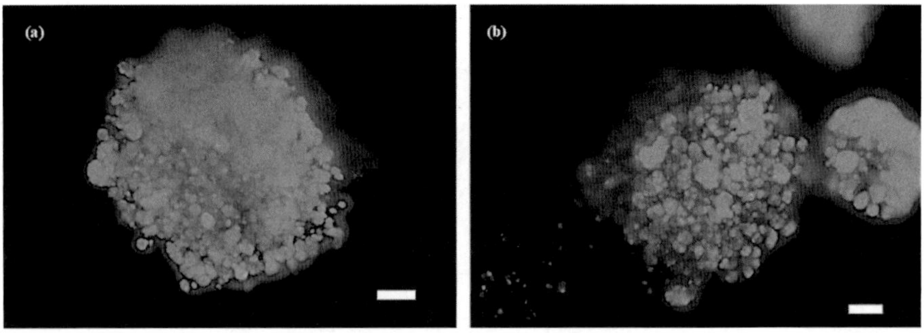

Figure 6.19 Fluorescence photographs of live and dead assay staining on (a) negative control and (b) live and dead cells in biofilament (Scale bar: 100 μm). (Othman et al. 2021). (Reproduced with permission of Springer Nature)

6.3.12 Previous Work of Extrusion-based 3D Bioprinting System

A comparison between the proposed extrusion-based 3D bioprinting system and previous development is summarised in Table 6.3. In this work, the proposed 3D bioprinter is much simpler and more economical than the previous development, which primarily uses commercial bioprinter. Moreover, the proposed bioprinter can print up to 10 to 50 mm/s, which consumes only a short fabrication time. The bigger diameter of the nozzle configured in the proposed bioprinter minimizes the shear stress and increases cell viability.

Table 6.3 Comparison of proposed 3D bioprinter with previous development

Technical Specification	Previous Development			Current Work
References	Webb and Doyle 2017	Freeman and Kelly 2017	Billiet et al. 2014	–
Type of printer	Inkredible + bioprinter	RegenHU's 3D Bioplotter	EnvisionTEC's 3D Bioplotter	Self-development microextrusion-based bioprinter
Type of bio-ink	Sodium alginate and gelatin (low concentration)	Sodium Alginate	Gelatin methacrylamide	Sodium alginate and gelatin (high concentration)
Nozzle diameter	0.3 mm, 0.4 mm and 0.5 mm	0.26 mm	0.15 and 0.2 mm	1 mm
Printing speed	1 to 6 mm/s	8 mm/s	1 to 17 mm/s	10 to 50 mm/s
Fabrication time	Long	Long	Medium	Short
Printing structure	Line pattern structure	Line pattern structure	Grid structure	Multilayer hexagonal structure
Thickness of single-layer printed biofilament	0.3–2.3 mm	0.3 mm	0.15–0.2 mm	1 mm

6.3.13 Comparison of 3D Bioprinting of Bioink Laden with Microtissues and Single Cells

Table 6.4 summarises the differences between the proposed encapsulated microtissues within 3D calcium alginate/gelatine bioink laden with previous single cells in bioink laden. The type of bioink used, the crosslinkers, 3D constructs, strengths, and weaknesses were compared and highlighted. The proposed method showed better cell positioning and excellent mechanical strength throughout the comparison. In addition, the microtissues are well-protected in the bioink and showed high cell survival.

Table 6.4 The comparison of proposed bioink-laden microtissues with previous development of bioink-laden single cells

	3D Bioprinting of Bioink-laden	
	Bioink-laden Single Cells (Wu et al. 2016)	**Proposed Bioink-laden Microtissues**
Type of bioink	Collagen with sodium alginate and gelatine solution	Composite bioink of sodium alginate and gelatine
The crosslinker	$CaCl_2$	$CaCl_2$
Type of cells	Human corneal epithelial cells (HCECs)	HeLa cells
Encapsulation method	Not encapsulated	Encapsulated in calcium alginate using a flicking technique
Shape for 3D constructs	Mesh shape	Multilayer of hexagonal shape
Strength	Collagen contains the most abundant component of extracellular matrix (ECM), which makes the hydrogel much better at mimicking tissue microenvironments.	Encapsulated tissues coated with bioink provide better immunoprotection to the cells. Better cell positioning.
Weakness	The position of far-scattered cells limits cell-to-cell interaction. The collagen cannot blend with sodium alginate and gelatine, reducing the precision of the printed bioinks.	Fast gelation, which makes the bioink hard to control.

6.4 CONCLUSIONS

Based on the detailed discussion and findings presented in this chapter on extrusion-based 3D bioprinting, it is evident that significant advancements have been made in the development of bioprinting systems and bioinks suitable for tissue engineering applications. This study successfully demonstrated the integration of a mechatronic extruder with adjustable flow rates into an existing 3D printer, effectively converting it into a functional bioprinter capable of producing complex tissue constructs with high precision. The optimization of critical parameters such as extrusion flow rates, printing speed, and nozzle diameter was crucial in achieving accurate and viable prints. The characterization of the sodium alginate-gelatin bioink provided insights into its suitable viscoelastic properties, which are essential for maintaining the structural integrity and functionality of the printed constructs. Moreover, the study's successful printing of encapsulated 3D microtissues highlights the potential of this bioprinting system in creating functional tissue constructs that can be used in various biomedical applications. The comparison with previous developments in the field underscored the advantages of the proposed bioprinter, including its simplicity, cost-effectiveness, and improved cell viability. These attributes make it a promising tool for advancing tissue engineering, drug testing, and regenerative medicine. Overall, this research contributes significantly to the extrusion-based 3D bioprinting technology field by providing a practical approach to fabricating complex tissue structures. The advancements made in this study bring us closer to the realization of functional artificial organs and tissues, offering new possibilities for medical treatments and applications. Further studies and developments based on these findings are expected to enhance the capabilities and applications of bioprinting technology.

REFERENCES

Ali, M., E. Pages, A. Ducom, A. Fontaine and F. Guillemot. 2014. Controlling laser-induced jet formation for bioprinting mesenchymal stem cells with high viability and high resolution. Biofabrication. 6(4): 0450001.

Anton, D., H. Burckel, E. Josset and G. Noel. 2015. Three-dimensional cell culture: a breakthrough *in vivo*. International Journal of Molecular Sciences, 16(3): 5517–5527.

Billiet, T., E. Gevaert, T.D. Schryver, M. Cornelissen and P. Dubruel. 2014. The 3D printing of gelatin methacrylamide cell-laden tissue-engineered constructs with high cell viability. Biomaterials 35(1): 49–62.

Breslin, S. and O'Driscoll, L. 2013. Three-dimensional cell culture: The missing link in drug discovery. Drug Discovery Today. 18(5–6): 240–249.

Chai, Q., Y. Jiao and X. Yu. 2017. Hydrogels for biomedical applications: their characteristics and the mechanism behind them. Gels. 3(6): 1–15.

Charbe, N., P.A. McCarron and M.M. Tambuwala. 2017. Three-dimensional bio-printing: a new frontier in oncology research. World Journal of Clinical Oncology. 8(1): 21.

Choudhary, B., S.R. Paul, S.K. Nayak, D. Qureshi and K. Pal. 2018a. Synthesis and biomedical applications of filled hydrogels. pp. 283–302. *In*: K. Pal and I. Banerjee (eds). Polymeric Gels. Elsevier.

Choudhury, D., S. Anand and M.W. Naing. 2018b. The arrival of commercial bioprinters—towards 3D bioprinting revolution! International Journal of Bioprinting. 4(2): 139

Ferris, C.J., K.J. Gilmore, S. Beirne, D. McCallum, G.G. Wallace and M. In. Het. Panhuis. 2013. Bioink for on-demand printing of living cells. Biomaterials Science. 1(2): 224–230.

Freeman, F.E. and D.J. Kelly. 2017. Tuning alginate bioink stiffness and composition for controlled growth factor delivery and spatially directing MSC fate within bioprinted tissues. Scientific Report. 7: 17042.

Gillentine, M.A., L.N. Berry, R.P. Goin-Kochel, M.A. Ali, J. Ge. D. Guffey, et al., 2017. The cognitive and behavioral phenotypes of individuals with CHRNA7 duplications. Journal of Autism and Developmental Disorders. 47(3): 549–562.

Gungor-Ozkerim, P.S., I. Inci, Y.S. Zhang, A. Khademhosseini, A. and M.R. Dokmeci. 2018. Bioinks for 3D bioprinting: an overview. Biomaterials Science. 6(5): 915–946.

Hoarau-Véchot, J., A. Rafii, C. Touboul and J. Pasquier. 2018. Halfway between 2D and animal models: are 3D cultures ideal for studying cancer-microenvironment interactions? International Journal of Molecular Sciences. 19(1): 181.

Jose, R.R., M.J. Rodriguez, T.A. Dixon, F. Omenetto and D.L. Kaplan. 2016. Evolution of bioinks and additive manufacturing technologies for 3D bioprinting. ACS Biomaterials Science and Engineering. 2(10): 1662–1678.

Kapałczyńska, M., T. Kolenda, W. Przybyła, M. Zajączkowska, A. Teresiak, V. Filas et al. 2018. 2D and 3D cell cultures—a comparison of different types of cancer cell cultures. Archives of Medical Science. 14(4): 910–919.

Keriquel, V., H. Oliveira, M. Rémy, S. Ziane, S. Delmond, B. Rousseau, et al. 2017. *In situ* printing of mesenchymal stromal cells by laser-assisted bioprinting for *in vivo* bone regeneration applications. Scientific Reports. 7(1): 1778.

Lee, H., J.J. Yoo, H.W. Kang and D.W. Cho. 2016. Investigation of thermal degradation with extrusion-based dispensing modules for 3D bioprinting technology. Biofabrication. 8(1): 15011.

Li, D., Z. Hawley and K. Schnier. 2013. Increasing organ donation via changes in the default choice or allocation rule. Journal of Health Economics. 32(6): 1117–1129.

Liu, W., M.A. Heinrich, Y. Zhou, A. Akpek, N. Hu, X. Liu, et. al. 2017. Extrusion bioprinting of shear-thinning gelatin methacryloyl bioinks. Advanced Healthcare Materials. 6(12): 1–11.

Loessner, D., K.S. Stok, M.P. Lutolf, D.W. Hutmacher, J.A. Clements and S.C. Rizzi. 2010. Bioengineered 3D platform to explore cell-ECM interactions and drug resistance of epithelial ovarian cancer cells. Biomaterials. 31(32): 8494–8506.

Morris, V.B., S. Nimbalkar, M. Younesi, P. McClellan and O. Akkus. 2017. Mechanical properties, cytocompatibility, and manufacturability of chitosan: PEGDA hybrid-Gel scaffolds by stereolithography. Annals of Biomedical Engineering. 45(1): 286–296.

Ng, W.L., W.Y. Yeong and M.W. Naing. 2017. Polyvinylpyrrolidone-based bioink improves cell viability and homogeneity during drop-on-demand printing. Materials. 10(2):190.

Othman, S.A., C.F. Soon, N.L. Ma, K.S. Tee, G.P. Lim, M.M. Morsin, et al. 2021 Alginate-gelatin bio ink for bioprinting of hela spheroids in alginate-gelatin hexagon shaped scaffolds. Polymer Bulletin. 78: 6115–6135.

Pantani, R. and L.S. Turng. 2015. Manufacturing of advanced biodegradable polymeric components. Journal of Applied Polymer Science. 132(48): 42889.

Pereira, F.D.A.S., V. Parfenov, Y.D. Khesuani, A. Ovsianikov and V. Mironov. 2018. Commercial 3D bioprinters. 3D Printing and Biofabrication. 535–549.

Raman, R. and R. Bashir. Stereolithographic 3D bioprinting for biomedical applications. pp. 89–121. *In*: A. Atala and J.J. Yoo (eds). Essentials of 3D Biofabrication and Translation. Elsevier Inc.

Shafiee, A. and A. Atala. 2016. Printing technologies for medical applications. Trends in Molecular Medicine. 22(3): 254–265.

Sundaramurthi, D., S. Rauf and C. Hauser. 2016. 3D bioprinting technology for regenerative medicine applications. International Journal of Bioprinting. 2(2): 9–26.

Ventola, C.L. 2014. Medical applications for 3D printing: current and projected uses. P & T: A Peer-Reviewed Journal for Formulary Management. 39(10): 704–711.

Webb, B. and B.J. Doyle. 2017. Parameter optimization for 3D bioprinting of hydrogels. Bioprinting. 8: 8–12.

Wu, Z., X. Su, Y. Xu, B. Kong, W. Sun and S. Mi. 2016. Bioprinting three-dimensional cell-laden tissue constructs with controllable degradation. Scientific Reports. 6: 1–10.

Chapter **7**

Cell Adhesion and Proliferation of Silica Nanocomposite

Nik Ahmad Nizam Nik Malek*[1,2]
Nor Suriani Sani[3], and Ahmad Taufiq[4]

[1]Department of Biosciences, Faculty of Science, Universiti Teknologi Malaysia, 81310 Skudai, Johor, Malaysia

[2]Centre for Sustainable Nanomaterials (CSNano), Ibnu Sina Institute for Scientific and Industrial Research (ISI-ISIR), Universiti Teknologi Malaysia, 81310 UTM Johor, Malaysia

[3]Department of Deputy Vice Chancellor (Research and Innovation), Universiti Teknologi Malaysia, 81310 UTM Skudai, Johor, Malaysia

[4]Department of Physics, Faculty of Mathematics and Natural Sciences, Universitas Negeri Malang, Indonesia

7.1 INTRODUCTION

The use of biomaterials for medical implants and regenerative medicine has gained popularity due to the ability of specific materials to repair and replace damaged or missing tissue and organs (Kareem et al. 2024, Todros et al. 2021). While metals are often utilized for skeletal implants due to their durability, they do not naturally integrate with cells and tissues at the implantation site (Thanigaivel et al. 2022). Additionally, metallic implants are prone to rusting when exposed to bodily fluids, which can release harmful metallic ions and cause damage to surrounding tissues (Li et al. 2024). Most of these materials also do not aid in forming new bone structures. Calcium phosphate materials, including tri-calcium phosphates, carbonated hydroxylapatite, and hydroxyapatite, have displayed great

*For Correspondence: niknizam@utm.my

potential as biomaterials (Chozhanathmisra et al. 2024, Idaszek et al. 2023). These materials are chemically similar to minerals found in calcified tissues of humans and vertebrates, such as tooth enamel, bone, and dentine. However, hydroxyapatite has some drawbacks, including poor mechanical properties, brittleness, and low fatigue resistance (Obada et al. 2023). Its stoichiometry also limits its bioactivity, which slows down its integration and reactivity with cells, leading to longer patient rehabilitation times (Ma and Liu 2009). Researchers have investigated immobilising or adding hydroxyapatite to support materials, such as those made of silica, to alleviate these concerns by improving the strength and mechanical rate as well as the quality of the bone tissue (Sani et al. 2019, 2020).

Silica, a compound comprised of the two most abundant elements on Earth, oxygen and silicon, can exist in either a "free" form or combined with other elements to form silicates. There are two main types of silica, namely crystalline and amorphous, and both can be found in natural or synthetic forms. Crystalline silica has been related to adverse health outcomes such as cancer and silicosis (Sato et al. 2018). Silicosis is a long-term health problem, a lung disease brought on by years of exposure to crystalline silica dust in the air (Leung et al. 2012). However, the effects of amorphous silica on human health are unclear. The stone industry dates to prehistoric times, and materials such as granite, sandstone, clay, limestone, and Portland cement all contain silicates. Furthermore, silicate glass and ceramics production is a long-standing, global industry. Silica, modified with hydroxyapatite, has recently caught the attention of many researchers due to its promising potential as an innovative biomaterial (Sani et al. 2019, 2020). This is primarily because of the beneficial effects of silica on cell growth. As a result, scientists are exploring the possibility of using modified silica with hydroxyapatite in various biomedical applications (Mabrouk et al. 2012). The combination of silicon, hydroxyapatite, and glycerol-hydrogel with different contents of hydroxyapatite (0.75 and 1.75 wt.%) that was obtained using a sol-gel method was studied by Khonina et al. (2020). They used this nanocomposite as a biomaterial for dental applications since the Si element could strengthen the remineralization of tooth enamel. In another recent work, Noviyanti and her colleague (2023) synthesized hydroxyapatite with silica via the hydrothermal method for bone graft material. They found enhanced mechanical stability after being immersed in simulated body fluid.

One kind of silica material is silica aerogel. Because of its extraordinary lightness and transparency, this solid substance is precious and genuinely astonishing. The name "frozen smoke" is frequently applied to it because of its hazy blue look. This material has a low density and is very porous. It comprises interconnected nanopores that combine to create a microstructure with various advantageous characteristics. With its high melting point, low bulk density, low thermal conductivity, and tuneable pore size and volume, silica aerogel is a superior insulator. It finds extensive application in biomedical imaging, pharmaceuticals, cosmetics, glass, cement, ceramics, and cosmetics. Silica aerogel is being investigated for its possible use in controlled drug release, biocompatible scaffolding for biological material immobilisation, biosensor matrix design, bone replacement, and drug-delivering particle carrier systems. Furthermore, silica aerogel's composition can be altered to mimic the bone bioactive qualities of traditional bioactive glasses, improving the material's biocompatibility.

A nanocomposite consisting of hydroxyapatite incorporated into silica aerogels has been developed as a promising biomaterial that possesses improved mechanical properties, biocompatibility, and bio-functionality (Sani et al. 2019, 2020). This hydroxyapatite-silica aerogel nanocomposite is found to have the potential to act as a cell delivery vehicle and carrier for implant and wound care applications. In this chapter, we discuss the results of assessing the efficiency and potential of the hydroxyapatite-silica aerogel nanocomposite against normal human osteoblast cells (NHOst) *in vitro*. An *in vitro* cell proliferation assay was used as it is a fast, easy, and cost-effective technique that allows for a clear understanding of cellular responses, as well as the biocompatibility and bio-functionality of the material. This type of assessment is sensitive enough to identify potentially toxic materials, enabling avoidance of such materials before animal testing. Conducting biocompatibility and bio-functionality assessments ensures their effectiveness when developing biomaterials. Biocompatibility refers to the ability of any material to integrate with host cells without harmful effects, such as cancer or developmental issues. A biocompatible material is composed of safe components that can facilitate the growth of cells and tissues. Meanwhile, bio-functionality measures how well the biomaterial can foster cell attachment and encourage cell migration, which is essential for its success.

To examine the biomaterial's ability to support attachment and promote osteoblastic growth, the hydroxyapatite-silica aerogel nanocomposite with a weight ratio of hydroxyapatite/silica at 0.5 was synthesized. This ratio was selected based on previous reports identifying it as resulting in the highest cell viability (Sani et al. 2019, 2020). Hydroxyapatite powder was uniformly dispersed to produce the nanocomposite in an extracted sodium silicate solution from rice husk ash. The gel was then prepared using concentrated H_2SO_4 in a Teflon beaker and left to age for 2 d at room temperature (24°C). After rinsing with double-distilled water to eliminate impurities and sodium sulphate salt, the gel was dried using an acetone solvent extraction technique for 2 d until completely dehydrated. Ambient-pressure drying (APD) technique was then used to remove any remaining water from the gel at 40°C. Finally, the hydroxyapatite-silica aerogel nanocomposite powder was kept in a sealed Teflon bottle at ambient conditions. For comparison purposes, silica aerogel was also produced using the same method but without hydroxyapatite.

The cells cultured on a hydroxyapatite-silica aerogel nanocomposite were observed for their morphology, and their proliferation was quantitatively evaluated using the MTT assay. The colourimetry MTT assay is a widely used method for assessing the metabolic activity of cells. It measures the cells' ability to change tetrazolium salt (MTT) into a purple formazan crystal. The quantity of purple formazan generated by the reaction directly relates to the number of living cells present. This can be measured by determining the solution's absorbance at a particular wavelength. The MTT assay is frequently used to assess cell proliferation, viability, and cytotoxicity in various disciplines, including toxicology, cancer biology, and drug development. Elemental analysis was conducted in the cell media after treatment with the samples to determine the impact of silicic acid and phosphate ions on cell proliferation and cell-material interaction. Comparative analysis was carried out with silica aerogel and free hydroxyapatite to understand the role of physical characteristics and chemical composition in cellular responses to materials, including attachment, growth, proliferation, and differentiation (Wei et al. 2008).

7.2 MORPHOLOGY ANALYSIS OF CELLS GROWN ON SILICA NANOCOMPOSITE

Cell culture involves growing cells in a sterile environment within a tissue culture lab. However, microbial contamination, such as yeast, bacteria, fungal spores, and mycoplasma, poses a significant challenge in tissue culture. This contamination can spread and compromise the entire experiment. All reagents, culture vessels, and consumables undergo sterilization in an autoclave at 121°C for 15 min to prevent this. This process effectively destroys all life forms, including bacterial spores, while avoiding denaturation of the media or other nutritious materials. All cell culture procedures were sterile using sterile techniques and a class II biosafety cabinet (Thermo 1300, series A2). Since 70% ethanol (Sigma-Aldrich; 99.8%) has the maximum germicidal efficacy, germs are kept out of the biosafety cabinet. The cell cultures were maintained in a controlled and operated environment simulating physiological conditions (37°C) similar to most mammalian organisms.

The OGM Bullet kit (OGMTM SingleQuotsTM) (Lonza) was used to cultivate normal human osteoblast cells (NHOst), which were then sub-cultured during the logarithmic (log) phase in osteoblast basal medium (OBM). The sub-cultured procedure was done by using 1 mL of trypsin-ethylenediaminetetraacetic acid, 0.25% (Lonza) to detach the cells, followed by the addition of roughly 2 mL of trypsin-neutralizing solution (Lonza) to neutralise the reaction and extract the detached cells. After centrifuging the cell suspension for 5 min at 1300 rpm, the entire media was added back, and the mixture was kept at 37°C in a humidified environment with 5% CO_2 until there were sufficient cells for analysis.

The ability of silica aerogel and hydroxyapatite-silica aerogel nanocomposite samples to promote the growth of NHOst can be analysed by understanding the cells' growth, cell orientation, and nutrient source of these treated cells in the sample. The morphology and nutrient supply within the culture were assessed to understand the function of chemical dispersion and consumption of the materials inside the cell-seeded sample. To obtain the cell-seeded samples, cell cultures were utilized, and the morphology of the samples and cultured cells' attachment was evaluated using Field Emission-Scanning Electron Microscopy (FESEM) and Energy Dispersive X-ray (EDX) analysis.

As part of the bio-functionality assay, a cell proliferation test was performed to evaluate the samples' ability to support the attachment of NHOst and promote cell growth. The study examined the impact of various parameters, such as the composition and topography of the materials, cell proliferation, apoptosis, and necrosis. Additionally, cell proliferation was used as an indicator of cell adhesion since better adhesion results in superior proliferation. To evaluate the proliferation and adhesion behaviour of NHOst, the cells were seeded immediately and directly onto discs (13 mm-diameter 0.1 mm-height) of the hydroxyapatite-silica aerogel nanocomposite. The cells on the discs were incubated and analyzed for several days, up to 21 d. The bio-functionality of the studied materials as a potential bone replacement material can be assessed by analysing the proliferation behaviour of the cells. The micrograph was captured using the FESEM (model JSM 6701F, JEOL) at an emission current of 2.00 kV, with a working distance of 3.0 mm and a probe current of 8 kV. At the same time, the EDX spectrometer (model EX-2300 BU, JEOL) was used for the elemental analysis of the samples.

Figure 7.1 shows the NHOst cell morphology after contact with the samples. Furthermore, the EDX analyser was utilised to confirm the presence of significant elements on the sample surfaces, listed in Table 7.1. Following a 5 d culture period, cell adhesion was microscopically observed on hydroxyapatite-silica aerogel nanocomposite, silica aerogel, and free hydroxyapatite samples using FESEM. The resulting micrographs in Figure 7.1 were carefully analyzed.

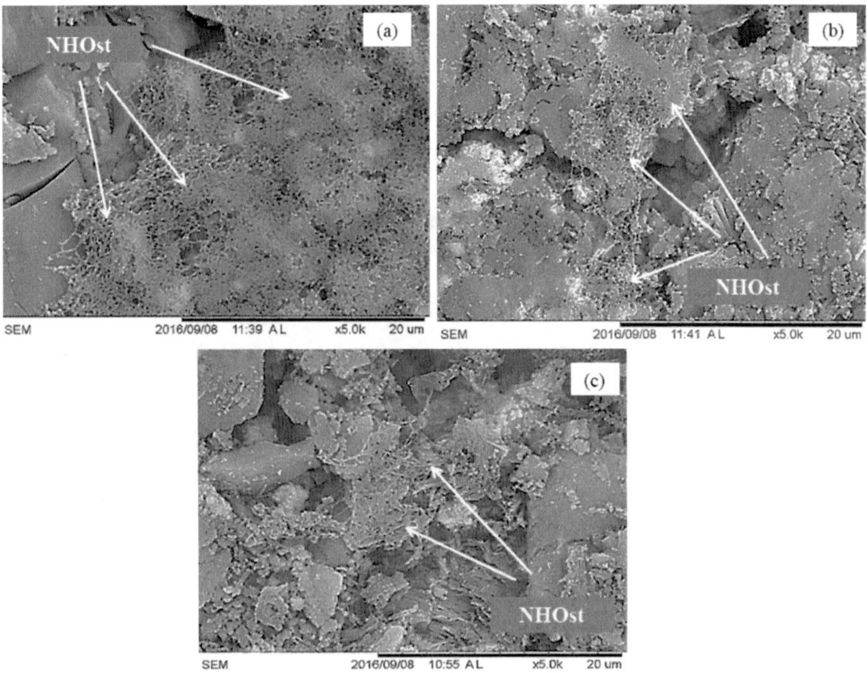

Figure 7.1 FESEM micrographs showing cell adhesion and proliferation of NHOst on (a) hydroxyapatite-silica aerogel nanocomposite, (b) silica aerogel and (c) free hydroxyapatite surfaces after 5 d of culturing indicate the presence of cells identified by arrows.

There are noticeable differences in the way cells respond to various surfaces. Several cells were found to adhere to the samples with polygonal elongated morphological shapes with multiple extensions. This appearance shows that the cells are connected and attached to the hydroxyapatite-silica aerogel

Table 7.1 Analysis of major elements (mass% SD) on hydroxyapatite-silica aerogel nanocomposite (HAESA), silica aerogel (SA), and free hydroxyapatite (HA) samples after several days of culturing with NHOst

Sample	Element (mass % ± SD)					
	C	O	Na	Si	P	Ca
SA-dy5	25.65 ± 0.82	49.02 ± 0.67	5.00 ± 0.00	14.32 ± 1.81	2.00 ± 1.00	4.01 ± 2.00
HA-dy5	16.28 ± 8.14	22.22 ± 1.11	nd	nd	23.48 ± 1.51	38.01 ± 1.01
HAESA-0.5-dy1	15.32 ± 7.97	43.70 ± 2.73	nd	35.49 ± 1.45	1.92 ± 1.00	3.56 ± 1.85
HAESA-0.5-dy3	19.29 ± 8.63	42.09 ± 1.82	nd	28.01 ± 1.53	3.98 ± 1.78	6.64 ± 2.97
HAESA-0.5-dy5	19.26 ± 9.63	22.31 ± 1.16	nd	24.53 ± 1.26	13.03 ± 6.52	20.87 ± 0.43
HAESA-0.5-dy7	21.35 ± 0.68	28.49 ± 1.25	nd	10.01 ± 5.01	14.94 ± 7.47	25.20 ± 1.60

*SA: silica aerogel. HA: Hydroxyapatite, dy: day

nanocomposite. EDX analysis confirmed that the surface of the hydroxyapatite-silica aerogel nanocomposite had higher percentages of P and Ca than that of silica aerogel and free hydroxyapatite. Conversely, cells on free hydroxyapatite were similarly dispersed to those cultured on silica aerogel after 5 d of culturing (see Figure 7.1 (b-c)). On the fifth day, cells cultured on silica aerogel and hydroxyapatite exhibited slower osteoblastic growth than on hydroxyapatite-silica aerogel nanocomposite. Most cells showed limited spreading with negligible prolongations and appeared slightly smaller. Only a little cell colony was seen on free hydroxyapatite and silica aerogel. Cells needed more time to culture to reach the confluent monolayer stage.

In Figure 7.2, the proliferation and adhesion of NHOst cells on a hydroxyapatite-silica aerogel nanocomposite are given throughout 1, 3, 5, and 7 d. On the surface of the studied materials, the cells are demonstrated to adhere and proliferate well; as the number of cultivation days increases, so does the colony of NHOst cells. The fact that the cells had adhered to the surface after only 3 d of culture suggests that the hydroxyapatite-silica aerogel nanocomposite is highly reactive and can promote cell adhesion. The number of cells on the surface rapidly rose as the days went by, and by day seven, a uniform layer of cells had colonised and covered most of the surface. The mass percentages of P and Ca in the hydroxyapatite-silica aerogel nanocomposite increased with this expansion. The NHOst cells firmly adhered to the surface by creating cell-cell connections between surface gaps.

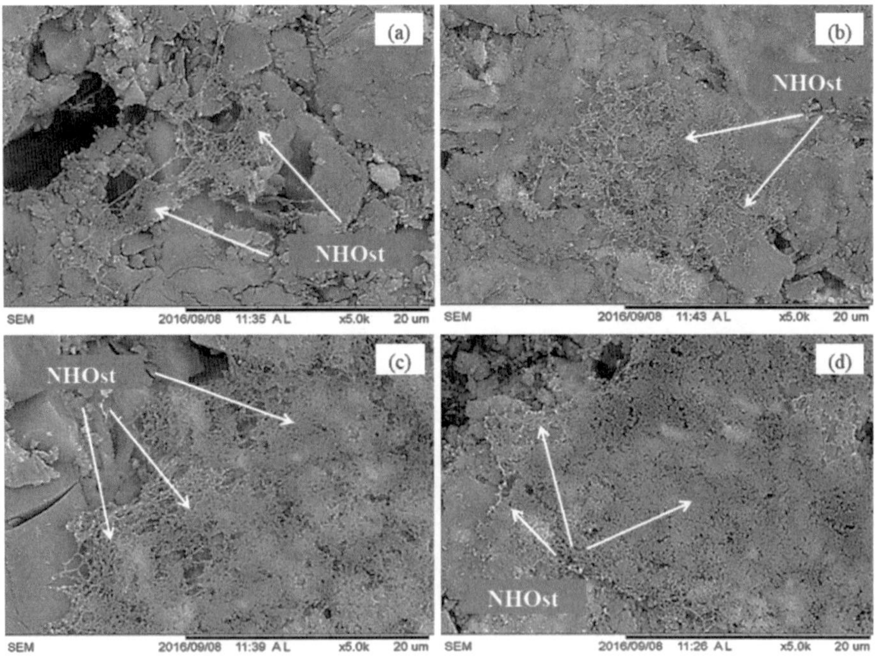

Figure 7.2 **FESEM** micrographs showing the NHOst cell adhesion on hydroxyapatite-silica aerogel nanocomposite at different days (a) 1, (b) 3, (c) 5 and (d) 7 d (arrow indicate the presence of cells).

The hydroxyapatite-silica aerogel nanocomposite's surface roughness, as seen in Figure 7.2, is essential for encouraging cell adhesion and proliferation. This is particularly important when developing materials for bone tissue engineering because many other inorganic materials need additional components and growth factors to improve their cell response and biocompatibility. The hydroxyapatite-infused nanocomposite will likely stop distant osteogenesis from happening. Osteogenesis is a phenomenon where the bone growth is congested due to the gap between the implant and the surrounding bone. Additionally, the hydroxyapatite-silica aerogel nanocomposite can promote the process of cell in-growth, ensuring that the cells can anchor themselves onto both the sample and the surface gaps. This feature can enable the nanocomposite to provide better integration and stability, prolonging the implant's lifespan. Eventually, this promotes cell migration, penetration, and strong adhesion, enabling successful cell integration into the material.

7.3 CELL PROLIFERATION ANALYSIS ON SILICA NANOCOMPOSITE

The proliferation of NHOst was assessed using an MTT-based colourimetry test. Following seeding onto an aerogel nanocomposite containing hydroxyapatite and silica, the cells were submerged in a complete media solution and incubated for 3, 5, 7 and 21 d at 37°C with 5% CO_2 in the atmosphere. The MTT test assessed the cell growth rate following each incubation period. The calculation for cell proliferation was conducted based on this equation (Rismanchian et al. 2013):

$$\text{Cell Proliferation (\%)} = \frac{\text{Mean OD of treated cells} - \text{Mean OD of blank}}{\text{Mean OD of untreated cells} - \text{Mean OD of blank}} \times 100$$

The hydroxyapatite-silica aerogel nanocomposite's bio-functionality was evaluated by subjecting cells to the material and examining the growth features of NHOsts. The MTT assay was used to quantify the proliferation of NHOst cells grown on free hydroxyapatite, hydroxyapatite-silica aerogel nanocomposite, and silica aerogel. MTT optical density (OD) measurements show how cells grow and proliferate on various surfaces. The proportion of cell proliferation following interaction with the substances is displayed in Figure 7.3.

After culture on each sample, the number of active cells increases, as seen in Figure 7.3. After culture for 2 to 7 d, the Tukey post hoc test, which came after the two-way ANOVA, revealed that the cell proliferation percentage for all materials under study was considerably ($p < 0.01$ – $p \leq 0.0001$) more significant than the control. Cells that proliferate after contact with hydroxyapatite-silica aerogel nanocomposite show a significant ($p \leq 0.0001$) growth with a longer culture time, and the highest growth can be seen on the 5th day. This indicates that the hydroxyapatite-silica aerogel nanocomposite significantly ($p \leq 0.0001$) affected the high metabolic rate on day five. The morphological analysis, as in Figure 7.2, supports this result in which the images showed that the rough surface of hydroxyapatite-silica aerogel nanocomposite had been adhered by the cells due to the cell proliferation activity for up to 7 d.

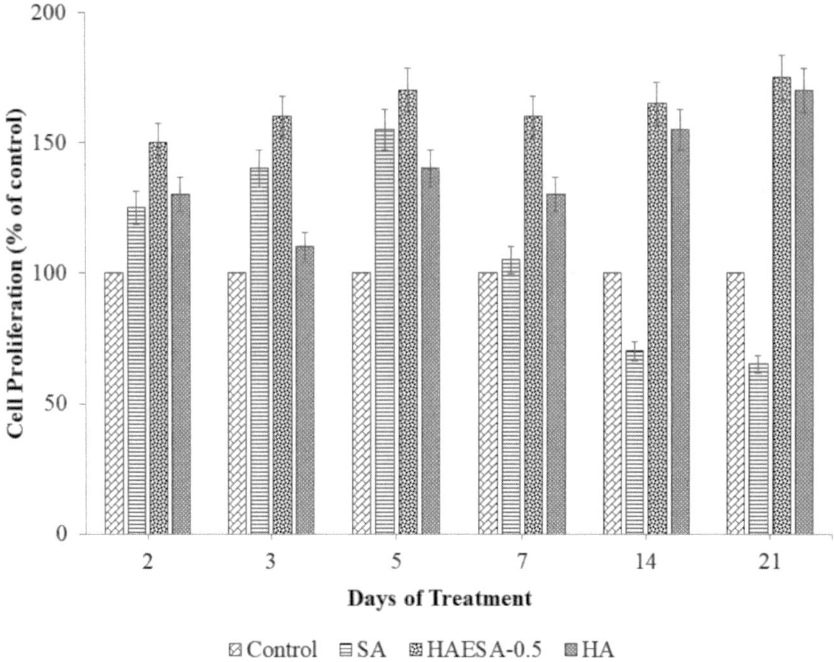

Figure 7.3 Percentage of NHOst cell proliferation after 3, 5, 7, 14 and 21 d cultured on silica aerogel (SA), hydroxyapatite-silica aerogel nanocomposite (HAESA0.5) and free hydroxyapatite (HA). The results are presented as the percentage of the control groups. Data are presented as three independent experiments' mean±standard deviation (SD). $^*P \leq 0.1$, $^{**}P \leq 0.01$, $^{***}P \leq 0.001$ and $^{****}P \leq 0.0001$ in comparison with other samples by two-way ANOVA.

The proliferation of NHOst on silica aerogel cultured samples shows a 10% to 25% increase compared to those on free hydroxyapatite. Figure 7.3 shows cell proliferation activity on silica aerogel and free hydroxyapatite after 2 to 5 d of culturing. However, the cell proliferation cultured on silica aerogel significantly decreased ($p \leq 0.01 - p \leq 0.0001$) after 7 to 21 d of treatment, whereas the cells cultured on free hydroxyapatite continued to increase. The early stages of cell proliferation are found to be influenced by the interaction between the cell and matrix, commonly called cell-matrix interaction. This interaction is crucial in promoting cell growth and proliferation activity. It is hypothesized that the silica (Si) element released into the media during this interaction could act as a growth promoter and contribute to the initial increase in cell proliferation. This early increase in proliferation activity is considered an essential factor in the overall development and growth of the cells. In addition, hydroxyapatite plays a vital role in maintaining osteoblastic functionality during extended mineralisation. As depicted in Figure 7.3, cells cultured on hydroxyapatite-silica aerogel nanocomposite grew by 10% to 50%, significantly higher than those on free hydroxyapatite ($p \leq 0.001 - p \leq 0.0001$). This validates that the hydroxyapatite-silica aerogel nanocomposite is an excellent biomaterial that could promote cell growth and proliferation because it has both hydroxyapatite and silica at optimal conditions.

7.4 PHOSPHATE AND SILICIC ACID IN THE MEDIA

Numerous tissue engineering studies have demonstrated the significant impact of chemical compositions on cell response. The ions released influence the interaction between cells and materials (Wei et al. 2008, Bohner 2009). Tissue regeneration at the implant site depends on material dissolution and subsequent resorption (Arora 2013). Essential ions are necessary for cellular differentiation, proliferation, apoptosis, and membrane integrity maintenance in mammalian cell culture. The effect of crucial ions from hydroxyapatite-silica aerogel nanocomposites on NHOst was examined by analysing the amounts of phosphate and silicic acid (mg/L) that remained in the media before and after different days of culture.

An in-depth elemental analysis was conducted to understand better the impact of phosphate and silicic acid ions on cell-material interaction. The study involved quantifying the presence of these ions in cell media after immersion with samples, focusing on how they affect cell growth and the biocompatibility of the materials. Supernatants from the remaining media were analyzed for phosphate and silicic acid, essential for hydroxyapatite and silica aerogel framework formation. The presence of these ions in the media may suggest possible framework dissolution.

The NANOCOLOR® UV-Vis spectrophotometer by Macherey-Nagel with Method 1–77 for phosphate and Method 1–48 for silicic acid (silica) analysis kits were used to measure the concentrations of both ions. Silica levels were determined using reagent kit set 91848. A 20 mL sample was mixed with reagent R1 and left for 3 min. Reagents R2 and R3 were added, and measurements were taken with the NANOCOLOR® Vis spectrophotometer at 690 nm wavelength in a 10 mm cuvette after 15 min. Method 1–77 from the NANOCOLOR Standard Test, developed by Macherey-Nagel in Germany, was used to calculate the phosphate levels. The reference number 91877 was utilised on a reagent kit. Pipetting 20 ml of the test sample into a 25 ml volumetric flask, one ml of R1 reagent (acid sulphuric, H_2SO_4, 30%) and one ml of R2 reagent (sodium disulphide, $Na_2O_5S_2$, <25%) were mixed. Deionized water was added to the test, and blank samples were placed at the 25 mL mark and then gently swirled. The mixture was put into a 10 mm cuvette and measured at 690 nm using the NANOCOLOR® Vis spectrophotometer after 10 min.

The calculation for the percentage loss of phosphate and silicic acid is given based on this formula (Sani et al. 2017):

$$\%\text{Element}_{\text{loss}} = \frac{\text{Mean } [E]_{\text{mgL}^{-1}} \text{ untreated solution} - \text{Mean } [E]_{\text{mgL}^{-1}} \text{ treated solution}}{\text{Mean } [E]_{\text{mgL}^{-1}} \text{ untreated solution}} \times 100$$

After the cells were cultured, the percentage of PO_4^{3-} loss in the media was determined to confirm that phosphate ions were utilised during cell proliferation on the materials under study. The concentration (mg/L) of phosphate ions in media after 21 d of being untreated or treated with silica aerogel, hydroxyapatite-silica aerogel nanocomposite, and free hydroxyapatite, as well as the percentage loss of phosphate ions in the media after 3, 5, 7, 14, and 21 d of culturing, are displayed on the graph in Figure 7.4.

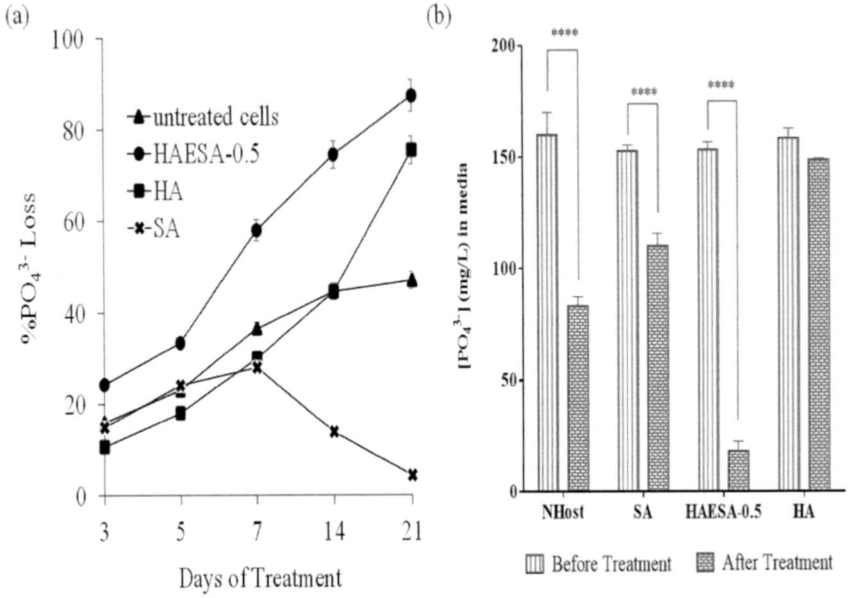

Figure 7.4 (a) Percentage loss of phosphate ions in media after 3, 5, 7, 14 and 21 d of culturing. (b) Concentration (mg/L) of phosphate ions in media after 21 d untreated and treated with silica aerogel (SA), hydroxyapatite-silica aerogel nanocomposite (HAESA-0.5) and free hydroxyapatite (HA).

Figure 7.4(a) demonstrates that during the initial 5 d of culturing, there was no apparent difference in the $\%PO_4^{3-}$ loss between the untreated cells, silica aerogel, hydroxyapatite-silica aerogel nanocomposite, and free hydroxyapatite. This aligns with the prior result that these materials are biocompatible with NHOst. However, as the culturing process continued, cellular growth patterns diverged. The concentration of phosphate ions in the media also changed over time, as depicted in Figure 7.4(a). Across all samples, there was a rapid increase in $\%PO_4^{3-}$ loss up to day 7, followed by distinct trends among the different materials. Compared to the other samples, free hydroxyapatite showed the highest concentration of phosphate ions in the media by the 21st day. The dissolution of free hydroxyapatite due to an under-saturated media environment (Ma et al., 2005) is further demonstrated in Figure 7.4(b), which releases many phosphate ions into the media. This may affect implant stability and result in an inadequate osteoblastic response. This result is corroborated by FESEM imaging (Figure 7.1), which demonstrates that osteoblast apoptosis is triggered by poor cell attachment density on free hydroxyapatite when exposed to concentrated phosphate ions (Laquerriere et al. 2003).

In contrast, the findings presented in Figure 7.4(a) demonstrate that the hydroxyapatite-silica aerogel nanocomposite is effective in reducing phosphate ions in media after 7 d of treatment, with nearly 60% of phosphate ions eliminated upon contact with the nanocomposite. Notably, the hydroxyapatite-silica aerogel nanocomposite exhibits a more remarkable ability to consume phosphate ions in the media compared to other samples, leading to increasing cell proliferation as evidenced by the MTT assay in Figure 7.3. Furthermore, Figure 7.4 shows that extending the culturing time from 14 to 21 d results in the complete removal of treated phosphate ions by the hydroxyapatite-silica aerogel nanocomposite, leading to cell confluence as shown in the morphological analysis in Figure 7.2. Phosphate ions treated with the nanocomposite had a concentration of only 18 mg/L after 21 d, which was significantly ($p < 0.0001$) less than that of the original media (153 mg/L) (Figure 7.4(b)). These findings validate that apoptosis and changes in cell membrane caused by excessive phosphate release from free hydroxyapatite can be avoided by covering the hydroxyapatite with a silica aerogel network on the surface of the hydroxyapatite-silica aerogel nanocomposite (Wang et al. 2015).

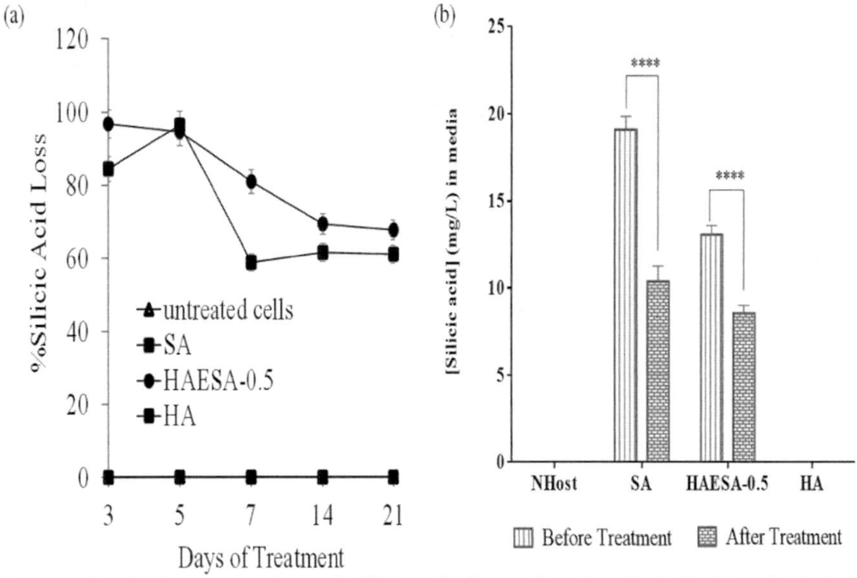

Figure 7.5 (a) Percentage loss of silicic acid in media after 3, 5, 7, 14 and 21 d of culturing. (b) Concentration (mg/L) of silicic acid in media after 21 d untreated and treated with silica aerogel (SA), hydroxyapatite-silica aerogel nanocomposite (HAESA-0.5) and free hydroxyapatite (HA).

Based on Figure 7.5(a), almost 85% of silicic acid was found to be released into the media by the samples containing silica (including the nanocomposite) in comparison to the free hydroxyapatite and untreated media. This happened in the early days of the incubation. This finding proved that the silica aerogel and the nanocomposite can be easily decomposed and release Si elements into the media. Figure 7.5(b) also shows that silica aerogel released more silicic acid in the media than hydroxyapatite-silica aerogel nanocomposite did. Interestingly, the results from the MTT assay (Figure 7.3) demonstrate that osteoblast cells cultured on hydroxyapatite-silica aerogel nanocomposite exhibited higher proliferation than those on silica aerogel. This could be due to the attachment of the osteoblast cells that could avoid the Si element release from hydroxyapatite-silica aerogel nanocomposite surfaces or prevent the uptake of Si by the cells (Quignard et al. 2012, Kalia et al. 2016). Hence, the hydroxyapatite-silica aerogel nanocomposite enhances osteoblast cell adhesion, proliferation, and new bone formation by stimulating osteoblasts through the ionic dissolution products that contain silicic acid.

An elevated level of silicic acid can lead to a slight reduction in cell proliferation. The graph in Figure 7.5(a) highlights that silicic acid concentration primarily affects the percentage of cells. This implies that the disruption of silica aerogel's networks caused by excessive dissolving may prevent cell adhesion and proliferation. Furthermore, the results consistently show that encouraging the growth of osteoblast-like cells and eliciting a biological response in NHOst cells depend on maintaining an ideal silicic acid concentration (Shie et al. 2011).

The hydroxyapatite-silica aerogel nanocomposite has a resorbable matrix of interconnected particles that can dissolve in a physical solution, promoting the proliferation of osteoblast cells. Osteoblasts cultured on the nanocomposite rapidly proliferated towards the confluence stage, more than silica aerogel and free hydroxyapatite. The finding also discovered that releasing a suitable quantity of silicic acid and phosphorus into the culture conditions greatly enhanced the early proliferation of cells on hydroxyapatite-silica aerogel nanocomposite. The hydroxyapatite-silica aerogel nanocomposite is ideal for the attachment and growth of NHOst cells, according to all biocompatibility and bio-functionality data. The nanocomposite's surface characteristics and dissolving behaviour can be changed by mixing hydroxyapatite and silica aerogel, accelerating bio-mineralization.

7.5 SUMMARY

The promise of the hydroxyapatite-silica aerogel nanocomposite as a novel bone implant material with excellent mechanical and biocompatibility performance was confirmed by cell proliferation experiments. After 21 d of exposure, the results demonstrated that the viability of cell growth was greater than that of the control on the hydroxyapatite-silica aerogel nanocomposite and statistically significant ($p \leq 0.01 - p < 0.0001$). The observation of elevated osteoblast cell adhesion and proliferation on the nanocomposite implies that the silicic acid-containing ionic dissolution products promote osteoblast proliferation, hence aiding in the expansion of osteoblast cells and the creation of new bone. Because of its high phosphate ion consumption capacity and excellent cell attachment surface, the hydroxyapatite-silica aerogel nanocomposite promoted NHOst adhesion at a higher density than silica aerogel and hydroxyapatite. According to these findings, the hydroxyapatite-silica aerogel nanocomposite, which was produced using the aqueous sol-gel ambient pressure drying (APD) method from rice husk ash, did not cause any harm to normal human osteoblast cells. Instead, this nanocomposite can enhance normal human osteoblast proliferation and adhesion, supporting or guiding the cells during bone implantation. This nanocomposite can be used as a novel biomaterial in the future to replace hard tissues in medical applications and as an alternative biomaterial for regenerative medicine.

ACKNOWLEDGEMENT

Authors would like to thanks Universiti Teknologi Malaysia and Ministry of Higher Education, Malaysia for financially supporting the project under Fundamental Research Grant Scheme (FRGS), Vot. 4F514.

REFERENCES

Arora, M. 2013. Cell culture media: a review. Mater. Methods. 3: 175–202.

Bohner, M. 2009. Silicon-substituted calcium phosphates—a critical view. Biomaterials. 30: 6403–6406.

Chozhanathmisra, M., A. Murugesan, P.S. Kumar, M. Loganathan, G. Sampath, N. Sedhu, et al., 2024. Fabrication of a novel biomaterial of mono-layer of halloysite nanotube-magnesium oxide/β-tricalcium phosphate composite on implant substrate: structural and biological properties. Ceramics International. 50(4): 6703–6712.

Idaszek, J., J. Jaroszewicz, E. Choińska, Z. Górecka, A. Hyc, A. Osiecka-Iwan, et al., 2023. Toward osteomimetic formation of calcium phosphate coatings with carbonated hydroxyapatite. Biomaterials Advances. 149: 213403.

Kalia, P., R.A. Brooks, S.D. Kinrade, D.J. Morgan, A.P. Brown, N. Rushton, et al., 2016. Adsorption of amorphous silica nanoparticles onto hydroxyapatite surfaces differentially alters surfaces properties and adhesion of human osteoblast cells. PLoS One. 11(2): 1–15.

Kareem, R.O., N. Bulut and O. Kaygili. 2024. Hydroxyapatite biomaterials: a comprehensive review of their properties, structures, medical applications, and fabrication methods. Journal of Chemical Reviews. 6(1): 1–26.

Khonina, T.G., O.N. Chupakhin, V.Y. Shur, A.P. Turygin, V.V. Sadovsky, Y.V. Mandra, et al., 2020. Silicon-hydroxyapatite-glycerohydrogel as a promising biomaterial for dental applications. Colloids and Surfaces B: Biointerfaces. 189: 110851.

Laquerriere, P., A. Grandjean-Laquerriere, E. Jallot, G. Balossier, P. Frayssinet and M. Guenounou. 2003. Importance of hydroxyapatite particles characteristics on cytokines production by human monocytes *in vitro*. Biomaterials. 24: 2739–2747.

Leung, C.C., I.T.S. Yu and W. Chen. 2012. Silicosis. The Lancet. 379(9830): 2008–2018.

Li, M., J. Wu, W. Geng, P. Gao, Y. Yang, X. Li, et al., 2024. Interaction pathways of implant metal localized corrosion and macrophage inflammatory reactions. Bioactive Materials. 31: 355–367.

Ma, G. and X.Y. Liu. 2009. Hydroxyapatite: hexagonal or monoclinic? Crystal Growth and Design. 9(7): 2991–2994.

Ma, S., T. Yang, D.L. Carnes, K. Kim, S. Park, S.H. Oh, et al., 2005. Effects of dissolved calcium and phosphorous on osteoblast responses. Journal of Oral Implantol. 31(2): 61–67.

Mabrouk, M., M.M. Selim, H. Beherei and M.I. El-Gohary. 2012. Effect of incorporation of nano bioactive silica into commercial glass ionomer cement (GIC). Journal of Genetic Engineering and Biotechnology. 10(1): 113–119.

Noviyanti, A.R., J. Juliandri, E.E. Ernawati, H. Haryono, S. Solihudin, D. Dwiyanti, et al., 2023. Hydrothermally synthesized hydroxyapatite-silica composites with enhanced mechanical properties for bone graft applications. Chemistry. 5(3): 1645–1655.

Obada, D.O., S.A. Osseni, H. Sina, A.N. Oyedeji, K.A. Salami, E. Okafor, et al., 2023. Hydroxyapatite materials-synthesis routes, mechanical behavior, theoretical insights, and artificial intelligence models: a review. Journal of the Australian Ceramic. 59(3): 565–596.

Quignard, S., G. Mosser, M. Boissière and T. Coradin. 2012. Long-term fate of silica nanoparticles interacting with human dermal fibroblasts. Biomaterials. 33(17): 4431–4442.

Rismanchian, M., N. Khodaeian, L. Bahramian, M. Fathi and H. Sadeghi-Aliabadi. 2013. *In-vitro* comparison of cytotoxicity of two bioactive glasses in micropowder and nanopowder forms. Iran Journal of Pharmacy Research. 12(3): 437–443.

Sani, N.S., N.A.N.N. Malek, K. Jemon, M.R.A. Kadir and H. Hamda. 2017. Effect of mass concentration on bioactivity and cell viability of calcined silica aerogel synthesized from rice husk ash as silica source. Journal of Sol-Gel Science and Technology. 82: 120–132.

Sani, N.S., N.A.N.N. Malek, K. Jemon, M.R.A. Kadir and H. Hamdan. 2019. Preparation and characterization of hydroxyapatite incorporated silica aerogel and its effect on normal human dermal fibroblast cells. Journal of Sol-Gel Science and Technology. 90: 422–433.

Sani, N.S., N.A.N.N. Malek, K. Jemon, M.R.A. Kadir and H. Hamdan. 2020. *In vitro* bioactivity and osteoblast cell viability studies of hydroxyapatite-incorporated silica aerogel. Journal of Sol-Gel Science and Technology. 96: 166–177.

Sato, T., T. Shimosato and D.M. Klinman. 2018. Silicosis and lung cancer: current perspectives. Lung Cancer (Auckl). 91–101.

Shie, M.Y., S.J. Ding and H.C. Chang. 2011. The role of silicon in osteoblast-like cell proliferation and apoptosis. Acta Biomaterialia. 7(6): 2604–2614.

Thanigaivel, S., A.K. Priya, D. Balakrishnan, K. Dutta, S. Rajendran and M. Soto-Moscoso. 2022. Insight on recent development in metallic biomaterials: Strategies involving synthesis, types and surface modification for advanced therapeutic and biomedical applications. Biochemical Engineering Journal. 187: 108522.

Todros, S., M. Todesco and A. Bagno. 2021. Biomaterials and their biomedical applications: From replacement to regeneration. Processes. 9(11): 1949.

Wei, J., S.J. Heo, D.H. Kim, S.E. Kim, Y.T. Hyun and J.-W. Shin. 2008. Comparison of physical, chemical and cellular responses to nano- and micro-sized calcium silicate/poly(epsilon-caprolactone) bioactive composites. Journal of the Royal Society, Interface. 5(23): 617–630.

Wang, X., N. Ahmed, N.G.S. Alvarez, M.V. Tuttolomondo, C. Hélary, M.F. Desimone, et al., 2015. Sol-gel encapsulation of biomolecules and cells for medicinal applications. Current Topics in Medicinal Chemistry. 15(3): 223–244.

Chapter 8

3D Bioprinting of Phenotypically Stable Articular Cartilage

Nilotpal Majumder and Sourabh Ghosh*
Regenerative Engineering Laboratory,
Department of Textile and Fibre Engineering,
Indian Institute of Technology Delhi, Hauz Khas, New Delhi, India

8.1 INTRODUCTION

Osteoarthritis (OA) is marked by the gradual degeneration of the articular cartilage with subsequent degradation of the underlying subchondral bone (Hunter and Felson 2006). A wide array of factors contributes to the onset and progression of OA, including age, physical trauma, heavy exercise or metabolic dysregulations like obesity, hyperlipidemia, elevated blood sugar, estrogen deficiency (menopausal women) etc. (Heidari 2011). Synovial inflammation and osteophyte formation are thought to be the main risk factors that aggravate the deterioration of normal functioning articular cartilage (de Lange-Brokaar et al. 2012, Sanchez-Lopez et al. 2022). The largest impact of OA is observed on the articular cartilage tissue, primarily composed of a single cell type called articular chondrocytes. It lacks nerves and blood vessels as a highly specialized connective tissue. The chondrocytes and collagen fibrils are spatially localized differently within the various anatomical layers of the articular cartilage (Sophia Fox et al. 2009). During inflammation, the cytokines activate the matrix-degrading enzymes like

*For Correspondence: Sourabh.Ghosh@textile.iitd.ac.in

matrix metalloproteinases (MMP-1/13) and a disintegrin-like metalloproteinase with thrombospondin motif (ADAMTS) that accelerate the degradation of the surrounding cartilage matrix and contribute to OA progression. The external manifestation of OA is severe pain and swelling around the joint area, causing permanent mechanical stiffness around the joint site, limiting the patient's locomotion. When clinicians are mainly focused on pain management, tissue engineers worldwide have taken the forefront in devising strategies to restore the normal functioning of articular cartilage.

Researchers have used natural polymers like chitosan (Shen et al. 2021), collagen, alginate (Chakraborty et al. 2022) and synthetic polymers like polyethylene glycol (Zhao et al. 2016), polylactic acid (Han et al. 2021), polyglycolic acid (Moran et al. 2003), either standalone or in blend, to engineer articular cartilage tissue. Various cell types like chondrocytes isolated from humans (Chameettachal et al. 2016), goats or pigs, stem cells derived from adipose tissue (Yoon et al. 2012), bone marrow (Majumder et al. 2024a) or nasal turbinate (Das et al. 2015) and different cell lines. They have also utilized different scaffold fabrication techniques like electrospinning (Ghosh et al. 2009), porogen leaching, solvent casting and 3D printing with top-seeding of respective cells to fabricate a three-dimensional hyaline cartilage construct. Additionally, a large population of pro-chondrogenic growth factors was also used during the differentiation or maturation period of the tissue constructs like Transforming Growth Factor-β (TGF-β) (Chawla et al. 2017), Growth Differentiation Factor-5 (GDF-5) (Takahata et al. 2022), Bone Morphogenetic Protein (BMP) (Yoon and Lyons 2004), etc. Our laboratory has also used silk fibroin and gelatin in combination with different chondrocytes and stem cells to 3D bioprint articular cartilage tissue (Majumder et al. 2024b). Although an enhanced expression of cartilage-specific gene expressions (COL-II, SOX-9, ACAN) has been observed post-maturation for 21–28 d, an unexpected consistent expression of osteogenic genes (COL-X, MMP-13) complicated the overall clinical outcome of successful articular cartilage graft formation (Chameettachal et al. 2016).

The largest obstacle in cartilage tissue engineering is maintaining the stable chondrogenic phenotype post-implantation *in vivo* (Majumder and Ghosh 2023). Although scientists have proposed different theories relative to the biomaterial, cell type and growth factors, the primary cause of this failure is the formation of a transient cartilage phenotype from the stable articular cartilage phenotype. This unusual phenotype conversion is termed "chondrocyte hypertrophy." Many characteristics differentiate hypertrophic cartilage from healthy articular cartilage-like calcification of the extracellular matrix (Fuerst et al. 2009), angiogenic growth throughout the cartilage matrix (Bonnet and Walsh 2005), and increased chondrocyte size. As a result, the implanted cartilage starts assuming an osteogenic phenotype post-maturation under the influence of the OA microenvironment, thereby losing its elastic shock-absorbing characteristics. Although the tissue engineers have mentioned in their reports the presence of chondrocyte hypertrophy following the traditional chondrogenic differentiation protocol during chondrogenesis, the exact reason for its occurrence is still a mystery.

A few reports claim that it is the innate characteristics of the stem cells to undergo hypertrophic differentiation during chondrogenesis following the endochondral ossification route (Chawla et al. 2018). In contrast, a few indicated that the abnormal regulation of the chondrogenic signaling pathways and the subsequent over- and under-expression of their downstream target genes during the differentiation protocol could contribute significantly to hypertrophic maturation. We hypothesize that the onset of chondrocyte hypertrophy is a major hindrance to the development of permanent cartilage grafts. A plausible reason behind this mystical scenario is ignorance concerning the difference between the embryonic development of transient and permanent articular cartilage and the lack of adequate protocols to characterize the developed cartilage graft at molecular levels. Hence, this chapter will majorly focus on the current platforms to engineer articular cartilage tissue and their futile outcomes, a brief introduction to chondrocyte hypertrophy, the probable reasons for the onset of chondrocyte hypertrophy and finally, the measures adopted to develop phenotypically stable articular cartilage combating chondrocyte hypertrophy.

8.2 THE EXISTING APPROACHES TO DEVELOPING ENGINEERED CARTILAGE TISSUE

Various tissue engineering approaches have been devised to engineer cartilage tissue. However, various challenges exist to replicate the native tissue microarchitecture and physiological function. Hence, there has been a shift in focus toward three-dimensional cell culture approaches like spheroids and other scaffold-based and scaffold-free tissue engineering strategies.

8.2.1 Spheroids

Culturing the cells in spheroids was considered the initial step to replicate the 3D microenvironment of cells that will facilitate the intra-cellular communication to stimulate chondrogenesis. Researchers have reported the advantages of spheroid culture over monolayer in attaining enhanced chondrogenic differentiation. The ADSC spheroids (3×10^6 cells) in a spinning bioreactor showed an enhanced expression of COL-II over the monolayer, indicating the superiority of the spheroid system over the monolayer in attaining chondrogenesis (Yoon et al. 2012). Similar results using ADSCs were also reported by Tu et al. showing successful chondrogenic differentiation post-21 d of culture (Vy et al. 2020). A comparison between micromass and pellet culture using human MSCs demonstrated the upper hand of micromass culture in mitigating chondrocyte hypertrophy while simultaneously promoting chondrogenic differentiation (Zhang et al. 2010). Researchers also conducted a comparative study emphasizing the advantages of aggregate and pellet culture over explant and monolayer systems in attaining successful chondrogenesis using equine articular chondrocytes (Stewart et al. 2000). However, the onset of hypertrophy could not be prevented even after successful chondrogenesis with the expression of collagen type-I (COL-I), collagen type-X (COL-X), osteogenic genes and matrix-degrading enzymes. This could be because the spheroid system only mimicked the cell-to-cell communication. Still, it could not simulate the cell-ECM interactions that play a key role in the phenotypic stability of articular chondrocytes.

8.2.2 3D Bioprinted Models

Our laboratory has been instrumental in utilizing the advantages of silk fibroin gelatin and 3D bioprinting technology to develop articular cartilage grafts utilizing an array of cell types and differentiation factors. To compare the chondrogenesis profile of human BMSCs and human chondrocytes within the 3D bioprinted SF-G constructs, Chameettachal et al. have shown that articular cartilage tissue developed using 3D bioprinting of human BMSC-loaded SF-G bioink displayed the highest chondrogenic traits compared to articular chondrocytes (Chameettachal et al. 2016). In an attempt to evaluate the effect of chondrogenic growth factor TGF-β1 on chondrogenesis, Chawla et al. used 3D bioprinted murine BMSCs-encapsulated SF-G bioink and cultured them with or without the TGF-β1 for a period of 21 d and compared with 3D aggregate and monolayer culture. The enhanced expression of cartilage marker genes like COL-II, ACAN, and SOX9 in the 3D bioprinted group without TGF-β1 supplementation compared to the aggregate and monolayer after 21 d of culture indicated successful chondrogenesis. However, the undesired increased levels of COL-X, IHH, and MMP-13 expression in the same 3D bioprinted group demonstrated the hypertrophic maturation of the BMSCs post-chondrogenesis (Chawla et al. 2017). A comparative analysis study of 3D bioprinted and 3D spheroid system composed of human articular chondrocytes and M2 macrophages also demonstrated the upper hand of the 3D bioprinting system in positively regulating chondrogenic phenotype with mitigation of chondrocyte hypertrophy. A detailed qPCR and Immunofluorescence analysis revealed an enhanced expression of COL-II along with the anti-inflammatory M2 marker CD 163. Therefore, the coculture strategy in the 3D bioprinted system could successfully mitigate OA progression while maintaining the stable anti-inflammatory M2 phenotype (Majumder et al. 2024b).

The same combination of SF-G bioink and human bone marrow stromal cells were 3D bioprinted and subjected to chondrogenesis for 28 d with TGF-β3 to analyze the protein level expressions during chondrogenesis. A detailed proteomic analysis implicated that it is not only due to the advantages of differentiation factors or 3D bioprinting but also due to the upregulation of the pro-chondrogenic Wnt/β-catenin pathway by the SF-G bioink itself that promotes chondrogenesis (Chawla et al. 2021, Trucco et al. 2021). These benefits of 3D bioprinting, SF-G bioink and hBMSCs were further reaped by Majumder et al. in studying the long-term maintenance of stable chondrogenic phenotype of BMSCs under OA-mimicking conditions (Majumder et al. 2024a). Our research group also explored another biomaterial blend of Alginate and Gellan gum encapsulated with silk nanoparticles and murine BMSCs, followed by 3D bioprinting and chondrogenic differentiation. It was observed that the silk nanoparticle-embedded group displayed enhanced chondrogenic traits with higher mechanical strength (Chakraborty et al. 2022). Researchers have also deployed a light-based 3D bioprinting technique by chemically modifying the existing biomaterial to allow photo-crosslinking using a photoinitiator without compromising cell viability and chondrogenic properties (Bandyopadhyay et al. 2022). Despite the tremendous efforts the cartilage biologists, tissue engineers, material scientists, and clinicians put forward in developing a phenotypically stable articular cartilage graft, a few could make the lab-to-clinic leap. This is because during the conventional chondrogenic differentiation protocol, although the tissue displayed adequate chondrogenic characteristics, parallel expression levels of chondrocyte hypertrophy markers made the entire protocol futile.

8.2.3 Spheroid 3D Bioprinting

Spheroids have been the building blocks of tissue engineering as the initial stepping stone to 3D culture systems initiating cell-to-cell communication (Zhuang et al. 2021). On the contrary, 3D bioprinting allows the spatio-temporal organization of the cells and biomaterials to fabricate a multi-layered 3D tissue construct to add the benefit of cell-ECM interaction. Thus, combining these technologies traces back to the embryonic development of tissues and organs from the process of cellular assembly (Banerjee et al. 2022). Chameettachal et al. encapsulated BMSC and chondrocyte spheroids within the SF-G bioink and 3D bioprinted, followed by chondrogenic differentiation of 28 d. Compared to the chondrocyte spheroid group, an enhanced expression of chondrogenic markers with downregulation of chondrocyte hypertrophy markers was observed in the hBMSC spheroid group. Overall, when compared to the respective dispersed groups, the encapsulated spheroids groups displayed better chondrogenic traits (Chameettachal et al. 2016). However, during the culture period, the spheroids tend to disengage and get distributed within the hydrogel instead of fusion, thereby hindering the macroscale tissue formation. A similar study compared the chondrogenic maturation of microtissues with the 3D bioprinted spheroid system composed of 42 d of chondrogenically differentiated BMSCs spheroids within the gelatin methacrylate bioink. Although the standalone spheroids displayed major chondrogenic traits, when encapsulated within the GelMA bioink, it demonstrated enhanced glycosaminoglycan deposition with reduced production of hypertrophic marker COL-I. This directed assembly approach of light-based 3D bioprinting using different photoinitiators (Irgacure 2959 and Li-TPO-L) successfully enhanced cell viability and maintained a stable chondrogenic phenotype (De Moor et al. 2020). When BMSC spheroids embedded within fibrin are 3D bioprinted over a coagulation bath of Polyethylene glycol (PEG) and alginate, the spheroids within the gel showed the highest chondrogenic properties with enhanced nutrient diffusion and did not adversely affect the mechanical properties of the overall system (Melo et al. 2019).

The cells post chondrogenic differentiation *in vitro* assume an intermediate unstable phenotype that further aggravates the development of fully transient hypertrophic cartilage. Cartilage biologists have termed this aberrant phenotype alteration as chondrocyte hypertrophy. Although scientists have coined the presence of chondrocyte hypertrophy in many cartilage tissue engineering protocols as an undesired outcome, there are limited resources that elucidate the plausible reasons for the onset of this aberrant phenotype during the gold-standard chondrogenesis procedure. This is of utmost importance because the knowledge will play a vital role in understanding the reasons for cartilage failure cases post-implantation.

8.3 PROBABLE REASONS FOR THE ONSET OF CHONDROCYTE HYPERTROPHY

Under the influence of chondrogenic stimulus, the cells strongly tend to drift towards an aberrant phenotype after chondrogenic differentiation through hypertrophy. In this state, the cells are called 'hypertrophic chondrocytes.' These chondrocytes display an increased proliferation capability but inhibited differentiation potential, allowing it to convert to bone-like tissue. This restricted differentiated state of the chondrocytes is termed 'terminally differentiated chondrocytes,' marked by increased cell size and volume due to the influx of water through the aquaporins. The developed bone-like tissue showed increased expression levels of COL-X/I, MMP-1/13 and several osteogenic genes like Osteocalcin (OCN), Osteopontin (OPN), Alkaline Phosphatase (ALP) and Runt related transcription factor 2 (RUNX2) (Sahu et al. 2015, Sharma et al. 2019). The hypertrophic chondrocytes constituting the transient cartilage are significantly marked by a mineralized (Dreier 2010) and vascularized (Gerber et al. 1999) matrix. The viscoelastic ECM of the articular cartilage transforms into a calcified matrix with enhanced hydroxyapatite deposition (Iannotti et al. 1994). This is followed by the degradation of the surrounding matrix and vascular sprouting from the underlying subchondral bone, regulated by the MMPs (Ortega et al. 2003, 2004). A few researchers have reported 'cellular senescence' or 'apoptosis' as the final fate of hypertrophic chondrocytes, termed terminally differentiated chondrocytes. The inorganic phosphate ions and the caspase molecules regulate cellular apoptosis post-hypertrophic maturation (Blanco et al. 1998). However, some studies report the osteogenic differentiation of the hypertrophic chondrocytes through an embryonic process called endochondral ossification that converts the hypertrophic cartilage into a longitudinal bone (Chawla et al. 2018, Daly et al. 2016).

Although the scientists have mentioned the onset of chondrocyte hypertrophy and also elaborately described the phenotypic alterations and molecular-level characteristics that are associated with the hypertrophic differentiation of chondrocytes, the exact reason or the cause of the onset is still a matter to be resolved. The following are the hypotheses that we have postulated as the major causes of chondrocyte hypertrophy.

8.3.1 The Innate Tendency of the Stem Cells towards Hypertrophic Differentiation

The stem cells (BMSCs and ADSCs) have been specifically widely exploited by tissue engineers to fabricate patient-specific tissue grafts. Their unlimited proliferation capability and multilineage differentiation make them advantageous over other primary cell types and cell lines (Wang et al. 2017). Several research groups, including ours, have utilized the benefits of stem cells to develop articular cartilage tissue grafts that show excellent cartilage-like characteristics, but hypertrophic differentiation was an inevitable outcome that came along with chondrogenesis. For, when murine BMSCs were encapsulated in SF-G bioink and 3D bioprinted, followed by chondrogenic differentiation for 21 d, it showed upregulated expression of hypertrophic genes (COL-X, MMP-1/13) along with the stable expression of cartilage-specific genes (COL-II, ACAN and SOX9) (Chawla et al. 2017). Similarly, when BMSCs and ADSCs were compared for chondrogenic differentiation, ADSCs showed a strong propensity towards hypertrophic differentiation marked by increased expression levels of COL-X/I and RUNX2 (Diekman et al. 2010). Therefore, following the cues from the developmental biology route of articular cartilage development, initial mesenchymal condensation finally leads to longitudinal bone formation through the endochondral ossification route. Due to this, the stem cells develop a memory to differentiate towards hypertrophic cartilage post-chondrogenesis.

8.3.2 The Functional Duality Role of Chondrogenic Signaling Pathways

The growth factors and the signaling morphogens used during the chondrogenesis protocol profoundly impact the onset of hypertrophic differentiation through activation or dysregulation of the chondrogenic signaling pathways.

The transforming growth factor beta (TGF-β) superfamily of proteins significantly promotes chondrogenesis and inhibits hypertrophic differentiation. The general pathway by which TGF-β1 and TGF-β3 work is the ALK-5 phosphorylation route followed by the activation of Smad 2/3 and ternary complex formation with Smad 4. This ternary complex translocates to the nucleus and aids in the transcription of chondrogenic genes, thereby mitigating hypertrophic maturation (Wang et al. 2014). However, the same TGF-β can also function via the ALK-1 route through Smad 1/5/8. The activated complex promotes the transcription of hypertrophic marker genes like RUNX2 and COL-X (Javed et al. 2008). Furthermore, it also stimulates the expression of hypertrophic marker genes by activating the JunB-associated MAPK signaling pathway (Li et al. 2005). These undesired activations of osteogenic lineage pathways could be responsible for the onset of chondrocyte hypertrophy during chondrogenesis, as observed in several reports from the research groups (Futrega et al. 2021, Narcisi et al. 2012). The Bone Morphogenetic Protein (BMP) signaling cascade also plays a dual role in chondrogenesis. Some reports claim that adding BMP-2 ligands to the culture media significantly increased the expression of SOX9, COL-II and ACAN in mesenchymal progenitor cells (Park et al. 2005). The same BMP ligand was used to induce OA traits in primary articular chondrocytes when added to the culture media to develop an OA microcartilage model with an upregulation of COL-X expression (Chawla et al. 2020). An *in vivo* study further confirmed the signaling crosstalk between TGF-β and the BMP pathway that induces the activation of hypertrophic maturation of chondrocytes (Keller et al. 2011).

Some studies also examined the ambiguous role of Wingless Protein (Wnt) ligands in promoting chondrogenic differentiation depending on the choice of Wnt ligands. Reports showed that Wnt 5a has been extensively observed in the cartilage zones (Usami et al. 2016). Our study also showed that activating the canonical Wnt/β-catenin signaling pathway by the SF-G bioink stimulates the chondrogenic differentiation of BMSCs (Chawla et al. 2021). However, other Wnt ligands like Wnt 1, 8a, and 4 inhibit chondrogenesis. Huang et al. have extensively studied the ambiguous role of the Wnt pathway in chondrogenesis using a Wnt pathway inducer and a suppressor. The Wnt inducer group observed a higher expression level of the matrix-degrading enzyme. In contrast, the expression levels of cartilage-specific genes were enhanced in the Wnt suppressor group (Huang et al. 2018). Therefore, it could be postulated that under the influence of chondrogenic supplements, the Wnt pathway functions via a hypertrophy-activating route through the mechanistic crosstalk between the other hypertrophy-activating pathways like BMP and Indian Hedgehog (IHH).

8.3.3 Role of Chondrogenic Culture Conditions

Mimicking the mechanical loading condition of the knee joint during *in vitro* chondrogenic differentiation has been an important choice for tissue engineers when developing articular cartilage grafts. The native articular cartilage is subjected to mechanical forces, including shear forces, compression and hydrostatic pressure. Frank et al. have shown the positive regulation of shear deformation on articular cartilage development. They observed that a sinusoidal shear force enhanced the secretion of cartilage ECM-specific proteoglycans (Frank et al. 2000). Hydrostatic pressure and shear force have also stimulated the GAG and collagen production, further inhibiting chondrocyte hypertrophy. This was observed in an agarose scaffold encapsulated with bovine chondrocytes and cultured for 21 d (Nazempour et al. 2017). However, some investigations also state that the dynamic compression microenvironment suppressed the expression of chondrogenic marker genes in human mesenchymal stem cells within agarose (Ge et al. 2021). It was also observed that a 30% compression was sufficient to induce hypertrophic differentiation of the cultured human chondrocytes (Occhetta et al. 2019). On the contrary, enhanced cell viability was observed for the dynamic condition when the same cyclic loading system was compared against the static loading platform (Lucchinetti et al. 2002). The dynamic culture condition also increased GAG production (nearly 20%) (Li et al. 2001). A deeper field analysis revealed the activation of the inflammatory NF-κB signaling pathway that accelerated the ECM degradation by the excessive secretion of ADAMTS5 and MMPs (Chang et al. 2019). Therefore, it could be hypothesized that mechanic loading (static and dynamic) is crucial in the hypertrophic differentiation of chondrocytes if not regulated.

Hypoxic conditions are crucial in maintaining a stable chondrogenic phenotype during chondrogenesis (Li et al. 2018). However, this hypoxia is also a prudent stimulator of hypertrophic differentiation of stem cells and chondrocytes. Although studies have reported HIF-1α as a chondroprotective agent, its role in promoting inflammation and matrix degradation by excessive production of inflammatory cytokine IL-1β has been equally validated (Scholz and Taylor 2013). HIF-1α also regulates cellular apoptosis during chondrocyte hypertrophy by governing the activities of Bcl-2 and procaspase-8 proteins, respectively (Zhang et al. 2015). Another hypoxia protein, HIF-2α, is also a positive regulator of chondrocyte hypertrophy through its association with the hypoxia-responsive elements and promoting the transcription of COL-X, MMP-13 and pro-angiogenic VEGFA (Saito et al. 2010). The HIF-2α is also a potent activator of pro-inflammatory cyclooxygenase-2 and prostaglandin E2, creating a favourable inflammatory niche to assist OA development and progression (Lee et al. 2010). Hypoxic conditions are also considered the ideal angiogenic microenvironment, activating several pro-angiogenic pathways and influencing vessel pattern formation (Krock et al. 2011), a key hallmark of hypertrophic cartilage from the underlying subchondral bone.

Therefore, an interconnection exists between the chondrogenic culture conditions, the activated signaling pathways and the molecular symptoms that play a crucial role in the hypertrophic maturation of chondrocytes. For instance, excessive hydrostatic pressure causes overexpression of the HIF-2α gene, which has been established as a potential transcriptional activator of hypertrophic genes and promotes inflammation (Inoue et al. 2015). Thus, the above-mentioned hypotheses indicate 'chondrocyte hypertrophy' as a major roadblock in cartilage tissue engineering.

8.4 ADVANCED STRATEGIES USING SMALL MOLECULES TO DEVELOP PHENOTYPICALLY STABLE ARTICULAR CARTILAGE

8.4.1 Exogenous Addition of Signaling Morphogens to the Media

The activation and de-activation of the chondrogenic cell signaling pathways are crucial in regulating the hypertrophic maturation of chondrocytes. This process is tightly regulated by several small molecule regulators that inhibit or activate the respective signaling cascades through various mechanisms like phosphorylation, de-phosphorylation, oxidation, etc. (Chawla et al. 2022). One of those signaling pathway modulators is LDN193189, a dorsomorphin family of compounds that inhibits the activation of the BMP signaling pathway by hindering the formation of the ternary complex between Smad 1/5/8 and Smad 4. Chawla et al. in their research have demonstrated the role of exogenous supplementation of BMP inhibitors in abrogating OA progression in an OA microcartilage model. The small molecule inhibitor caused a sharp decrease in expression levels of hypertrophy markers like COL-X, MMP-13, IHH, etc. in the OA model (Chawla et al. 2020). A similar strategy was also adopted by Doenges et al. in which a human BMSC-led OA model was developed using an inflammatory cocktail comprising IL-1β, TNF-α and IL-6. Upon screening with the anti-OA compounds like TGF-β, LDN and IL1Ra, the regular chondrogenic characteristics of the 3D microcartilage model were restored with a reduction of hypertrophy traits (Dönges et al. 2024). Although successful chondrogenesis has been observed in all these strategies, the shorter half-life of these small molecules restricts their long-term usage in the culture media. As a result, frequent replenishment is required, making the overall protocol expensive. Hence, researchers have been exploring a far more advanced approach to overcome the mentioned challenges and fabricate a stable hyaline cartilage tissue.

8.4.2 Chemical Tethering of Small Molecules to the Bioink and 3d Bioprinting

The onset of chondrocyte hypertrophy was an inevitable hindrance to the success of the tissue graft. The signaling morphogens were covalently conjugated to the biomaterial instead of exogenously adding them to the differentiation media. The covalent attachment enhanced their bioavailability and bioactivity by exhibiting sustained release characteristics, solving the shorter half-span challenge (Majumder et al. 2024c) (Figure 8.1). In that regard, Majumder et al. have reported the advantage of covalent conjugation over the exogenous addition of TGF-β3 into the SF matrix to demonstrate enhanced chondrogenesis with human BMSCs and chondrocytes. It was observed that BMSCs cultured over the TGF-β3-conjugated SF matrix displayed higher levels of chondrogenic gene expression than where TGF-β3 was externally added to the media. The research group further used this cue to develop a hyaline cartilage graft that could maintain its chondrogenic phenotype under OA-mimicking conditions. The study aimed at investigating the pro-chondrogenic and anti-hypertrophic role of the selected small molecule inhibitors and growth factors: Low dose naltrexone (LDN193189, BMP pathway inhibitor), Transforming growth factor β (TGF-β3) and Interleukin 1 receptor antagonist (IL1Ra, IL-1β pathway inhibitor), conjugated to our silk fibroin gelatin (SF-G) bioink. The IL1Ra molecule displayed the maximum chondrogenic traits with simultaneous inhibition of chondrocyte hypertrophy irrespective of the donor cells, which is further validated using the GeneMANIA tool. Therefore, for the first time, this study underlined the significance of maintaining a stable chondrogenic phenotype under OA-mimicking conditions rather than simply maturing the tissue construct under a conventional chondrogenic medium (Majumder et al. 2024a). Thus, we could successfully fabricate a stable hyaline cartilage tissue as a regenerative therapy for OA.

Therefore, we anticipate developing phenotypically stable articular cartilage tissue, mimicking the developmental biology route of organoid formation. A common point of intersection was 'chondrocyte hypertrophy' during *in vitro* chondrogenesis, which every model encountered as the biggest threat to the

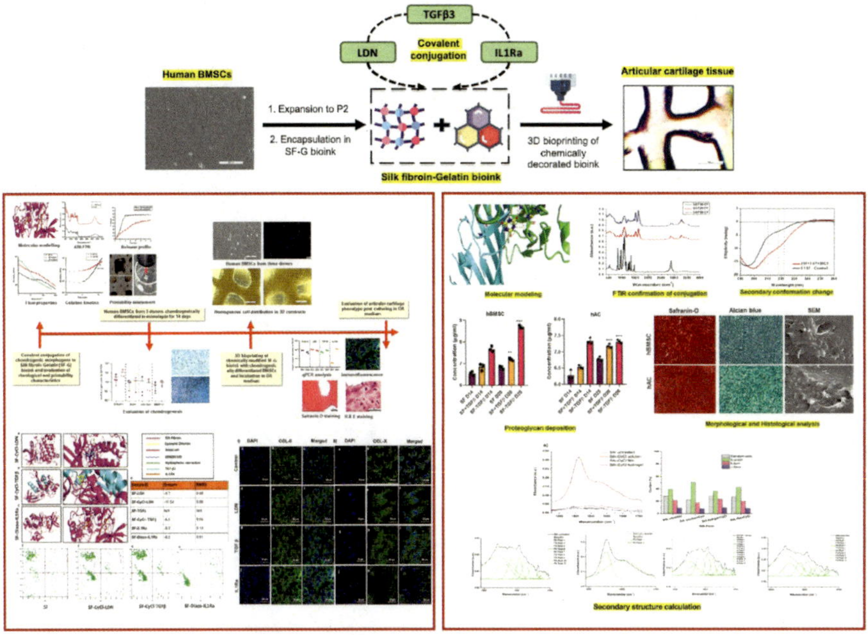

Figure 8.1 An advanced tissue engineering strategy by covalently tethering pro-chondrogenic and anti-hypertrophic signaling pathway modulators to the SF-G bioink to combat chondrocyte hypertrophy during in vitro chondrogenesis. Reprinted with permission Majumder et al. 2024a, "Covalent Conjugation of Small Molecule Inhibitors and Growth Factors to a Silk Fibroin-Derived Bioink to Develop Phenotypically Stable 3D Bioprinted Cartilage". ACS Applied Materials and Interfaces. 16(8): 9925–9943 and Majumder et al. 2024c, "An Advanced Bioconjugation Strategy for Covalent Tethering of TGFβ3 with Silk Fibroin Matrices and its Implications in the Chondrogenesis Profile of Human BMSCs and Human Chondrocytes: A Paradigm Shift in Cartilage Tissue Engineering". Advanced Healthcare Materials. 13(10): e2303513.

hyaline cartilage phenotype. Many studies have reported the occurrence of hypertrophic maturation of chondrocytes or the molecular symptoms associated with hypertrophy, but none of them mentioned the etiology of such occurrence. Understanding the root cause is paramount in devising strategies to mitigate them. Hence, we next pondered the possible theories regarding the onset of hypertrophic differentiation of chondrocytes and stem cells during chondrogenesis, citing evidence from the literature. Once the cause is known and the associated molecular parameters are examined, the next target is to optimize measures to mitigate hypertrophy. Different strategies on how small molecule regulators of chondrogenic signaling pathways can be implemented, either conjugated into the biomaterial or directly to the media, were utilized to 3D bioprint articular cartilage tissue that could successfully maintain its chondrogenicity even under inflammatory conditions. Thus, suppressing chondrocyte hypertrophy can be used as a principle to develop phenotypically stable articular cartilage tissue.

REFERENCES

Bandyopadhyay, A., B.B. Mandal and N. Bhardwaj. 2022. 3D bioprinting of photo-crosslinkable silk methacrylate (SilMA)-polyethylene glycol diacrylate (PEGDA) bioink for cartilage tissue engineering. Journal of Biomedical Materials Research—Part A. 110(4): 884–898.

Banerjee, D., Y.P. Singh, P. Datta, V. Ozbolat, A. O'Donnell, M. Yeo, et al., 2022. Strategies for 3D bioprinting of spheroids: a comprehensive review. Biomaterials. 291: 121881.

Blanco, F.J., R. Guitian, E. Vázquez-Martul, F.J. de Toro and F. Galdo. 1998. Osteoarthritis chondrocytes die by apoptosis: a possible pathway for osteoarthritis pathology. Arthritis and Rheumatism: Official Journal of the American College of Rheumatology. 41(2): 284–289.

Bonnet, C.S. and D.A. Walsh. 2005. Osteoarthritis, angiogenesis and inflammation. Rheumatology. 44(1): 7–16.

Chakraborty, J., N. Majumder, A. Sharma, S. Prasad and S. Ghosh. 2022. 3D bioprinted silk-reinforced alginate-gellan gum constructs for cartilage regeneration. Bioprinting. 28: e00232.

Chameettachal, S., S. Midha and S. Ghosh. 2016. Regulation of chondrogenesis and hypertrophy in silk fibroin-gelatin-based 3D bioprinted constructs. ACS Biomaterials Science and Engineering. 2(9): 1450–1463.

Chang, S.H., D. Mori, H. Kobayashi, Y. Mori, H. Nakamoto, K. Okada, et al., 2019. Excessive mechanical loading promotes osteoarthritis through the gremlin-1-NF-κB pathway. Nature Communications. 10(1): 1442.

Chawla, S., A. Kumar, P. Admane, A. Bandyopadhyay and S. Ghosh. 2017. Elucidating role of silk-gelatin bioink to recapitulate articular cartilage differentiation in 3D bioprinted constructs. Bioprinting. 7: 1–13.

Chawla, S., A. Sharma, A. Bandyopadhyay and S. Ghosh. 2018. Developmental biology-inspired strategies to engineer 3D bioprinted bone construct. ACS Biomaterials Science and Engineering. 4(10): 3545–3560.

Chawla, S., M.H.M. Berkelaar, B. Dasen, C. Halleux, S. Guth-Gundel, I. Kramer, et al., 2020. Blockage of bone morphogenetic protein signalling counteracts hypertrophy in a human osteoarthritic micro-cartilage model. Journal of Cell Science. 133(23): jcs249094.

Chawla, S., G. Desando, E. Gabusi, A. Sharma, D. Trucco, J. Chakraborty, et al., 2021. The effect of silk–gelatin bioink and TGF-β3 on mesenchymal stromal cells in 3D bioprinted chondrogenic constructs: a proteomic study. Journal of Materials Research. 36(19): 4051–4067.

Chawla, S., A. Mainardi, N. Majumder, L. Dönges, B. Kumar, P. Occhetta, et al., 2022. Chondrocyte hypertrophy in osteoarthritis: mechanistic studies and models for the identification of new therapeutic strategies. Cells. 11(24): 4034.

Daly, A.C., G.M. Cunniffe, B.N. Sathy, O. Jeon, E. Alsberg and D.J. Kelly. 2016. 3D bioprinting of developmentally inspired templates for whole bone organ engineering. Advanced Healthcare Materials. 5(18): 2353–2362.

Das, S., F. Pati, Y.J. Choi, G. Rijal, J.H. Shim, S.W. Kim, et al., 2015. Bioprintable, cell-laden silk fibroin-gelatin hydrogel supporting multilineage differentiation of stem cells for fabrication of three-dimensional tissue constructs. Acta Biomaterialia. 11(1): 233–246.

de Lange-Brokaar, B.J.E., A. Ioan-Facsinay, G.J.V.M. van Osch, A.M. Zuurmond, J. Schoones, et al., 2012. Synovial inflammation, immune cells and their cytokines in osteoarthritis: a review. Osteoarthritis and Cartilage. 20(12): 1484–1499.

De Moor, L., S. Fernandez, C. Vercruysse, L. Tytgat, M. Asadian, N. De Geyter, et al., 2020. Hybrid bioprinting of chondrogenically induced human mesenchymal stem cell spheroids. Frontiers in Bioengineering and Biotechnology. 8: 484.

Diekman, B.O., C.R., Rowland, D.P. Lennon, A.I. Caplan and F. Guilak. 2010. Chondrogenesis of adult stem cells from adipose tissue and bone marrow: induction by growth factors and cartilage-derived matrix. Tissue Engineering. Part A. 16(2), 523–533.

Dönges, L., A. Damle, A. Mainardi, T. Bock, M. Schönenberger, I. Martin, et al., 2024. Engineered human osteoarthritic cartilage organoids. Biomaterials. 308: 122549.

Dreier, R. 2010. Hypertrophic differentiation of chondrocytes in osteoarthritis: The developmental aspect of degenerative joint disorders. Arthritis Research and Therapy. 12(5): 216.

Frank, E.H., M., Jin, A.M., Loening, M.E. Levenston and A.J. Grodzinsky. 2000. A versatile shear and compression apparatus for mechanical stimulation of tissue culture explants. Journal of Biomechanics. 33(11): 1523–1527.

Fuerst, M., Bertrand, J., Lammers, L., Dreier, R., Echtermeyer, F., Nitschke, Y., et al., 2009. Calcification of articular cartilage in human osteoarthritis. Arthritis and Rheumatism. 60(9): 2694–2703.

Futrega, K., P.G., Robey, T.J., Klein, R.W. Crawford and M.R. Doran. 2021. A single day of TGF-β1 exposure activates chondrogenic and hypertrophic differentiation pathways in bone marrow-derived stromal cells. Communications Biology. 4(1): 29.

Ge, Y., Y. Li, Z. Wang, L. Li, H. Teng and Q. Jiang. 2021. Effects of mechanical compression on chondrogenesis of human synovium-derived mesenchymal stem cells in agarose hydrogel. Frontiers in Bioengineering and Biotechnology. 9: 697281.

Gerber, H.P., T.H., Vu, A.M., Ryan, J. Kowalski, Z. Werb and N. Ferrara. 1999. VEGF couples hypertrophic cartilage remodeling, ossification and angiogenesis during endochondral bone formation. Nature Medicine. 5(6): 623–628.

Ghosh, S., M., Laha, S. Mondal, S. Sengupta and D.L. Kaplan. 2009. *In vitro* model of mesenchymal condensation during chondrogenic development. Biomaterials. 30(33): 6530–6540.

Han, X., S. Chang, M. Zhang, X. Bian, C. Li and D. Li. 2021. Advances of hydrogel-based bioprinting for cartilage tissue engineering. Frontiers in Bioengineering and Biotechnology. 9: 746564.

Heidari, B. 2011. Knee osteoarthritis prevalence, risk factors, pathogenesis and features: Part I. Caspian Journal of Internal Medicine. 2(2): 205–212.

Huang, X., L., Zhong, J. Hendriks, J.N. Post and M. Karperien. 2018. The effects of the WNT-signaling modulators BIO and PKF118-310 on the chondrogenic differentiation of human mesenchymal stem cells. International Journal of Molecular Sciences. 19(2): 561.

Hunter, D.J. and D.T. Felson. 2006. Osteoarthritis. BMJ (Clinical Research (ed.). 332(7542): 639–642.

Iannotti, J.P., S. Naidu, Y. Noguchi, R.M. Hunt and C.T. Brighton. 1994. Growth plate matrix vesicle biogenesis. The role of intracellular calcium. Clinical Orthopaedics and Related Research. 306: 222–229.

Inoue, H., Y. Arai, T. Kishida, R. Terauchi, K. Honjo, S. Nakagawa, et al., 2015. Hydrostatic pressure influences HIF-2 alpha expression in chondrocytes. International Journal of Molecular Sciences. 16(1): 1043–1050.

Javed, A., J.-S. Bae, F. Afzal, S. Gutierrez, J. Pratap, S.K. Zaidi, et al., 2008. Structural coupling of Smad and Runx2 for execution of the BMP2 osteogenic signal. Journal of Biological Chemistry. 283(13): 8412–8422.

Keller, B., T. Yang, Y. Chen, E. Munivez, T. Bertin, B. Zabel, et al., 2011. Interaction of TGFβ and BMP signaling pathways during chondrogenesis. PloS One. 6(1): e16421.

Krock, B.L., N. Skuli and M.C. Simon. 2011. Hypoxia-induced angiogenesis: good and evil. Genes and Cancer. 2(12): 1117–1133.

Lee, J.J., M. Natsuizaka, S. Ohashi, G.S. Wong, M. Takaoka, C.Z. Michaylira, et al., 2010. Hypoxia activates the cyclooxygenase-2-prostaglandin E synthase axis. Carcinogenesis. 31(3): 427–434.

Li, K., A. Williamson, A. Wang and R. Sah. 2001. Growth responses of cartilage to static and dynamic compression. Clinical Orthopaedics and Related Research. 391: S34–S48.

Li, T.-F., R.J. O'Keefe and D. Chen. 2005. TGF-β signaling in chondrocytes. Frontiers in Bioscience: A Journal and Virtual Library. 10: 681–688.

Li, H., X. Li, X. Jing, M. Li, Y. Ren, J. Chen, et al., 2018. Hypoxia promotes maintenance of the chondrogenic phenotype in rat growth plate chondrocytes through the HIF-1α/YAP signaling pathway. International Journal of Molecular Medicine. 42(6): 3181–3192.

Lucchinetti, E., C.S. Adams, W.E.J. Horton and P.A. Torzilli. 2002. Cartilage viability after repetitive loading: a preliminary report. Osteoarthritis and Cartilage. 10(1): 71–81.

Majumder, N. and S. Ghosh. 2023. Unfolding the mystery behind the onset of chondrocyte hypertrophy during chondrogenesis: toward designing advanced permanent cartilage-mimetic biomaterials. Advanced Functional Materials. 33(35): 2300651.

Majumder, N., C. Roy, L. Doenges, I. Martin, A. Barbero and S. Ghosh. 2024a. Covalent conjugation of small molecule inhibitors and growth factors to a silk fibroin-derived bioink to develop phenotypically stable 3D bioprinted cartilage. ACS Applied Materials and Interfaces. 16(8): 9925–9943.

Majumder, N., S. Roy, A. Sharma, S. Arora, R. Vaishya, A. Bandyopadhyay, et al., 2024b. Assessing the advantages of 3D bioprinting and 3D spheroids in deciphering the osteoarthritis healing mechanism using human chondrocytes and polarized macrophages. Biomedical Materials (Bristol). 19(2): 025005.

Majumder, N., S. Seit, N.S. Bhabesh and S. Ghosh. 2024c. An advanced bioconjugation strategy for covalent tethering of TGFβ3 with silk fibroin matrices and its implications in the chondrogenesis profile of human BMSCs and human chondrocytes: a paradigm shift in cartilage tissue engineering. Advanced Healthcare Materials. 13(10): e2303513.

Melo, B.A.G. De., Y.A. Jodat, S. Mehrotra, M.A. Calabrese, T. Kamperman, B.B. Mandal, et al., 2019. 3D printed cartilage-like tissue constructs with spatially controlled mechanical properties. Advanced Functional Materials. 29(51): 1906330.

Moran, J.M., D. Pazzano and L.J. Bonassar. 2003. Characterization of polylactic acid-polyglycolic acid composites for cartilage tissue engineering. Tissue Engineering. 9(1): 63–70.

Narcisi, R., R. Quarto, V. Ulivi, A. Muraglia, L. Molfetta and P. Giannoni. 2012. TGFβ-1 administration during Ex vivo expansion of human articular chondrocytes in a serum-free medium redirects the cell phenotype toward hypertrophy. Journal of Cellular Physiology. 227(9): 3282–3290.

Nazempour, A., C.R. Quisenberry, N.I. Abu-Lail and B.J. Van Wie. 2017. Combined effects of oscillating hydrostatic pressure, perfusion and encapsulation in a novel bioreactor for enhancing extracellular matrix synthesis by bovine chondrocytes. Cell and Tissue Research. 370(1): 179–193.

Occhetta, P., A. Mainardi, E. Votta, Q. Vallmajo-Martin, M. Ehrbar, I. Martin, et al., 2019. Hyperphysiological compression of articular cartilage induces an osteoarthritic phenotype in a cartilage-on-a-chip model. Nature Biomedical Engineering. 3(7): 545–557.

Ortega, N., D. Behonick, D. Stickens and Z. Werb. 2003. How proteases regulate bone morphogenesis. Annals of the New York Academy of Sciences. 995(1): 109–116.

Ortega, N., D.J. Behonick and Z. Werb. 2004. Matrix remodeling during endochondral ossification. Trends in Cell Biology. 14(2): 86–93.

Park, Y., M. Sugimoto, A. Watrin, M. Chiquet and E.B. Hunziker. 2005. BMP-2 induces the expression of chondrocyte-specific genes in bovine synovium-derived progenitor cells cultured in three-dimensional alginate hydrogel. Osteoarthritis and Cartilage. 13(6): 527–536.

Sahu, N., P. Baligar, S. Midha, B. Kundu, M. Bhattacharjee, S. Mukherjee, et al., 2015. Nonmulberry silk fibroin scaffold shows superior osteoconductivity than mulberry silk fibroin in calvarial bone regeneration. Advanced Healthcare Materials. 4(11): 1709–1721.

Saito, T., A., Fukai, A. Mabuchi, T. Ikeda, F. Yano, S Ohba, et al., 2010. Transcriptional regulation of endochondral ossification by HIF-2α during skeletal growth and osteoarthritis development. Nature Medicine. 16(6): 678–686.

Sanchez-Lopez, E., R. Coras, A. Torres, N.E. Lane and M. Guma. 2022. Synovial inflammation in osteoarthritis progression. Nature Reviews Rheumatology. 18(5): 258–275.

Scholz, C.C. and C.T. Taylor. 2013. Targeting the HIF pathway in inflammation and immunity. Current Opinion in Pharmacology. 13(4): 646–653. https://doi.org/10.1016/j.coph.2013.04.009.

Sharma, A., G. Desando, M. Petretta, S. Chawla, I. Bartolotti, C. Manferdini, et al., 2019. Investigating the role of sustained calcium release in silk-gelatin-based three-dimensional bioprinted constructs for enhancing the osteogenic differentiation of human bone marrow derived mesenchymal stromal cells. ACS Biomaterials Science and Engineering. 5(3): 1518–1533.

Shen, Y., Y. Xu, B. Yi, X. Wang, H. Tang, C. Chen, et al., 2021. Engineering a highly biomimetic chitosan-based cartilage scaffold by using short fibers and a cartilage-decellularized matrix. Biomacromolecules. 22(5): 2284–2297.

Sophia Fox, A.J., A. Bedi and S.A. Rodeo. 2009. The basic science of articular cartilage: structure, composition, and function. Sports Health. 1(6): 461–468.

Stewart, M.C., K.M. Saunders, N. Burton-Wurster and J.N. Macleod. 2000. Phenotypic stability of articular chondrocytes *in vitro:* the effects of culture models, bone morphogenetic protein 2, and serum supplementation. Journal of Bone and Mineral Research: The Official Journal of the American Society for Bone and Mineral Research. 15(1): 166–174.

Takahata, Y., H. Hagino, A. Kimura, M. Urushizaki, S. Yamamoto, K. Wakamori, et al., 2022. Regulatory mechanisms of Prg4 and Gdf5 expression in articular cartilage and functions in osteoarthritis. International Journal of Molecular Sciences. 23(9): 4672.

Trucco, D., A. Sharma, C. Manferdini, E. Gabusi, M. Petretta, G. Desando, et al., (2021). Modeling and fabrication of silk fibroin–gelatin-based constructs using extrusion-based three-dimensional bioprinting. ACS Biomaterials Science and Engineering, 7(7), 3306–3320.

Usami, Y., A.T. Gunawardena, M. Iwamoto and M. Enomoto-Iwamoto. 2016. Wnt signaling in cartilage development and diseases: lessons from animal studies. Laboratory Investigation. 96(2): 186–196.

Vy, T., H. Le, X. To, P. Nguyen, P. Huynh, T. Le, et al., 2020. Method for *in vitro* production of cartilage microtissues from scaffold-free spheroids composed of human adipose-derived stem cells. Biomedical Research and Therapy. 7: 3697–3708.

Wang, W., D. Rigueur and K.M. Lyons. 2014. TGFβ signaling in cartilage development and maintenance. Birth Defects Research. Part C, Embryo Today: Reviews. 102(1): 37–51.

Wang, M., Z. Yuan, N. Ma, C. Hao, W. Guo, G. Zou, et al., 2017. Advances and prospects in stem cells for cartilage regeneration. Stem Cells International. 2017: 4130607.

Yoon, B.S. and K.M. Lyons. 2004. Multiple functions of BMPs in chondrogenesis. Journal of Cellular Biochemistry. 93(1): 93–103.

Yoon, H.H., S.H. Bhang, J.-Y. Shin, J. Shin and B.-S. Kim. 2012. Enhanced cartilage formation via three-dimensional cell engineering of human adipose-derived stem cells. Tissue Engineering. Part A. 18(19–20): 1949–1956.

Zhang, L., P. Su, C. Xu, J. Yang, W. Yu and D. Huang. 2010. Chondrogenic differentiation of human mesenchymal stem cells: a comparison between micromass and pellet culture systems. Biotechnology Letters. 32(9): 1339–1346.

Zhang, F., J. Wang, J. Chu, C. Yang, H. Xiao, C. Zhao, et al., 2015. MicroRNA-146a induced by hypoxia promotes chondrocyte autophagy through Bcl-2. Cellular Physiology and Biochemistry. 37(4): 1442–1453. https://doi.org/10.1159/000438513.

Zhao, X., A. Papadopoulos, S. Ibusuki, D.A. Bichara, D.B. Saris, J. Malda, et al., 2016. Articular cartilage generation applying PEG-LA-DM/PEGDM copolymer hydrogels. BMC Musculoskeletal Disorders. 17(1): 245.

Zhuang, P., Y.H. Chiang, M.S. Fernanda and M. He. 2021. Using spheroids as building blocks towards 3D bioprinting of tumor microenvironment. International Journal of Bioprinting. 7(4): 444.

Chapter **9**

Biodegradation Behaviour of Bioactive Gallium Based Nanosilver Doped Alloy in Simulated Body Fluid

Yap Yean Weiy[1], Mahmood Anwar*[2], Nabisab Mujawar Mubarak[3,4], Sumaiya Islam[5] and Sujan Debnath[1]

[1] Department of Mechanical Engineering, Curtin University, CDT 250, 98009 Miri, Sarawak, Malysia

[2]Department of Mechanical Engineering, SCEBE, Glagow Caledonian University, Glasgow G4 0BA, Scotland, United Kingdom

[3]Petroleum and Chemical Engineering, Faculty of Engineering, Universiti Teknologi Brunei, Bandar Seri Begawan, BE1410, Brunei Darussalam

[4]Department of Biosciences, Saveetha School of Engineering, Saveetha Institute of Medical and Technical Sciences, Chennai, India

[5]School of Engineering, Chisholm Institute, Melbourne, Australia

9.1 INTRODUCTION

Gallium alloys have generated tremendous interest in orthopaedic and dental implants due to their unique biocompatible and mechanical properties. Approved by the Food and Drug Administration (FDA), studies have shown that gallium

*For Correspondence: mahmood.anwar@gcu.ac.uk

could serve as an anti-tumour and anti-hypercalcemia drug (Dong et al. 2019). Gallium shares specific chemical characteristics with Iron (III) ions regarding its chemical valence and ionic radius (Wang et al. 2017). Gallium also contains high affinity binding to transferrin, an iron transport protein in the human body's circulation. In addition to the unique properties of gallium, it is also a feasible candidate to develop flexible devices such as sensors due to its electrical conductivity and extremely deformable properties (Kim et al. 2018). Gallium liquid metal is conductive in its liquid state. It flows naturally at room temperature due to its low melting point. Compared to the most commonly known liquid metal – mercury—gallium could form shapes more easily than mercury. The toxicity of mercury was also an issue when compared to gallium, as pure gallium is non-toxic to the human body.

The flexibility and biocompatibility of Gallium make it a feasible candidate for the fabrication of biosensors. However, biosensors require suitable electrical conductivity materials, which is a property that gallium requires to enhance. Silver is a promising material with the highest electrical conductivity and antimicrobial properties. Silver was used in many healthcare products due to its antimicrobial activity and low toxicity to human cells (Ahmed et al. 2006). Hence, doping nanosilver and gallium may lead to a new flexible material with high electrical conductivity and biocompatibility to the human body for future biomedical applications. Despite the unique properties that may be developed by doping nanosilver with gallium, it is essential to determine the biodegradation behaviour necessary for many bio-related applications and wearable devices. Limited studies have been reported on the doping of gallium with silver nanoparticles. However, various studies have been conducted on gallium doping with other materials (Liu et al. 2024). Studies have shown multiple experiments of gallium-doped alloys incubated in simulated body fluid to determine their degradation behaviour.

Interestingly, the biodegradation behaviour of nanosilver doped Gallium could be determined by incubating the alloy in simulated body fluid. Simulated body fluid (SBF) mimics blood plasma conditions for *in vitro* research experiments. In this study, the types of simulated body fluid used were deionized water, DMEM (Dulbecco's Modified Eagle's Medium) high glucose, and DMEM low glucose concentration. After time-dependent incubation, the biodegradation behaviour was determined by analysing the alloy's mass loss and surface morphology. In addition, the electrical conductivity of nanosilver-doped gallium was also determined.

9.2 CRITICAL REVIEW

Liquid metal is an emerging engineering material in various industries, the most common being applications in electronics. However, a new side of liquid metal was discovered and researched to be applied in the medicine industry, typically in developing medical tools, devices, and implants to improve the diagnosis of diseases. Liquid metals, similar to biometals characteristics (Sudin et al. 2011), have good biomaterial characterstics, electrical and thermal conductivity. The ductility of liquid metals allowed inventions such as flexible tubes and flexible biosensors to exist. With that, the liquid metal required in the medicine industry must be equipped with antimicrobial properties to prevent bacterial infection in open wounds, and such liquid metal is gallium.

A recent study was conducted with gallium-doped titania nanotubes to determine its effects on gram-positive bacteria (Dong et al. 2019). Casting gallium ions on titania nanotubes has proven to inhibit bacterial activity. The degradation study in this research was to analyze gallium ion release with PDLLA on titania nanotube bone plates. The gallium-containing bone plates were immersed in 50 ml of PBS and kept in a water bath at 37°C for 14 days. Then, 10 ml of the solution was sucked out from the sample after 1 day, 3 days, 7 days, 10 days and 14 days. ICP-OES analyzed gallium ion release. Overall, the immersion test on the first three days had a large amount of gallium release, followed by a gradual increase at a lower rate. The presence of gallium ions inhibited the growth of bacteria cells, significantly improving the antibacterial ability of titania nanotubes.

According to Sandra (Sanchez-Salcedo et al. 2018), gallium ions can be beneficial to bone cell growth. The research experiment was conducted by enriching highly bio reactive silica-based mesoporous bioactive glass with gallium ions. Four compositions of mesoporous bioactive glasses were formed, three enriched with gallium, to determine the effect of gallium ions on the glass. The glasses were soaked in SBF at 37°C for 1, 3 and 7 days. The composition of glasses and gallium with phosphorus content exhibits hydroxyapatite formation from converting amorphous calcium phosphate. However, glass composition with no phosphorus content does not exhibit hydroxyapatite formation. Hence, it was reported that the hydroxyapatite formation is dependent strictly on P_2O_5 content in the glasses. Besides that, higher P_2O_5 content would result in higher amount of gallium and calcium ions release. Regardless, the amount of gallium released was inside the non-cytotoxic level and was in the range to combat against *P. aeruginosa* cells. It was also reported that the first gallium-enriched mesoporous bioactive glass was first formed in this article, which exhibits antibacterial properties and *in vitro* bioactivity with a non-cytotoxic effect.

In another study, a series of gallium-doped bioactive glasses have been tested on osteosarcoma cells, a type of cancer cell in the bone, to determine their potential in bone cancer applications (Rana et al. 2017). A kind of SBF was prepared by referring to the method of Saravanapavan (Saravanapavan et al. 2003) and it is used as an incubation medium for the gallium bioactive glasses to test for *in vivo* hydroxyapatite formation. The samples were prepared in powder form ranging from 40–60 μm, and sealed in SBF for 7 days at 37°C and as reported, the bioactive glasses, when soaked in SBF after 7 days, will form an apatite layer on the surface. Bioactive glasses with the presence of gallium showed no significant difference or reduction in apatite formation. Gallium ion released was observed with inductively coupled mass spectrometry (ICP), and the results showed rapid gallium ion release in the first few hours, then stabilized after 24 hours. Minimal gallium ion release was observed at 48 and 72 hours. Gallium antimicrobial activity was observed in osteosarcoma cell viability. Sarcoma cells were significantly decreased when the gallium oxide content increased. At 72 hours and 3% gallium condition, the osteosarcoma cell viability was less than 50%. Gallium showed antimicrobial properties in osteosarcoma cells with no significant cytotoxicity effect, resulting in a positive control of osteosarcoma cells. Gallium-doped bioactive glass exhibits not only antimicrobial properties, but the biodegradable behaviour of the bioactive glass will dissolve over time, which is beneficial in the long term. Hence, gallium-doped bioactive glass shows significant potential in bone cancer applications.

In recent research, gallium has been reported to possess therapeutic properties. It is a compatible dopant for inorganic materials that exhibit promising results for bone implants or bone substitute materials. When released from bioactive glasses, it is said that gallium ions could treat autoimmune diseases and fight against gram-positive and gram-negative bacteria. However, gallium also promotes angiogenesis, which is the process of blood vessel growth from the existing vasculature. According to literature by Aleksandra, unlike iron, gallium cannot change its oxidation number, affecting the metabolic process (Wajda et al. 2018). The gallium ions released from the mesoporous glass surface can quickly enter the microbe cells through its iron transport mechanism and block its metabolic pathways, leading to cell death. In addition to that, gallium can also cause cell membrane and protein structure damage by interfering with bacterial DNA and protein synthesis. Bioactivity and ion release tests were conducted incubating in SBF at 37°C for 24 hours, 7 days, 14 days and 21 days. An *in vitro* test was conducted with osteoblasts-like cells incubated in DMEM at 37°C for 48 hours to test the biocompatibility. Nonetheless, antibacterial studies of gallium-containing-glasses were conducted on gram-negative and gram-positive bacteria. The results reported that the presence of gallium deprives the glasses of their bioactive properties. Gallium addition inhibited the growth of hydroxyapatite layer formation by reducing the degradation rate. Gallium-doped glasses promote higher cell viability and mitochondrial activity when tested on osteoblast-like cells. Statistics showed that the level of mitochondrial activity in gallium-doped glasses is >110%, compared to without gallium, which is between the range of 80% to 100%. Thus, the biocompatibility test of gallium-doped glasses showed a more favourable result for cell proliferation and non-cytotoxicity. However, gallium ions may potentially enhance neovascularization as vascular endothelial growth factor secretion increases proportionally with gallium ion release. Regardless, a significant antibacterial effect was shown towards gram-positive and gram-negative bacteria after 24 hours of incubation. It was also the time with the highest release of gallium ions (at about 6 ppm). Therefore, this study showed that gallium hinders hydroxyapatite layer formation on glasses, depriving them of their bioactive properties. However, gallium ions showed no negative influence on bioactivity and cytotoxicity. Gallium ions could improve the glasses' metabolic activity, angiogenesis and antibacterial properties.

In addition to the antibacterial characteristics of gallium ions, Yang et al. (Yang et al. 2017) discussed the inhibition of microbial growth on Pseudomonas aeruginosa bacteria cells with gallium-doped amorphous calcium apatite (ACP) nanoparticles. The antibacterial effect measurement was conducted based on disk-diffusion methodology, whereby small pressed pieces of gallium-doped ACP nanoparticles were placed on bacteria-inoculated plates and incubated overnight in the air at 37°C. Ion release measurements of gallium ions were conducted with ICP-MS (inductively coupled plasma mass spectrometry) at 24 hours, 48 hours, 72 hours and 96 hours. It was reported that the ion release concentration increases at a faster rate within 24 hours when the (Ca + Ga)/P molar ratio is high (>2.7). Thus, the concentration of (Ca + Ga)/P promotes the release of gallium ions. However, a higher (Ca + Ga)/P molar ratio increases the growth inhibition of *Pseudomonas aeruginosa*. These results showed a close relationship between the antibacterial effect and gallium ion concentration. Hence, a higher concentration of gallium ions leads to a higher antibacterial effect efficiency.

Gallium doped with phosphate-based glasses could develop a promising new therapeutic agent as it has been reported to demonstrate antimicrobial activity (Valappil et al. 2009). The research indicated that applying gallium-doped phosphate-based glasses could be helpful in applications such as controlled drug delivery to deliver gallium ions to combat *P. aeruginosa* bacteria. A degradation study of the gallium-doped phosphate-based glass was conducted by incubating the doped product into 50 ml of deionized water at 37°C for 6 hours, 24 hours, 48 hours, 72 hours, and 120 hours. The mass loss rate was taken per unit area to determine the degradation of the doped product. An ion release study of the glass was also conducted by using inductively coupled mass spectrometry (ICPMS) for the previously said time points. The degradation behaviour of gallium-doped phosphate-based glasses showed an inversely proportional relationship with CaO content—gallium oxide in the glasses dissolved entirely after 72 hours of incubation.

Besides gallium, silver has been also proven to exhibit antimicrobial properties, and such studies were conducted by incorporating silver with phosphate-based glasses. Silver-doped phosphate-based glasses could counter the effect of bacteria cells and give excellent long-term silver ions release into the medium (Ahmed et al. 2006). A total of 8 silver-doped phosphate-based glass compositions were formed with increasing mol% of silver to determine the bacteria cell viability via disk diffusion methodology. The disks were placed onto inoculated plates of *P. aeruginosa*, *E. coli*, *Bacillus cereus*, and *C. albicans* cells and incubated overnight at 37°C in air. A degradation study is performed by measuring the mass loss of the disk after incubation. The disks were removed from the inoculated cell plates, and the excess moisture was dried from the surface. Then, the disks were incubated in nutrient broth at 37°C and agitated at 200 rpm for up to 120 hours. The following equation measured the weight loss of the disks:

$$\text{Weight loss} = (M_0 - M_t)/A \tag{9.1}$$

As reported from the results, the disks agitated at 200 rpm showed a decreased dissolution rate with increasing silver concentration. Also, the growth rate of bacterial cells was inhibited due to the presence of silver, which proved that silver has bacterial growth inhibition properties and exerts a bactericidal effect. However, inhibiting silver's bacterial growth is more efficient in inhibiting the growth rate of *P. aeruginosa*. In the case of the other bacterial cells, silver only lowers the growth rate to a minimum but has not exhibited any complete growth inhibition behaviour. A decrease in pH value is observed as the dissolution rate increases, thus depicting that an increase in silver ions would lower the medium's pH. It was concluded that the most efficient mol% of silver to combat bacterial cell growth is 3 mol% of silver doped composition. This composition also showed excellent long-term silver ion release, which would be beneficial in tackling biomaterial-related infections.

A study on the application of silver as an antimicrobial agent in medical devices was also conducted in 2010. Silver could be absorbed by the human body through ingestion, inhalation, or insertion through medical devices and skin contact (Lansdown 2010). Excessive absorption of silver in the human body would cause argyria, which was caused by excessive absorption of silver that exceeds the capacity of the liver and kidney to eliminate the metal. However, the amount of silver that was considered excessive was not known. It was stated that nanosilver particles exhibited a solubility 100-fold greater than metallic silver. Still, it also showed specific biological action through its influence on physical and chemical properties. Tested on *P. aeruginosa* bacteria cells, antibacterial effects were present in microcrystalline silver. A wound treated with nanocrystalline silver also demonstrated success in infection control. Excessive exposure to silver would lead to adverse health effects, but as long as the silver exposure is maintained within the safety standards of exposure limits of 0.01 mg/m^3 ionisable silver compounds in air, and 0.10 mg/L in drinking water, which was set by National Institute for Occupational Safety and Health and the American Conference on Governmental Industrial Hygienists, then silver would not pose any negative impact on the human body.

To determine the compatibility of nanosilver doped gallium alloy as an implant material, it would be necessary to first determine the effects of the metal, especially the degradation behaviour, through *in vitro* experiments. *In vitro* experiments could be conducted using a non-living cell culture medium such as Dulbecco's Modified Eagle's Medium (DMEM). Besides that, the presence of amino acids in DMEM composition made it a better candidate to mimic biological conditions. Hence, DMEM was used in this study. Interestingly, most of the research studies were conducted experiments with DMEM (Kokubo and Takadama 2006). DMEM, however, comes in low glucose and high glucose concentrations, where the L-glutamine composition was different. In the case of this research study, the alloy was incubated in low glucose-concentration medium to identify the degradation behaviour in normal blood conditions. No living cells were cultured in this research study.

An essential factor to consider when doping gallium with nanosilver alloy was to determine the changes in electrical conductivity. It was assumed that silver, the highest electrical conductor, would effectively enhance its electrical conductivity when doped with gallium. Besides that, after the alloy was incubated, the ideal result was to obtain no significant difference or, even better, a decrease in electrical resistivity. The electrical conductivity could be measured by referring to the method introduced by Pouladrazeh et. al. (Pouladzadeh et al. 2018), in the study of the electrical conductivity of carbon nanotubes. The equations related to determining the electrical conductivity were as shown in Equation 9.2 and Equation 9.3.

$$\rho = \frac{R_c A}{h} \tag{9.2}$$

$$\sigma = \frac{1}{\rho} \tag{9.3}$$

A study of electrical conductivity variation due to incubation in a cell culture medium was also conducted by Kaklamani et. al. which was a study on the electrical conductivity of hydrogel in a DMEM cell culture medium where the hydrogels were mixed with DMEM solution and incubated at 37°C (Kaklamani et al. 2018). After incubation in a cell culture medium, the conductance of the hydrogels increased significantly. This is due to the DMEM medium's mobile ions dissociated with the proteins and carbohydrates. The ions dissociated were Na^+, K^+, SO_4^{2-}, Ca^{2+}, Cl^- and Mg^{2+}. Such ions could contribute to the increase of electrical conductivity.

As the alloy was incubated in two different mediums, namely deionized water and DMEM low glucose concentration medium, the stability of the alloy could be disrupted due to the reaction with the incubation medium. Horasawa et al. conducted a research study on the effects of pure gallium in water and other saline solutions. It was observed that immersing gallium in deionized water would result in gallium release, and changes in appearance would only occur after seven days of incubation (Horasawa et al. 1997). After 7 days of incubation in deionized water, the sample's surface had dark brown or greyish-brown pigment yet maintained its metallic luster. However, a layer of greyish-white products was formed after 28 days of incubation. Incubation of pure gallium in deionized water would also result in gallium release. Two types of reactions were considered on the surface of the specimen, as gallium and water reaction gave the following chemical equation:

$$2Ga^+ + 2(OH)^- \rightarrow 2GaO(OH) + H_2O$$

From the chemical reaction, gallium oxide mono hydroxide GaO(OH) would give white colour. The white precipitates that was formed in this research study could be depicted as the existence of gallium hydroxide in the molecular disperse and partly in the colloidal state. As for silver, it was reported that incubating silver at 38°C would not result in any dissolution (Johnson 1970).

9.3 EXPERIMENTAL DESIGN

9.3.1 Preparation of Nanosilver Doped Gallium Alloy

Nanosilver doped gallium alloy was prepared by melting solid gallium (Ga) at 30°C, then mixing with nanosilver (nAg) powder in Ga/nAg ratios of 1:1, 1:3 and 1:9. The mixed product was placed in a furnace and heat treated at a rate (heating and cooling) of 4°C/min and 5°C/min at 1000°C for 12 hours under a controlled argon gas environment. Heating nanosilver powder with liquid gallium in an inert gas environment allows the doping of novel nanosilver doped gallium alloy without the influence of oxygen to prevent oxidation of metals. Before incubation, the samples were grouped. Each group was split into three sub-groupings in terms of incubation period of 5 days, 10 days, and 20 days. The surface morphology of the samples was studied by Olympus BX53M Upright Metallurgical optical microscope for each incubated sample.

9.3.2 *In Vitro* Degradation Test

The degradation study of the samples was based on mass loss. Hence, to obtain the mass loss, the initial mass (M_0) of the samples was measured before incubation, and the final mass (M_f) was measured after incubation to fulfill Equation 9.4 (Ahmed et al. 2006). The mass loss of the samples was plotted in graphs with a comparison of 4°C/min and 5°C/min heating rate samples at various incubation periods.

$$\text{Mass loss} = M_0 - M_f \tag{9.4}$$

9.3.3 Electrical Conductivity

To study the effect of SBF incubation on the electrical behaviour of the doped alloy, the samples were subjected to a direct current, and the resistance measurements were recorded using a multimeter. Average resistance values were calculated by taking 3 resistance readings in a sample. The surface area of the samples was calculated by determining the samples' length and width to determine the samples' resistivity. Samples' resistivity was calculated using the following method (Pouladzadeh et al. 2018), fulfilling the following equations:

$$\text{Resistivity} = \frac{\rho L}{A}$$

$$\text{Electrical conductivity} = \frac{1}{\text{Resistivity}}$$

9.4 SURFACE MORPHOLOGY STUDY

An optical microscope was used to determine the surface morphology of the samples before and after incubation. Upon removal of the sample after incubation in deionized water and low glucose concentration, a silver clump was observed on the surface of the sample, which looked very similar to gallium. Under microscope observation, brown and red colour spots were observed on the surface of the samples. Gallium reaction with deionized water would show a dark brown or greyish brown surface after immersion in 7 days (Horasawa et al. 1997). Reaction with carbohydrate chains could also affect the colour change of the metal, according to literature (Konwarh et al. 2011).

Another hypothesis on the surface's colour changes could result from the occurrence of complex composites. Metals, when oxidized, would exhibit typical colours; for example, copper minerals would form green, blue, or sometimes red shades. The pigment formed on the sample's surface from light reflection (Pracejus 2014). The colour observed in a complex composite is known as absorption colour, and the variation of the colour depends on the crystal's orientation concerning plane of light. This phenomenon is known as pleochroism (MacKenzie and Adams 2007). Light etching also causes colour variation, especially in silver-containing complex composites (Pracejus 2014).

9.4.1 Incubation in Deionized Water

After 5 days of incubation, a 1:1 4°C/min sample exhibited brown pigment stained on a large sample surface with a small red dot. The surface roughness appeared to have improved slightly. However, silver oxide formation covered a larger area after incubation. 1:3 4°C/min sample demonstrated the grain formation after incubation. Besides that, no significant changes were observed. 1:9 4°C/min sample had yet to exhibit any change in surface morphology. Moving on to 5°C/min heating rate samples, 1:1 and 1:3 Ga/nAg ratio samples exhibited pigmentation on the surface. 1:1 5°C/min demonstrated light brown pigment stained on the surface with apparent grain size observed next to the pigmentation. In contrast, 1:3 5°C/min sample exhibited a large red pigment on the surface. Meanwhile, 1:9 5°C/min sample did not present any significant change except the surface was slightly stained in brown pigment.

After 10 days of incubation, 1:1 4°C/min sample did not show any significant difference. 1:3 4°C/min as well did not demonstrate a change in surface morphology. However, 1:9 4°C/min revealed a shift in grain structure. Grain formation was seen on the surface after the sample was incubated; silver oxide formation spread across the sample's surface. As for 5°C/min heating rate samples, 1:1 5°C/min exhibited a less shiny metallic surface than before incubation. The surface was observed to be dull and grainy. However, significant changes in surface morphology were observed in 1:3 and 1:9 5°C/min samples. As observed from 1:3 5°C/min sample, crystallization was detected on the surface, resembling the shape of tetrahedral geometry. The crystal formations observed were spread across the surface of the sample. As for 1:9 5°C/min sample, grain formation was observed after incubation. The grain formation spread more comprehensively than the formation observed in 1:9 4°C/min. These grain formations were deduced as fracturing and deformation. Fracturing and deformation are the results of hydrothermal recrystallization. Due to elevated temperatures, the sample may be subjected to recrystallization to readjust to the environment.

After 20 days of incubation, 4°C/min heating rate samples demonstrated noticeable surface morphology changes. Observation of 1:1 4°C/min sample presented brown pigment stained on the surface with a longitudinal formation of shiny metallic surface. 1:3 4°C/min sample revealed less nucleation growth, but brown pigment was also observed on the surface. As for 1:9 4°C/min sample, the surface area was also affected by the brown pigment with cracked lines observed on the shiny metallic surface. From 5°C/min heating rate samples,

1:1 5°C/min sample exhibited grain formation clearly. 1:3 and 1:9 5°C/min samples both exhibited pigmentation on the surface. However, the pigment on 1:3 5°C/min sample revealed a colourful and shiny shade that sides towards a pinkish purple colour. As for 1:9 5°C/min, the surface was affected by brown stains. All 5°C/min samples after incubation revealed a dull shade, not as shiny as before. The highlighted results of samples i.e. incubated for 20 days are shown in Figure 9.1.

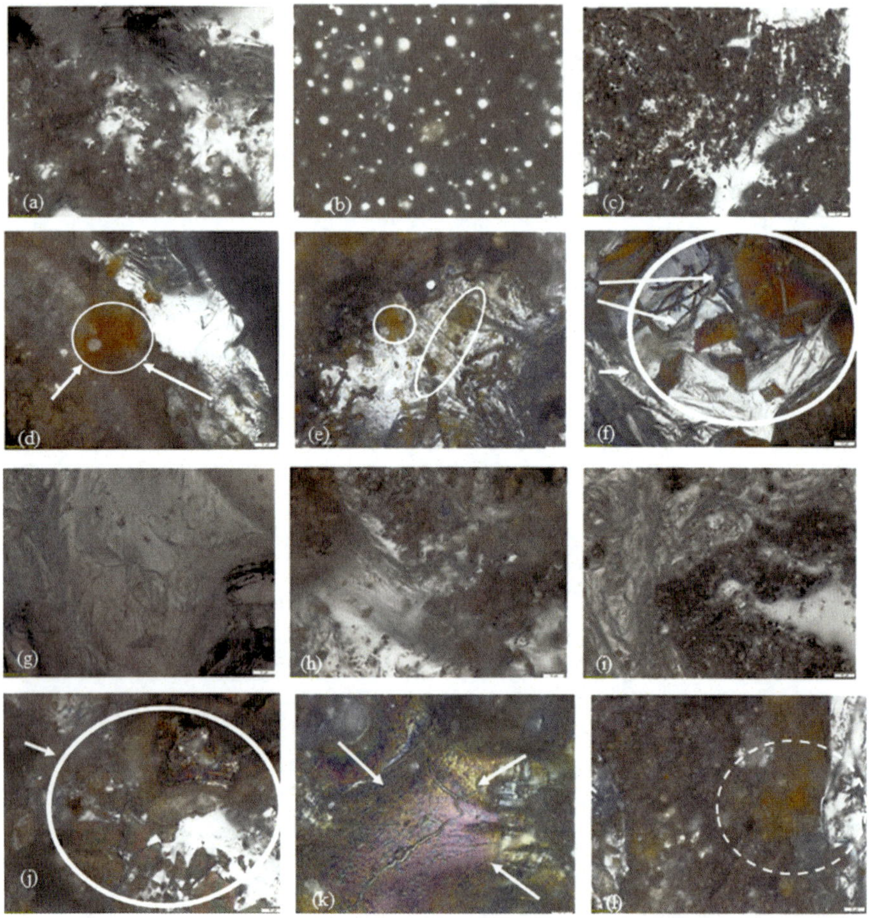

Figure 9.1 (a) 1:1, (b) 1:3, (c) 1:9, 4°C/min, before 20 days incubation in deionized water; (d) 1:1, (e) 1:3, (f) 1:9, 4°C/min, after 20 days incubation in deionized water; (g)1:1, (h) 1:3, (i) 1:9, 5°C/min, before 20 days incubation in deionized water; (j) 1:1, (k) 1:3, (l) 1:9, 5°C/min, after 20 days incubation in deionized water.

Based on observations of 4°C/min heating rate samples, the most significant change in surface morphology was exhibited after 20 days of incubation. The 1:9 sample exhibited the most changes, demonstrating grain formation after 10 days of incubation and brown stains on the surface after 20 days.

Meanwhile, 5°C/min solidification rate samples demonstrated significant surface morphology differences in 5 days of incubation. Changes were observed at the beginning of 5 days up to 20 days. This gave a stable degradation behaviour of the alloy along the incubation period, unlike 4°C/min solidification rate samples. As observed, pigmentations were found on the surface. 1:1 5°C/min sample showed a large red spot after 5 days of incubation. The red spot could be hypothesized as a different silver thickness on light exposure. Besides colour changes, crystal formation was also observed after 10 days of incubation. The crystal resembles the structure of a tetrahedral crystal. The silver content in the samples could have reacted with deionized water to form silver hydride, a complex composition of silver that would generate tetracapped tetrahedral geometry (Liu 2010).

9.4.2 Incubation in DMEM Low Glucose Concentration Medium

After 5 days of incubation, 1:1 4°C/min sample showed no significant difference in surface morphology. 1:3 4°C/min sample exhibited nucleation growth on the surface, similar to before, except after incubation it resulted in more nucleation growth. The grain structure observed from the sample significantly resembled the images from before incubation. A reaction rim (darker grey) on the surface may extend into the supergene conditions. The white shiny surface was deduced as native silver. Reaction rims are standard in forming hydrothermal replacements, as the texture was assumed to be the result of exsolution. Hydrothermal replacement in post-magmatic environments could occur in elevated temperatures (Pracejus 2014). As for 1:9 4°C/min sample, no significant changes were observed. Moving on to 5°C/min heating rate samples, 1:1 Ga/nAg ratio demonstrated a formation of yellow pigment. The formation of yellow pigment was assumed to be the formation of composites of the sample. With light exposure, the silver-containing composite would develop different coatings that exhibit various colours concerning thickness. The colour generated could be the surface breakdown of a silver-rich composite, chalcopyrite (Pracejus 2014). As for 1:3 5°C/min sample, the grain size changed significantly after incubation, which was observed from the magnified surface of a shiny metallic clump. Before incubation, a large sample area was covered by black silver oxide formation. However, it was not shown on the sample after incubation. The observation deduced that the sample experienced dehydration. The opening cracks usually would be filled with secondary complex chemical compounds, resulting in the observed variation of colours. The folded paper-like appearance demonstrated twinning, which could be the occurrence of lamellar pressure associated with deformation (Pracejus 2014). Nonetheless, 1:9 5°C/min showed no significant difference in surface morphology after incubation. A reaction rim was observed on the surface, similar to 1:3 4°C/min sample, which could result from hydrothermal replacement.

After 10 days of incubation, from 1:1 4°C/min sample observation, the surface revealed a significant firmness difference. 1:3 4°C/min exhibited crystalline formation that resembled a tetragonal crystalline structure, with a hint of black and brown pigmentation on the surface of the crystal structure. Besides that, 1:9 4°C/min sample demonstrated brown pigmentation on a large surface area of the sample with a hint of black silver oxide formation. The surface resembled fracturing and deformation of the sample and brown pigments, which could result from light etching of silver-rich chalcopyrite composite formation (Pracejus 2014). Moving on to 5°C/min samples, 1:1 Ga/nAg ratio had shown brown pigmentation on the surface, to 1:3 ratio. The pigmentation observed was similar to the results in 4°C/min heating rate samples. The nucleation formed before incubation was not observed on the surface after incubation. 1:3 Ga/nAg ratio exhibited supergene replacement, where regular linear features were observed. Supergene weathering is the occurrence of oxidation and hydration, hence deducing that the sample may have demonstrated silver oxidation or silver hydration during incubation (Pracejus 2014). Crystalline structure formation was found on 1:9 5°C/min sample, where at 1:9 Ga/nAg ratio, the crystal structure observed was thin and long.

After 20 days of incubation, all 4°C/min samples have brown colour pigmentation observed on the surface. 1:1 4°C/min had a wide surface stained in brown pigmentation. The large surface invaded by black silver oxide observed before incubation had subsided and was no longer observed on the sample after incubation. In addition to that, the sample was not as shiny as before incubation. Hydrothermal replacement was observed on the surface, which may result from exsolution. Crack lines were also observed on the hydrothermal surface, possibly resulting from supergene replacement. Besides that, the 1:3 4°C/min sample revealed less nucleation growth and a small surface stained in brown pigment after incubation. Moreover, the grain structure was seen more clearly after incubation, demonstrating cumulus formation. A cumulus is formed when high melting point grains start to crystallize and form euhedral to subhedral grains, which separates oxides and melts the low melting point. As a result, edge disintegration was formed. The surface also resembled the formation of chloroargyrite, which is formed by the chemical composition of AgCl. As for 1:9 4°C/min sample, a small red spot on the surface was observed. The surface of 1:9 4°C/min resembled a composite formation, miargyrite. Miargyrite had a strong bluish-to-brown anisotropy and had red internal reflections. Besides that, firework-like sparks were observed, which could result from silver-rich light etching of canfieldite (Pracejus 2014). For 5°C/min samples, 1:1 Ga/nAg ratio improved surface roughness. Changes in surface roughness were also observed in 1:3 5°C/min samples, which showed smoother yet less shiny surfaces and the formation of cracked lines, which could be resulted from crystallization of the surface later than the surrounding matrix. However, 1:9 5°C/min sample exhibited brown pigment stained on the surface. The highlighted results of samples incubated for 20 days were shown in Figure 9.2.

Based on the results obtained for 4°C/min heating rate samples, no significant change was observed until 10 days of incubation. After 10 days of incubation, brown pigment was observed to be stained on the surface of samples 1:3 4°C/min and 1:9 4°C/min. 1:1 4°/min only exhibited brown pigment after 20 days of incubation. The pigmentation formed was brown, seen in samples 1:3 4°C/min and 1:9 4°C/min after 10 days of incubation and 1:1 4°C/min after 20 days. However, after 20 days, the brown pigment in 1:3 and 1:9 4°C/min samples had subsided. Interestingly, 5°C/min heating rate samples exhibited a relatively stable and predicted relationship of reaction behaviour. Although after 5 days of

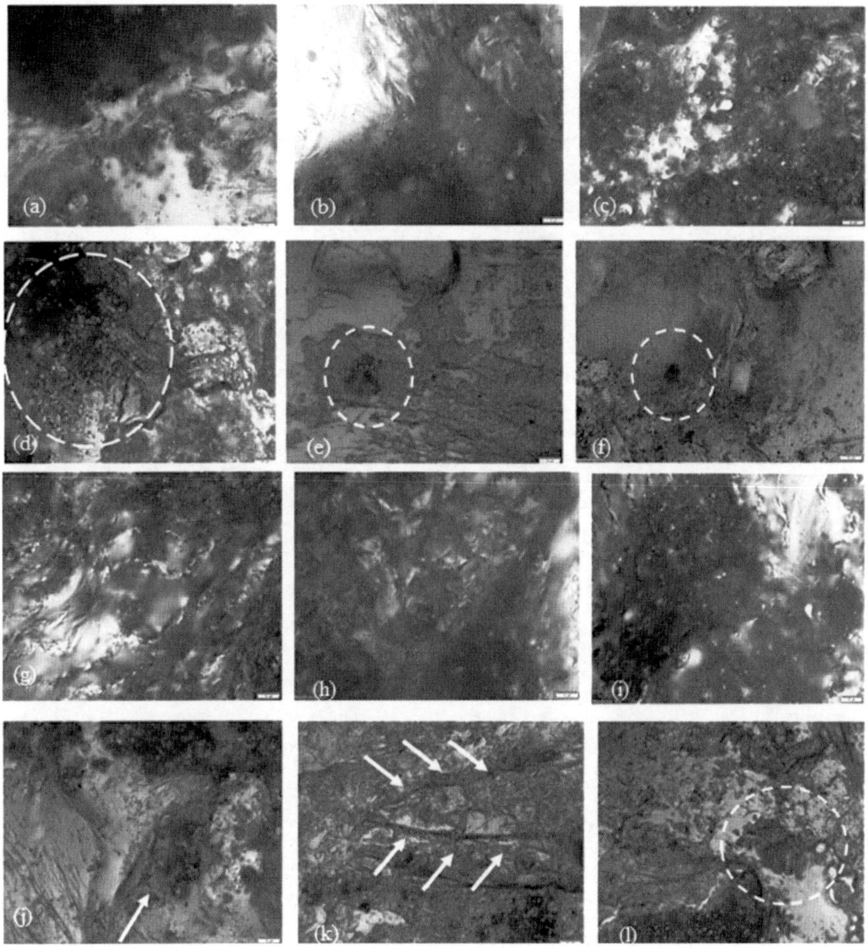

Figure 9.2 (a) 1:1, (b) 1:3, (c) 1:9, 4°C/min, before 20 days incubation in DMEM low glucose concentration medium; (d) 1:1, (e) 1:3, (f) 1:9, 4°C/min, after 20 days incubation in DMEM low glucose concentration medium; (g) 1:1, (h) 1:3, (i) 1:9, 5°C/min, before 20 days incubation in DMEM low glucose concentration medium; (j) 1:1, (k) 1:3, (l) 1:9, 5°C/min, after 20 days incubation in DMEM low glucose concentration medium.

incubation, the surface morphology changes were very significant, the changes gradually subsided and stabilized in the extended course of incubation periods. The observed modifications were mainly related to the surface roughness after incubation and a minority of colour pigmentation. The colour pigment formation demonstrated behaviour similar to the observation discussed for 4°C/min heating rate samples. As for surface roughness, the samples showed improvement in surface roughness throughout incubation. The crystalline formation was also observed, especially after 10 days of incubation; in the case of 1:3 5°C/min, the crystalline formation persisted until after 20 days. Based on these observations, the most stable sample after incubation in DMEM low glucose concentration medium was demonstrated by a 1:3 5°C/min sample.

9.4.3 Degradation Study for Incubation in Deionized Water

Referring to the plotted results of samples incubated in deionized water, despite the broken samples, the mass loss increments for 4°C/min heating rate samples were observed to be 0.0019 g, 0 g and 0.0005 g at 5 days incubation for Ga/nAg ratio of 1:1, 1:3 and 1:9, respectively. At 10 days, the mass loss increments obtained for the respective ratios are 0g, 0.0025 g, and 0.0066 g. Meanwhile, at 20 days, the mass loss increments obtained for the respective ratios are 0 g, 0 g, and 0.0006 g, respectively. Based on the received data, the mass loss of 4°C/min heating rate samples had an abrupt relationship with the concentration of nanosilver with significant changes of mass loss in the samples. On the contrary, the mass loss of 5°C/min heating rate samples varies in the minimal range of 0 ~ 0.0002 g for all Ga/nAg 1:3 ratio samples after 5 days, 10 days and 20 days. The mass loss of the samples was very close to zero, demonstrating a stable degradation behaviour. The plotted results could be observed based on Figure 9.3. Based on the results plotted on the graph, the ideal mass loss or degradation behaviour was observed at 5°C/min samples. The sample demonstrating the most stable and negligible degradation was 1:3 5°C/min. As observed, the mass loss obtained throughout the incubation was very close to zero. 1:1 and 1:9 5°C/min samples presented similar results except 1:1 5°C/min in 20 days and 1:9 5°C/min in 10 days. However, the samples had weak surface firmness, which resulted in breaking. Degradation of the metal alloy could be affected by the factor of ion release. As nanosilver doped gallium alloy was immersed and soaked in deionized water for up to 20 days incubation at 37°C, a small percentage of ions, either silver or gallium, was bound to release from the alloy, resulting in mass loss of the sample.

9.4.4 Degradation Study for Incubation in DMEM Low Glucose Concentration Medium

Based on the degradation behaviour of nanosilver doped gallium in DMEM low glucose concentration medium, the mass loss of the samples after 5 days ranged between 0 ~ 0.0003 g, which showed minimum signs of degradation. At 10 days, no signs of degradation were observed as the mass loss for all Ga/nAg ratios is almost 0 g. However, after 20 days, the samples for ratio 1:1 and 1:3 showed mass increments of –0.0005 g and –0.0002 g, respectively, except 1:3 5°C/min sample. The negative sign refers to the mass increment calculated from Equation 9.1. For 5°C/min heating rate samples, the mass loss of the 1:3 and 1:9 samples insignificantly increased, demonstrating stability. Therefore, the mass loss observed at 20 days was relatively stable; however, there was a mass loss in the 1:9 ratio sample. Based on the results, 5°C/min heating rate samples although exhibited insignificant abrupt behaviour after 5 and 10 days of incubation, they gradually stabilized when it reached 20 days of incubation. Generally, the samples exhibited minimum mass loss after incubating in a medium of DMEM low glucose concentration. The results were plotted in graphs and illustrated in Figure 9.4.

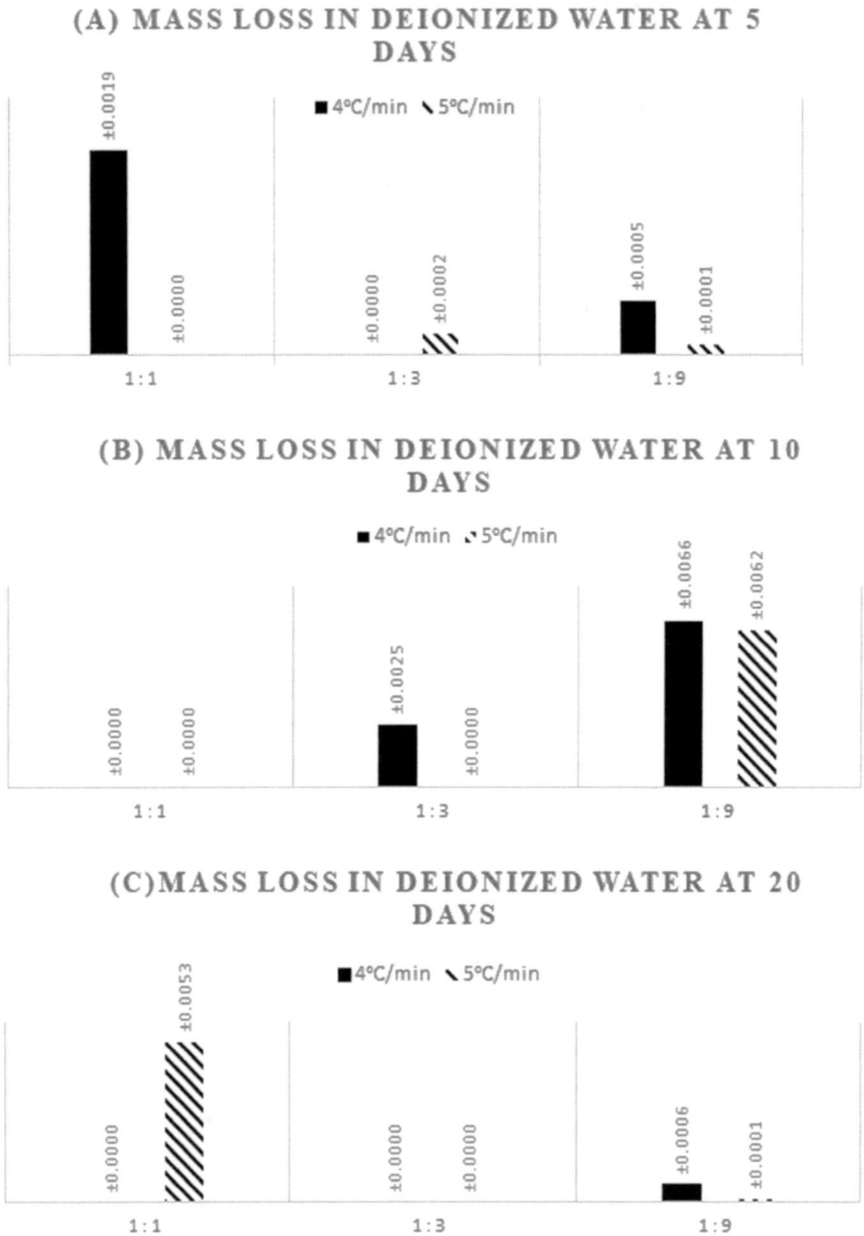

Figure 9.3 (A) mass loss of samples incubated in deionized water at 5 days; (B) Mass loss of samples incubated in deionized water at 10 days; (C) Mass loss of samples incubated in deionized water at 20 days.

Incubation of metal alloy in DMEM would result in gallium ion release (Wadge et al. 2018). The higher the percentage of gallium, the higher the gallium ion leaching. The degradation behaviour of gallium depended on the concentration of calcium and phosphorus ions, whereby an increase in calcium and phosphorus ions would result in more gallium leaching. As the concentration of calcium and

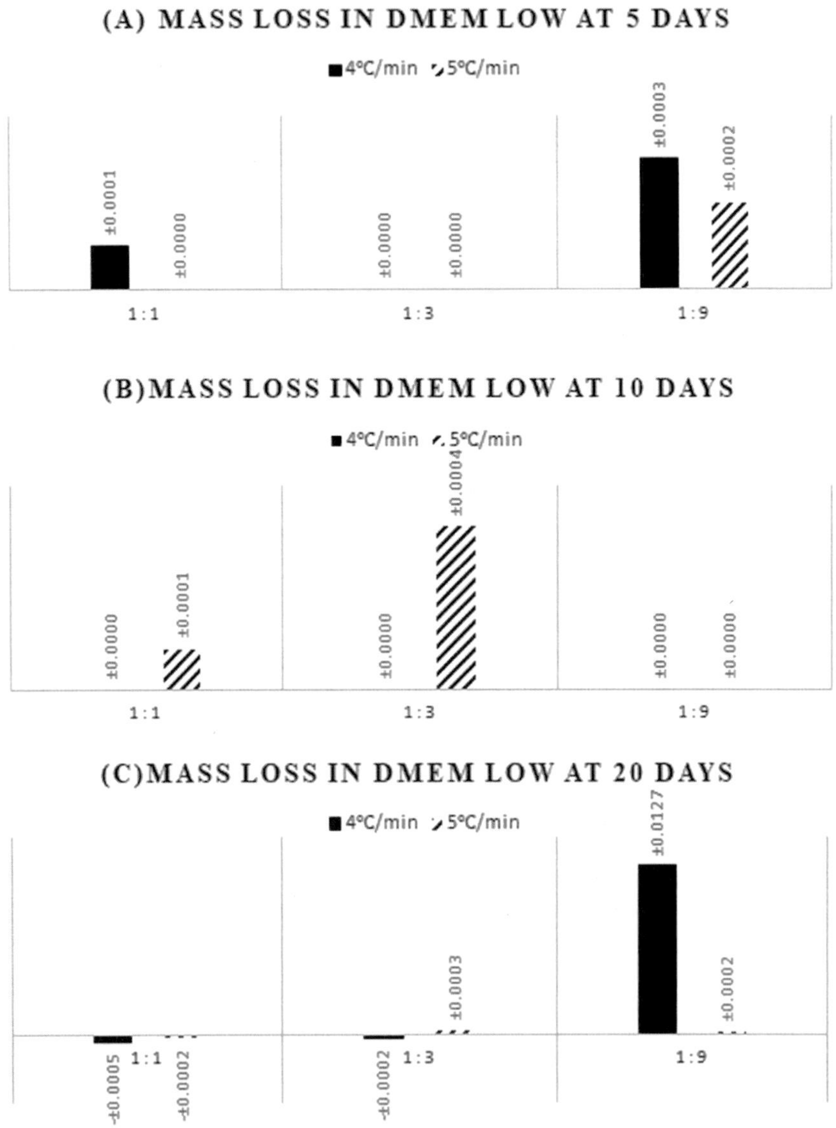

Figure 9.4 (A) Mass loss of samples incubated in dmem low glucose concentration medium at 5 days; (B) Mass loss of samples incubated in dmem low glucose concentration medium at 10 days; (C) Mass loss of samples incubated in DMEM low glucose concentration medium at 20 days.

phosphorus ions decreased, it reflected in the release of gallium ions. Interestingly, in this study, the working factor was the concentration of silver ions. Hence, the mass loss observed was due to the ratio of Ga/nAg and its metallic bond. However, the most effective degradation behaviour was 1:3 ratio throughout incubation with steady surface firmness where 1:9 ratio showed weak surface firmness that resulted in breaking. Nonetheless, 5°C/min heating rate samples showed better degradation behaviour compared to 4°C/min heating rate samples.

9.4.5 Electrical Conductivity Analysis for Incubation in Deionized Water

The samples in deionized water showed no significant difference in resistivity for 4°C/min samples. Hence, this explains that the electrical conductivity of the samples was not influenced by incubating them in deionized water. However, this was very different behaviour for 1:1 5°C/min samples which showed significant decrease in resistivity after incubation for 5 days, 10 days, and 20 days. The relationship of electrical resistivity was inversely proportional to electrical conductivity, where a reduction of electrical resistivity increased electrical conductivity. Therefore, 1:1 5°C/min samples exhibited an increase in electrical conductivity after incubating in deionized water. In contrast, samples 1:1 4°C/min, 1:3 4°C/min, 1:3 5°C/min, 1:9 4°C/min, and 1:9 5°C/min showed no significant difference in electrical conductivity after incubation. The graphs are provided in Figure 9.5. Based on the results obtained for electrical conductivity after incubating in deionized water, the most stable electrical conductivity behaviour was demonstrated at the 1:3 and 1:9 Ga/nAg ratio samples. However, 1:1 5°C/min demonstrated enhanced electrical conductivity after incubation as it's resistiviy reduced throughout incubation from 0.00006 ohm.m after 5 days incubation to 0.00004 ohm.m after 20 days incubation.

9.4.6 Electrical Conductivity Analysis for Incubation in DMEM Low Glucose Concentration Medium

After 5 days of incubation, 1:1 4°C/min and 1:3 5°C/min samples experienced a reduction in electrical resistivity, while the remaining samples i.e. 1:1 5°C/min, 1:3 4°C/min, 1:9 4°C/min, and 1:9 5°C/min did not demonstrate any significant differences in terms of electrical resistivity. As the samples reached 10 days incubation, 1:1 4°C/min, and 1:3 5°C/min samples presented stabilized behaviour and showed no changes in electrical resistivity. However, a reduction in electrical resistivity was observed in 1:1 5°C/min and 1:3 4°C/min samples, respectively. No changes were observed for 1:9 Ga/nAg ratio samples indicating stable electrical conductivity. The resistivity reduction for 1:1 5°C/min sample was maintained at 20 days of incubation except 1:3 4°C/min sample. Interestingly, reduction in electrical resistivity of 1:3 5°C/min sample was notable after 5 days incubation. According to (Kaklamani et al. 2018), the salts existing in the medium could be a contributing factor to the improvement of electrical conductivity. The protein and carbohydrates, along with the salts in the medium, would be dissociated,

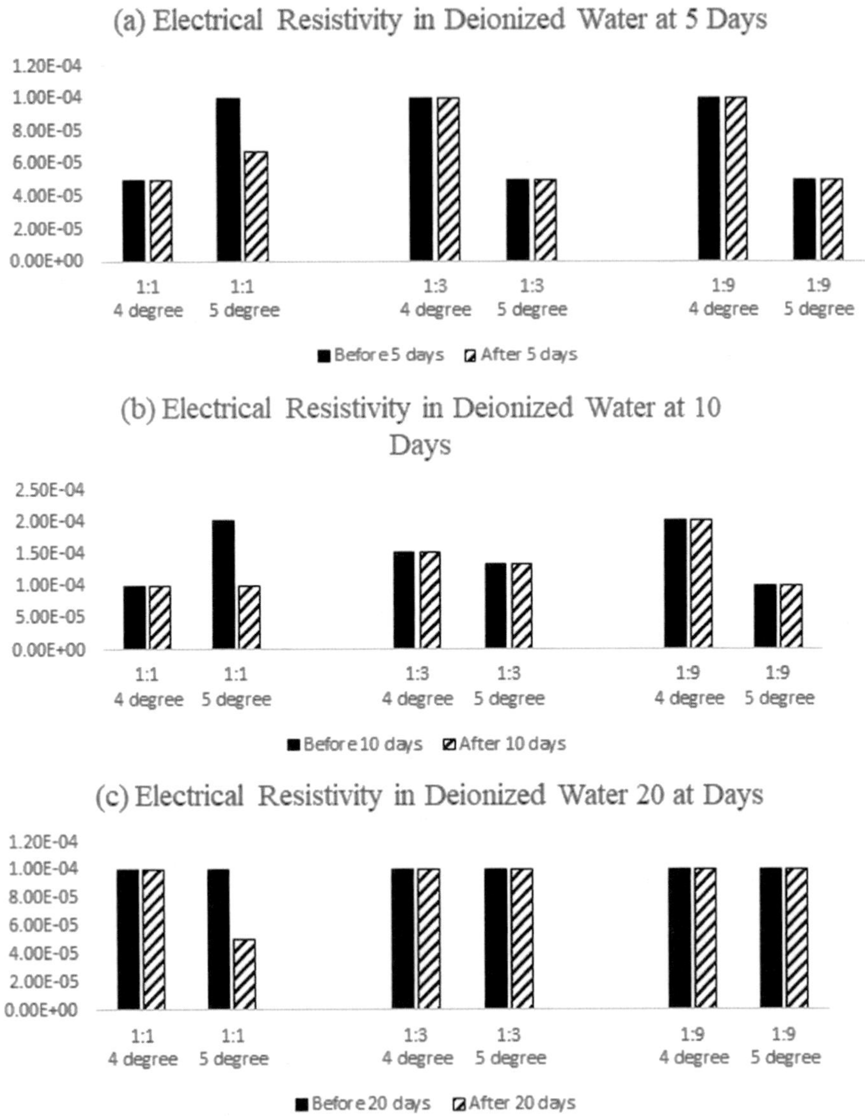

Figure 9.5 (a) Electrical resistivity of samples in deionized water at 5 days; (b) Electrical resistivity of samples in deionized water at 10 Days; (c) Electrical resistivity of samples in deionized water at 20 days.

which increased the existence of free ions such as Ca^{2+}, K^{+}, Na^{+}, SO_4^{2-}, Cl^{-} and Mg^{2+}. The mobility of the ions released from the medium could be the reason for improved electrical conductivity. Hence, this explained the reduction in resistivity as observed from the data obtained for samples incubated in a DMEM low glucose concentration medium. Generally, incubation in DMEM low glucose medium was proven effective for the conductance of nanosilver doped gallium alloy, except for a 1:3 4°C/min sample after 20 days of incubation. Graphs are illustrated in Figure 9.6. According to the observations obtained from the graphs, the most stable electrical conductivity behaviour was presented by 1:9 ratio samples, where no significant differences were observed throughout the incubation periods. Despite that, the most effective electrical behaviour

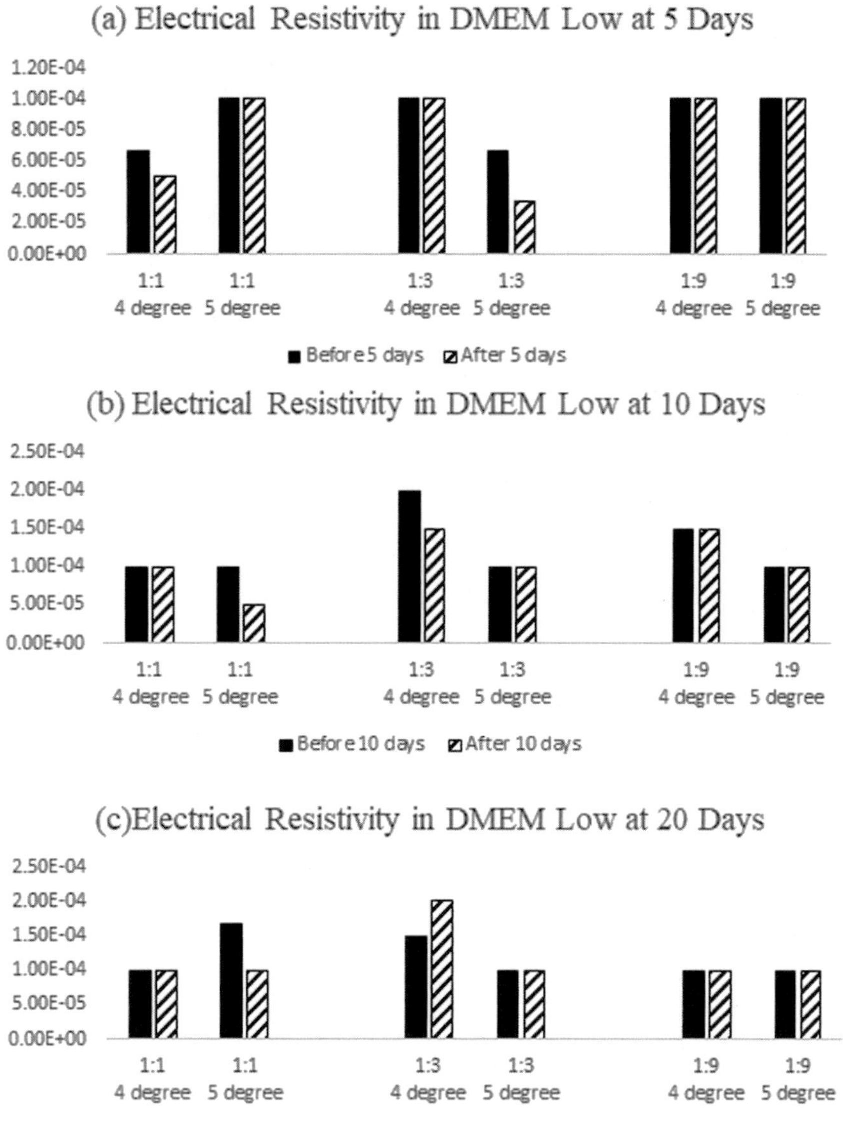

Figure 9.6 (a) Electrical resistivity of samples in DMEM low glucose concentration medium at 5 days; (b) Electrical resistivity of samples in DMEM low glucose concentration medium at 10 days; (c) Electrical resistivity of samples in DMEM low glucose concentration medium at 20 days.

of the sample was demonstrated at 1:3 5°C/min sample, as 1:3 Ga/nAg ratio showed a notable decrease in electrical resistivity except only 1:3 4°C/min sample which changed after 20 days of incubation, making it less effective compared to 1:3 5°C/min sample.

9.5 CONCLUSION

This research has been resolved through analysing factors with determining the effective concentration ratio and the solidification rate of nanosilver-doped gallium. Based on the overall results and discussion of this research, it is revealed that 1:3 gallium to nanosilver ratio (1:3 Ga/nAg) as the effective concentration ratio and 5°C/min as the effective solidification rate among all fluid conditions (Deionized water and DMEM Low Glucose concentration medium), demonstrating stable degradation behaviours including steady surface morphology and effective electrical behaviour. It is noticed that solidification rate contributed dominantly to degradation study and electrical conductivity analysis. Interestingly, it is also observed that a complex composition of silver was formed when the alloy was incubated in fluids resulting in pigmentation on the surface, especially in silver-containing complex compounds. It is expected such findings would bring light into enhancing the electrical conductivity of medical devices with metal alloys that would not be harmful to the human body and vulnerable to corrosive environments.

ACKNOWLEDGEMENT

The authors would like to express their gratitude to Curtin University Malaysia, Glasgow Caledonian University, Universiti Teknologi Brunei, Chisholm Institute and Saveetha Institute of Medical & Technical Sciences for their further support.

REFERENCES

Ahmed, I., D. Ready, M. Wilson and J.C. Knowles. 2006. Antimicrobial effect of silver-doped phosphate-based glasses. Journal of Biomedical Materials Research Part A. 79A: 618–626.

Dong, J., D. Fang, L. Zhang, Q. Shan and Y. Huang. 2019. Gallium-doped titania nanotubes elicit anti-bacterial efficacy *in vivo* against Escherichia coli and Staphylococcus aureus biofilm. Materialia. 5: 100209.

Horasawa, N., H. Nakajima, S. Takahashi and T. Okabe.1997. Behavior of pure gallium in water and various saline solutions. Dental Materials Journal. 16: 200–208.

Johnson, B.A. 1970. The dissolution of metal corrosion products in deionized water at 38°C (100°F). Nuclear Applications and Technology. 8(1): 58–61. https://doi.org/10.13182/NT70-A28634.

Kaklamani, G., D. Kazaryan, J. Bowen, F. Iacovella, S.H. Anastasiadis and G. Deligeorgis. 2018. On the electrical conductivity of alginate hydrogels. Regenerative Biomaterials. 5: 293–301.

Kim, J.-H., S. Kim, J.-H. So, K. Kim and H.-J. Koo. 2018. Cytotoxicity of gallium–indium liquid metal in an aqueous environment. ACS Applied Materials and Interfaces. 10: 17448–17454.

Kokubo, T. and H. Takadama. 2006. How useful is SBF in predicting *in vivo* bone bioactivity? Biomaterials. 27: 2907–2915.

Konwarh, R., N. Karak, C.E. Sawian, S. Baruah and M. Mandal. 2011. Effect of sonication and aging on the templating attribute of starch for "green" silver nanoparticles and their interactions at bio-interface. Carbohydrate Polymers. 83: 1245–1252.

Lansdown, A.B.G. 2010. A pharmacological and toxicological profile of silver as an antimicrobial agent in medical devices. Advances in Pharmacological Sciences. 2010: 910686. doi: 10.1155/2010/910686.

Liu, C.W., H.-W. Chang, B. Sarkar, J.-Y. Saillard, S. Kahlal and Y.-Y. Wu. 2010. Stable silver(I) hydride complexes supported by diselenophosphate ligands. Inorganic Chemistry. 49: 468–475.

Liu, Q., Y. Si, X. Cao, J. Cheng, S. Zhu and R. Xu. 2024. Effects of metal element doping on the lubrication behaviours and mechanisms of gallium-based liquid metals. Wear. 538–539: 1–17.

MacKenzie, W.S. and A.E. Adams. 2007. A Color Atlas of Rocks and Minerals in Thin Section, 9th Ed. Manson Publishing.

Pouladzadeh, F., A.A. Katbab, N. Haghighipour and E. Kashi. 2018. Carbon nanotube loaded electrospun scaffolds based on thermoplastic urethane (TPU) with enhanced proliferation and neural differentiation of rat mesenchymal stem cells: The role of state of electrical conductivity. European Polymer Journal. 105: 286–296.

Pracejus, B. 2014. The Ore Minerals Under the Microscope: An Optical Guide, 2nd Ed. Elsevier Science.

Rana, K.S., L.P.d. Souza, M.A. Isaacs, F.N.S. Raja, A.P. Morrell and R.A. Martin. 2017. Development and characterization of gallium-doped bioactive glasses for potential bone cancer applications. ACS Biomaterials Science and Engineering. 3: 3425–3432.

Sanchez-Salcedo, S., G. Malavasi, J.A. Salinas, G. Lusvardi, L. Rigamonti and L. Menabue. 2018. Highly-bioreactive silica-based mesoporous bioactive glasses enriched with gallium(III). Materials. 11.

Saravanapavan, P., J.R. Jones, R.S. Pryce and L.L. Hench. 2003. Bioactivity of gel–glass powders in the CaO–SiO_2 system: A comparison with ternary (CaO–P_2P_5–SiO_2) and quaternary glasses (SiO_2–CaO–P_2O_5–Na_2O). Journal of Biomedical Materials Research Part A. 66A: 110–119.

Smerker, H. 2014. Weaning and adaptation of SH-SY5Y Cells to Low glucose media for manganese exposure studies. Proceedings of the National Conference on Undergraduate Research (NCUR), University of Kentucky, Lexington, KY April 3–5.

Sudin, I., A. Kadir, M. Anwar, E.M.N. Engku Abu Bakar, A. Nalisa and M. Konneh. 2011. Effect of carburization process on adhesion strength of Ti carbide layer on titanium alloy substrate. Advanced Materials Research. 197–198: 219–224.

Valappil, S.P., D. Ready, E.A. Abou Neel, D.M. Pickup, L.A. O'Dell and W. Chrzanowski. 2009. Controlled delivery of antimicrobial gallium ions from phosphate-based glasses. Acta Biomaterialia. 5: 1198–1210.

Wadge M.D., B.W. Stuart, K.G. Thomas and D.M. Grant. 2018. Generation and characterisation of gallium titanate surfaces through hydrothermal ion-exchange processes. Materials and Design. 155: 264–277.

Wang, H., Y. Zheng, J. Liu, C. Jiang and Y. Li. 2017. *In vitro* corrosion properties and cytocompatibility of Fe–Ga alloys as potential biodegradable metallic materials. Materials Science and Engineering C. 71: 60–66.

Yang, M., J. Ren and R. Zhang. 2017. Novel gallium-doped amorphous calcium phosphate nanoparticles: preparation, application and structure study. Journal of Non-Crystalline Solids. 466–467: 15–20.

Chapter **10**

Characterizing Physicochemical Properties of Alginate-Gelatin-Cholesteryl Ester Liquid Crystals Scaffold for Microtissue Regeneration

Alyaa Idrees Abdulmaged[1] and Chin Fhong Soon*[2]

[1]Human Leukocyte Antigens Typing Laboratory,
Al-Hadbaa Hospital for Hematology and Bone Marrow Transplantation,
Nineveh Health Department, Iraqi Ministry of Health, 41002 Nineveh, Iraq

[2]Microelectronics and Nanotechnology-Shamsuddin Research Centre,
Institute for Integrated Engineering, Universiti Tun Hussein Onn Malaysia,
Parit Raja, 86400 Batu Pahat, Johor, Malaysia

10.1 INTRODUCTION

Tissue engineering (TE) integrates three-dimensional (3D) matrices, scaffolds, cells, bioactive molecules, and growth factors to create functional tissue-like constructs. These 3D matrices offer structural and mechanical support for tissue regeneration, aiming to restore, replace, maintain, or enhance the function of damaged tissues or failing organs (Sefat et al. 2016, Zhang et al. 2017, Zhu et al. 2016). Designing an effective 3D matrix is crucial for guiding the formation of new functional tissue (Ashammakhi et al. 2019). An optimal 3D scaffold should

*For Correspondence: soon@uthm.edu.my

closely mimic the physicochemical properties of native tissue *in vitro*, facilitating construct integration into the host tissue. It should also provide the necessary cues to influence cell behavior and differentiation toward the desired phenotypes (Jiang et al. 2019, Jang et al. 2018).

3D bioprinting is a widely used technology in creating 3D tissue and organ models for various tissue engineering applications. Advances in this field have enabled the replication of the complexity and heterogeneity of native tissues (Murphy and Atala 2014, Tabriz et al. 2015, Ashammakhi et al. 2019). Bioinks are crucial in 3D bioprinting, providing structural and mechanical support to enhance cell attachment, proliferation, and differentiation into specific tissue types (Liu et al. 2020, Berg et al. 2018, Sun et al. 2020). Selecting the appropriate materials for the scaffold is essential for engineering a 3D microenvironment that closely mimics the complexity of native tissues. Both natural and synthetic polymers, such as alginate, poly(ethylene glycol) (PEG), gelatin, hyaluronic acid, collagen, chitosan, cellulose, and polyvinyl alcohol (PVA), have been used in the development of bioinks to create biomimetic structures (Skardal and Atala 2015). A single biomaterial in bioinks cannot meet all mechanical and functional requirements necessary for producing tissue-like constructs, leading to the development of multi-material bioinks. These combine multiple types of biomaterials, cells, and bioactive molecules (Chimene et al. 2016, Rutz et al. 2015, Ashammakhi et al. 2019). Currently, multi-material bioinks are gaining increasing interest in 3D bioprinting research for their potential to enhance the performance of biomimetic structures.

Alginate is a linear, hydrophilic polysaccharide naturally derived from brown seaweed. It is an anionic polymer composed of (1–4)-linked β-D-mannuronic acid (M) and α-L-guluronic acid (G) monomers (Axpe and Oyen 2016, Andersen et al. 2015, Park et al. 2017). Alginate-based bioinks have been extensively studied and utilized for creating 3D cell matrices (Valot et al. 2019). This is because alginate offers desirable features for biomedical applications, such as excellent biocompatibility, non-immunogenicity, shear-thinning properties, and ease of gelation with divalent cations like Ca^{2+} (Lee and Mooney 2012, Jia et al. 2014, Panwar and Tan 2016). However, a significant challenge with alginate-based bioinks is their limited cell adhesion and interaction (Chimene et al. 2016, Wu et al. 2020, Kajave et al. 2020). To address this, modifying alginate-based bioinks by adding polymers with cell-binding sites, such as gelatin, has proven effective in enhancing the biological activity of printed structures (Chung et al. 2013, Shaw et al. 2015). Gelatin, a natural, biocompatible, and biodegradable protein derived from the partial hydrolysis of collagen (the most abundant protein in the mammalian extracellular matrix), supports cell growth and function, making it suitable for bioprinting applications (Chung et al. 2013, Di Giuseppe et al. 2018, Duan et al. 2013). Blending alginate with gelatin significantly improves cell viability and printing criteria. (Sarker et al. 2014a) demonstrated that adding gelatin to alginate results in more viable cells with intact nuclei and cell membranes within the alginate-gelatin scaffold than pure alginate, with the number of adherent cells increasing with higher gelatin content.

Liquid Crystals (LCs) form a distinct class of synthetic biomaterials, complementing the existing range of synthetic biopolymeric materials. They exhibit intermediate phases between liquid and solid states, known as liquid-crystalline or mesophases, and have a variety of chemical compositions and properties (Soule and Rey 2012, Urbanski et al. 2017). LCs are widely used in modern technologies and consumer products, including electronic display devices, biosensors, lubricants, optical devices, and drug delivery systems (Acharya 2011, Prévôt et al. 2018). The unique characteristics of LCs, such as rapid molecular reorganization, viscoelasticity, sensitivity to external stimuli, and flexibility, have attracted the attention of researchers in tissue engineering, prompting their combination with other materials (Iwabata et al. 2013, Sidiq and Pal 2016, Traugutt et al. 2020). Furthermore, many biological molecules, including nucleic acids (DNA), cell membranes, proteins, phospholipids, and cholesterols, exhibit liquid crystalline-like phases (Sung and Kim 2018, Soon et al. 2013, Mitov 2017). Consequently, significant attention has been given to liquid crystal polymers for designing novel 3D biomimetic microenvironments for tissue engineering applications. For instance, (Du et al. 2017) developed a chitosan/cholesterol hydroxypropyl cellulose ester liquid crystal (CS/LC) composite hydrogel. They found that LC provides consistent spatial orientations that enhance initial cell adhesion. LCs also offer active sites at substrate interfaces, promoting cell attachment, spreading, and extracellular matrix (ECM) secretion. (Wu et al. 2017) reported that the cellular behavior of human umbilical cord-derived mesenchymal stem cells (hUC-MSCs) cultured on PU/LC composite substrates is influenced by variations in substrate microstructure. Their study demonstrated that incorporating different LC contents into polyurethane (LC-10/PU, LC-30/PU, and LC-50/PU) enables cells to sense the bioactive patterns and viscoelastic properties of LC domains, responding to topographical and mechanical cues. This modulation of cellular behavior includes adhesion, proliferation, and differentiation.

Our previous study (Abdulmaged et al. 2020) explored the printability of an Alginate-Gelatin-Cholesteryl Ester Liquid Crystal (AGLC) bioink and its polymers using a microextrusion-based 3D printer. Alginate was chosen for its excellent biocompatibility and mild crosslinking with divalent ions, while gelatin was included to enhance cell attachment, spreading, and proliferation. Adding cholesteryl ester liquid crystal (CELC) introduced lipid moieties into the bioink. Our findings demonstrated that incorporating CELC significantly improved the printability of the AGLC bioinks. Specifically, the printability increased by approximately 27.8% compared to alginate alone, achieving a minimum width of 1.3 mm at a 1 mL/min flow rate when the CELC concentration was raised to 40%. This formulation combined the beneficial properties of alginate-gelatin hydrogels with CELC, which contains cholesterol moieties known for their universal affinity for cell membranes.

This study presents the characterization of the physicochemical properties of multi-material AGLC bioinks designed to mimic the complexity of natural tissues. The AGLC bioinks are composed of alginate, gelatin, and CELC. Their physicochemical properties were analyzed using Fourier Transform Infrared Spectroscopy (FTIR), Differential Scanning Calorimetry (DSC), Field Emission Scanning Electron Microscopy (FE-SEM), contact angle measurement, and MTT assay and were compared with AG1050 bioink. This paper contributes to the field by providing a detailed characterization of new bioinks, examining their physicochemical properties and biological activity. The results indicate that the flexible chains of LCs and the structural similarity of the Alginate-Gelatin (AG) hydrogel composite substrate closely mimic the characteristics of natural cell models, demonstrating favorable soft matter properties.

10.2 MATERIALS AND METHOD

10.2.1 Materials

Sodium alginate and gelatin powders were purchased from Sigma-Aldrich, USA. Cholesteryl Perlagonate ($C_{36}H_{62}O_2$), Cholesteryl Oleyl Carbonate ($C_{46}H_{80}O$) and Cholesteryl Chloride ($C_{27}H_{45}Cl$) and Calcium chloride ($CaCl_2$) of anhydrous, granular, ≤7.0 mm, ≥93.0% were purchased from Sigma-Aldrich, USA.

10.2.2 Synthesis of Alginate-Gelatin and Alginate-Gelatin-CELC bioink

Alginate-gelatin (AG1050) and CELC were prepared to construct the AGLC bioinks. The AG1050 consisted of 10% w/v alginate and 50% w/v gelatin dissolved in distilled water. The CELC biomaterial sample was made up of 37.5% Cholesteryl Oleyl Carbonate, 37.5% Cholesteryl Pelargonate, and 25.0% Cholesteryl Chloride, heated to its isotropic phase at 115–120°C before being mixed into the AG gel. Various volumes of CELC were added to AG1050 and mixed to obtain a homogeneous gel labeled AGLC105001, AGLC105005, AGLC105010, AGLC105020, AGLC105040, and AGLC105060. The last two digits of the sample label represent the percentage of CELC added. Table 10.1 shows the AGLC bioinks prepared with different concentrations of CELC.

Table 10.1 The concentration of the AGLC bioinks. Adapted with permission from Abdulmaged et al. 2022, by permission of the *Polymers* under the Creative Commons Attribution License.

No.	Sample Name	Alginate-Gelatin-CELC
1	AGLC105001	AG1050–1%
2	AGLC105005	AG1050–5%
3	AGLC105010	AG1050–10%
4	AGLC105020	AG1050–20%
5	AGLC105040	AG1050–40%

10.2.3 Calcium Chloride Preparation

A 1% Calcium Chloride ($CaCl_2$) solution was prepared by dissolving 0.6 g of calcium chloride in 60 mL of distilled water. The solution was thoroughly mixed for 1 minute to ensure complete dissolution of the $CaCl_2$ granules. This $CaCl_2$

solution was then used as a crosslinker in the polymerization of AGLC bioinks, helping to stabilize the printed structures.

10.2.4 Printing and Polymerization of AGLC Bioinks

Stabilization of the AGLC bioink structures post-printing was achieved using Ca^{2+} ions. The printed bioink structures were immersed in a 1% $CaCl_2$ solution for 15 minutes to facilitate polymerization. Simple square models with 3.5 × 3.5 cm dimensions were printed using a custom-built 3D micro-extrusion bioprinting system. A bioink-loaded syringe, equipped with a disposable tip with an inner diameter of 1 mm, was mounted on the plastic holder. All bioink structures were printed with a single layer at a speed of 30 mm/s, at room temperature (25°C), and under sterile conditions.

10.2.5 Fourier Transform Infrared Spectroscopy

Fourier Transform Infrared Spectroscopy (FTIR, PerkinElmer Spectrum 100, USA) was utilized to assess the chemical composition of the AGLC bioink and its constituent polymers. Following polymerization in a 1% CaCl solution, all AGLC bioink samples were cut into small pieces and placed onto the sample stage under the universal attenuated total reflectance (UATR) before scanning. A pressure arm was then positioned over the samples. Spectra were recorded in the wavelength range of 4000–600 cm^{-1} with a scanning resolution of 4 cm^{-1} and a scan rate of 32 scans per second. Measurements were taken in ten different areas of each sample.

10.2.6 Differential Scanning Calorimetry

The thermal properties of AG1050 and AGLC bioinks were investigated using Differential Scanning Calorimetry (DSC, Q20, USA). Approximately 8–12 mg of each sample was weighed, placed in an aluminum pan, and sealed with an aluminum lid before being heated at 10°C/min. An empty aluminum pan with its lid served as the reference for the DSC analysis. All measurements were performed under a nitrogen flow within a temperature range of 0 to 250°C.

10.2.7 Contact Angle Measurement

The surface wettability (hydrophilicity) properties of the AG1050 and AGLC bioinks were evaluated by the sessile drop technique in which the water contact angle was determined for each sample fixed on a glass slide and 10 μl of deionized water dropped onto the surface of the samples. A VCA-optima instrument (AST Inc., USA) was used for contact angle measurement. The images of the droplet were captured to record the contact angles in three different areas of each sample.

10.2.8 Field Emission Scanning Electron Microscopy

Surface morphology analysis of AG1050 and AGLC bioinks was carried out using a field emission scanning electron microscope (FESEM, JEOL JSM-7600F, Japan) at an accelerating voltage of 5.0 kV. Before the examination, a sample of AG1050 and AGLC bioinks (AGLC105001, AGLC105005, AGLC105010, AGLC105020, AGLC105040 and AGLC105060) were air-dried at room temperature (25°C) and the surface of the samples were coated with a thin layer of gold using a JFC-1600 auto fine coater (JEOL. Tokyo, Japan).

10.2.9 Cell Culture Preparation

HeLa cells used in this research were obtained from ATCC (Cell Line, US). The cells were cultured in a 25 cm^2 tissue-treated flask with Gibco® Dulbecco's Modified Eagle's Medium with High Glucose (Sigma Aldrich, UK), supplemented with 10 ml of L-glutamine, 50 ml of fetal bovine serum (Sigma Aldrich, UK), 5 ml of penicillin-streptomycin (Sigma Aldrich, UK), and 1 ml of Amphotericin (Sigma Aldrich, UK). When the cells reached approximately 80% confluency, 1 ml of trypsin was added to the culture flask, and the cells were incubated in a CO_2 incubator for 5 minutes. After trypsinization, the flask was inspected using a Nikon Eclipse TS100 phase-contrast microscope with a Go-5 charged-coupled device (CCD) camera to confirm that most cells had detached from the flask surface. To halt trypsinization, 5 ml of supplemented culture media was added. The cells were then transferred to a 15 ml tube and centrifuged at 1000 rpm for 5 minutes. Following centrifugation, the old media was discarded, and the cell pellet was resuspended in 5 ml of supplemented culture media, preparing the cells for the MTT assay.

10.2.10 MTT Assay Assessment

HeLa cells used in this study were obtained from ATCC (Cell Line, US). The cells were cultured in a 25 cm^2 tissue-treated culture flask with Gibco® Dulbecco's Modified Eagle's Medium with High-glucose (Sigma Aldrich, UK), supplemented with 10 ml of L-glutamine, 50 ml of fetal bovine serum (Sigma Aldrich, UK), 5 ml of penicillin-streptomycin (Sigma Aldrich, UK), and 1 ml of Amphotericin (Sigma Aldrich, UK). Once the cells reached 80% confluency, 1 ml of trypsin was added to detach them from the flask. The cytotoxicity of the AGLC bioinks was assessed using the 3-(4, 5-dimethylthiazol-2-yl)-2,5-diphenyltetrazolium bromide (MTT) assay. The cells were transferred to fresh culture medium and seeded into 96-well plates at a density of 10^4 cells per well, then incubated for 24 hours at 37°C in a humidified atmosphere with 5% CO_2 and 95% air. Following incubation, the medium was replaced with 100 μl of fresh media containing various concentrations of AGLC bioinks and incubated for an additional day. Afterward, the AGLC bioinks were removed, and the cells were rinsed with 100 μl of Hank's Balanced Salt Solution (HBSS). To assess cytotoxicity, 10 μl of MTT solution was added to each well and incubated for 2 hours. The resulting formazan crystals were dissolved with 100 μl of dimethyl sulfoxide (DMSO), and the absorbance of each well was measured at 540 nm using a microplate reader (Multiskan GO, Germany). The cells were seeded into a fresh culture medium (control) with the same seeding conditions. The HeLa cell viability (%) was calculated using the following equation:

$$\text{Viability}\,(\%) = \left(\frac{[\text{abs}]\,\text{sample} - [\text{abs}]\,\text{blank}}{[\text{abs}]\,\text{control} - [\text{abs}]\,\text{blank}}\right) \times 100 \qquad (10.1)$$

where abs sample, abs control, and abs blank represent the absorbance values for the cells treated with materials, HeLa cells treated with culture medium, and absorbance of blank solution, respectively.

10.2.11 Statistical Analysis

The results were presented as mean ± standard deviation (SD). The data obtained from FTIR and DSC was interpreted using the original Lab Pro software. A statistical comparison of MTT assay data was analyzed using a student's t-test. $P < 0.05$ was considered statistically significant.

10.3 RESULTS AND DISCUSSION

10.3.1 Polymerization Analysis

Figure 10.1 shows the printed structures of the AGLC bioink immersed in the DMEM culture medium. The AGLC printed structure in varying CELC concentrations was effectively stabilized with Ca^{2+} ions after being immersed in a 1% $CaCl_2$ solution for 15 minutes. This stabilization is attributed to the ability of Ca^{2+} ions to bind to the G-block units in the alginate, as noted by (Valot et al. 2019). The cross-linking process successfully restricted the mobility of CELC chains and captured them during network stabilization (Prévôt et al. 2018). The presence of CELC, particularly at high concentrations (40% AGLC bioink), did not affect the crosslinking between Ca^{2+} ions and the G-block units in alginate. This is likely due to the high solubility of $CaCl_2$ in aqueous solutions, which facilitates the diffusion of Ca^{2+} ions to the entangled alginate chains even in the presence of high CELC concentrations (AGLC105020 and AGLC105040).

10.3.2 Chemical Properties Analysis

Figure 10.2 displays the FTIR spectra of polymerized AGLC bioinks alongside those of AG hydrogel and CELC. The spectra for AGLC bioinks, formulated with 1, 5, 10, 20, and 40% CELC, exhibit features characteristic of both AG hydrogel and CELC. The IR spectrum of AG hydrogel shows distinct peaks at 3318, 1635, 1560, 1456, and 1033 cm^{-1}, while CELC peaks appear at 2928, 2852, 1737, 1466, 1376, and 1252 cm^{-1}. For the AGLC bioinks, the prominent peaks are observed

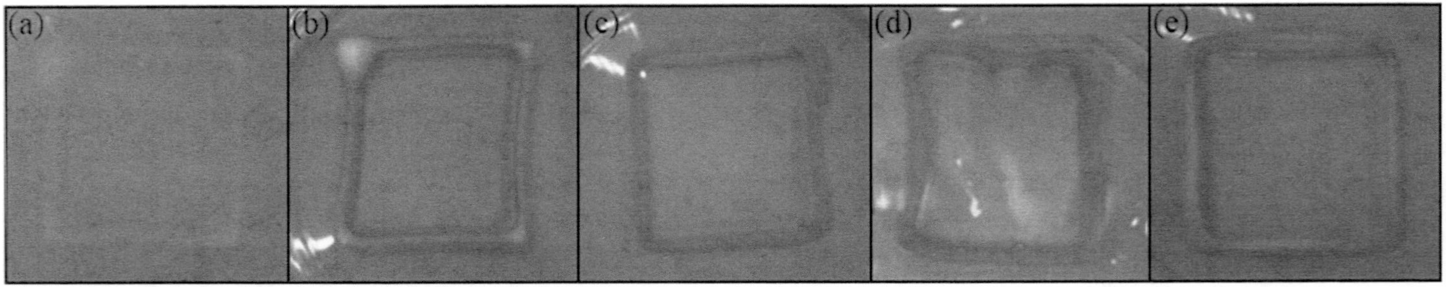

Figure 10.1 The polymerized structures of (a) AGLC105001, (b) AGLC105005, (c) AGLC105010, (d) AGLC105020 and (e) AGLC105040. Adapted with permission from Abdulmaged et al. 2022, by permission of the *Polymers* under the Creative Commons Attribution License.

around 3318, 2928, 2852, 1737, 1635, 1560, 1456, 1252, and 1033 cm^{-1}, as depicted in Figure 10.2.

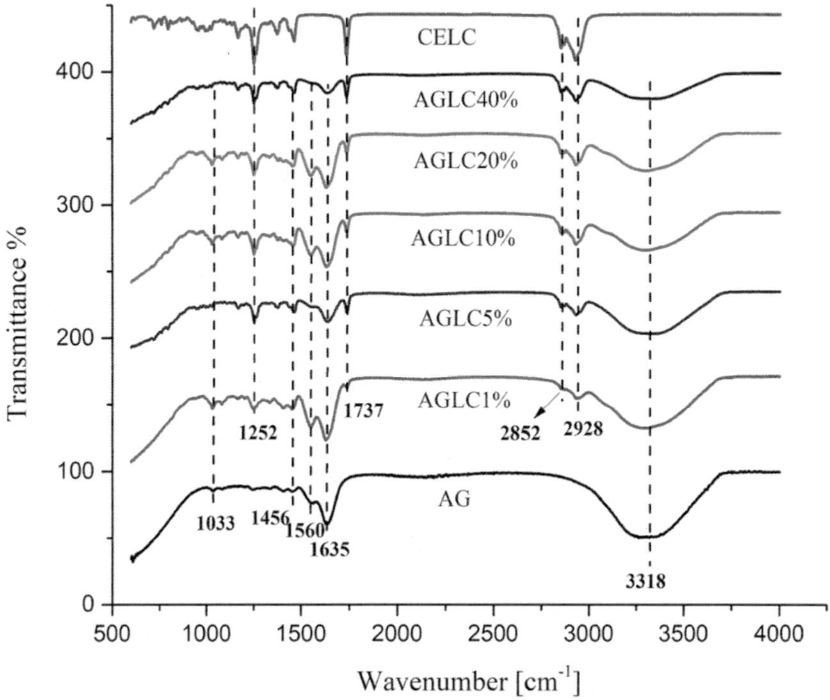

Figure 10.2 FTIR spectra of the AG hydrogel, AGLC bioinks, and CELC. Adapted with permission from Abdulmaged et al. 2022, by permission of the *Polymers* under the Creative Commons Attribution License.

The absorption band at 3318 cm^{-1} is attributed to the stretching vibration of the O–H functional group. This peak broadens, and its intensity diminishes due to water repulsion caused by the hydrophobic CELC in the bioinks. Two adjacent small peaks at 2928 cm^{-1} and 2852 cm^{-1} correspond to the C–H stretch from CELC. The carbonyl group (C=O) related to lipid ester is observed at 1737 cm^{-1} in CELC. The strong peak at 1635 cm^{-1} is associated with the asymmetric stretching of the COO^- bond from carboxylic acid in AG hydrogel.

The peak at 1560 cm^{-1} is linked to the electrostatic interaction between the negative charges of alginate carboxyl groups and the positive charges of gelatin amino groups. The distinctive peak for the polyelectrolyte complex ($CONH_2$) due to electrostatic reactions is typically found between 1500–1650 cm^{-1} (Devi and Kakati 2013, Wang et al. 2016). The absorption band at 1456 cm^{-1} is related to the symmetric and asymmetric bending vibrations of the methyl group in amino acid residues of AG hydrogel.

The bands at 1252 cm^{-1} and 1033 cm^{-1} correspond to C–O stretching vibrations in CELC and the C–H group of guluronic units in AG hydrogel, respectively. Additionally, the intensity of CELC peaks in AGLC bioink samples increases with the weight fraction of CELC. The spectra of AGLC bioinks show that combining AG hydrogel with CELC does not alter the positions of the characteristic peaks of the individual components. The C=O absorption peak at 1737 cm^{-1}, indicative of lipid ester, is present in all AGLC bioink spectra, potentially enhancing the biological activity of the bioinks.

CELC contains cholesteryl moieties with a general affinity for cell membranes, which may closely simulate *in vivo* conditions (Soon et al. 2014). These moieties are essential for cell membrane formation and stabilization of the bilayer lipid membrane (Hwang et al. 2002, Zhao et al. 2016). Therefore, modifying the AG1050 bioink with various concentrations of CELC could create a more favorable microenvironment for cellular activities such as adhesion, migration, proliferation, and differentiation.

10.3.3 Thermal Properties Analysis

Figure 10.3 (a–g) illustrates the thermal properties of polymerized AG and AGLC bioinks. The DSC curves for AG and AGLC bioink samples reveal endothermic peaks at 126, 140, 127, 121, 131, 143, and 111°C, respectively. For the AG sample without CELC (AG1050), two distinct endothermic peaks were observed at 127 and 140°C, as shown in Figure 10.3 (a). The first peak at 127°C corresponds to the evaporation of water molecules from the hydrogel, while the second peak at 140°C indicates the decomposition of more thermally stable structures in the alginate and gelatin hydrogels (Sarker et al. 2014b, Lopes et al. 2017).

For AGLC bioink samples containing 1% and 5% CELC, the DSC curves displayed a reduction in the melting temperature (Tm) for AGLC105001 and AGLC105005, as depicted in Figure 10.3 (b and c). This decrease suggests a reduction in water content, likely due to the hydrophobic nature of CELC (Lai et al. 2013, Wu and Gong 2011). Conversely, AGLC samples with 10% and 20% CELC (AGLC105010 and AGLC105020) exhibited increased thermal stability, with higher Tm values than samples with 1% and 5% CELC. This increase is attributed to the higher energy required to evaporate water molecules in these higher CELC concentration samples, leading to more excellent thermal stability and increased thermal degradation, as seen in Figures 10.3 (d and e). However, the sample with 40% CELC (AGLC105040) demonstrated a significant decrease in thermal stability, as shown in Figure 10.3 (f), which can be attributed to the lower melting point and higher CELC concentration. Overall, the DSC curves indicate that AGLC bioinks remain stable at application temperatures of room temperature (25°C) and incubation temperature (37°C).

10.3.4 Contact Angle Analysis

Figures 10.4 and 10.5 present the wettability measurements of various AGLC bioinks compared to AG hydrogel. It was observed that the contact angle values increased in AGLC bioinks with different CELC concentrations compared to

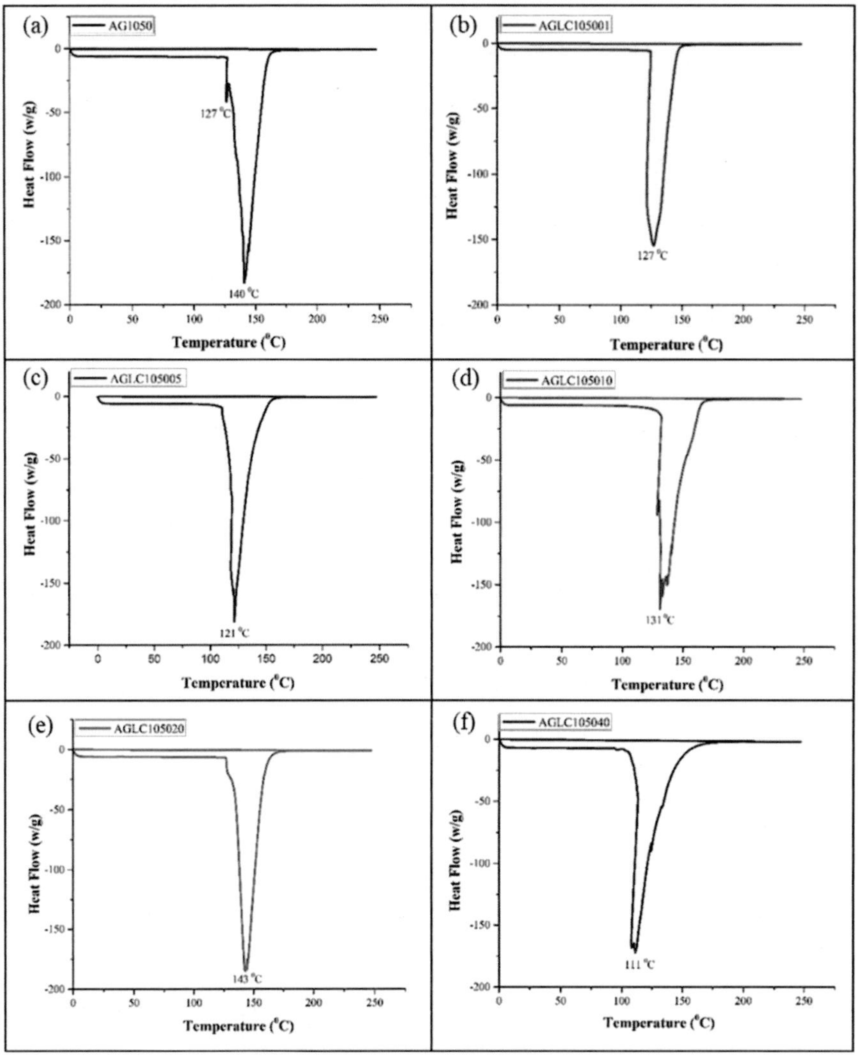

Figure 10.3 The DSC curves of (a) AG1050, (b) AGLC105001, (c) AGLC105005, (d) AGLC105010, (e) AGLC105020 and (f) AGLC105040. Adapted with permission from Abdulmaged et al. 2022, by permission of the *Polymers* under the Creative Commons Attribution License.

AG1050. The AGLC105001 bioink exhibited the highest hydrophilicity with a mean contact angle of 76.13°, while the AGLC105040 bioink had the highest contact angle of 88.6°. These results indicate that varying CELC concentrations lead to the formation of more compact structures with reduced hydrophilicity. CELC compounds feature flexible side chains chemically bonded to a rigid backbone. Figure 10.4 (a–f) displays images of AG1050 and AGLC bioinks at different CELC concentrations.

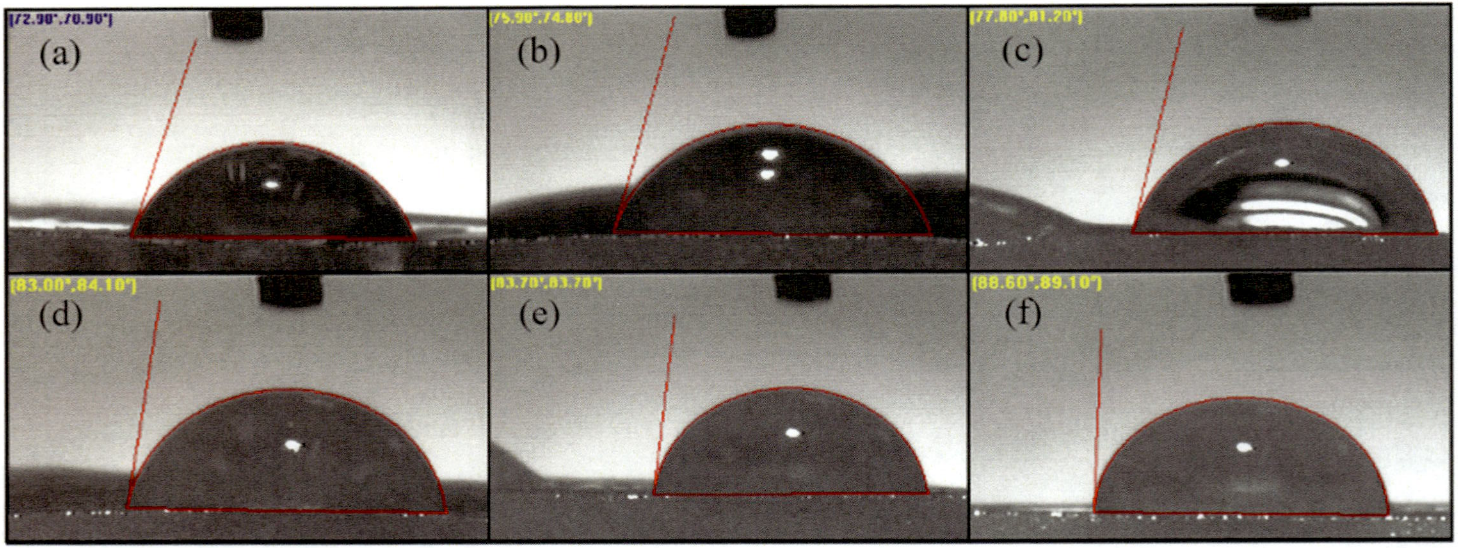

Figure 10.4 The contact angles of different concentrations of (a) AG1050, (b) AGLC105001, (c) AGLC105005, (d) AGLC105010 (e) AGLC105020 and (f) AGLC105040. Adapted with permission from Abdulmaged et al. (2022), by permission of the *Polymers* under the Creative Commons Attribution License.

In different AGLC bioinks, the side chains (esters) of CELC are shorter and less entangled with the soft components of AG1050 hydrogel. As a result, during the crosslinking of AGLC structures, the rigid backbone arranges itself in an orderly way while the softer side chains remain more mobile. This causes the hydrophobic aliphatic side chains to migrate to the surface of the bioinks, gradually making the AGLC bioink surface more hydrophobic as CELC content increases. Cellular behaviors, such as cell adhesion and protein absorption, are influenced by the hydrophilicity of the bioink composite substrate (Ferrari et al. 2019, Gilarska et al. 2018). However, as shown in Figure 10.5, all the bioinks exhibited moderate hydrophobicity, with contact angle values below 90 degrees, making them suitable for cell adhesion and proliferation.

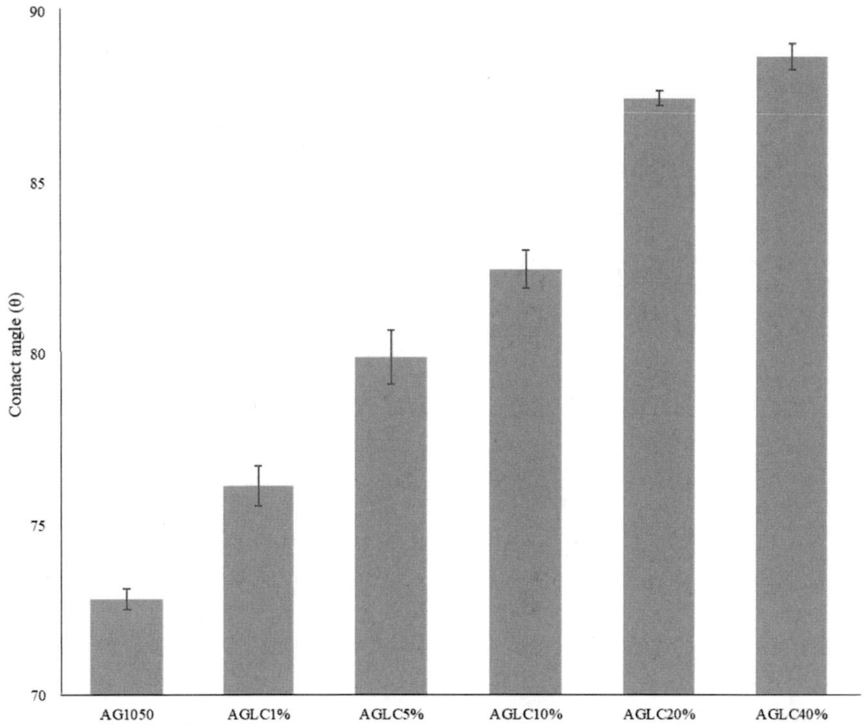

Figure 10.5 The contact angle values of AGLC bioinks. Adapted with permission from Abdulmaged et al. 2022, by permission of the *Polymers* under the Creative Commons Attribution License.

10.3.5 Surface Morphology of the Bioink

Figure 10.6 (a–g) presents the surface morphologies of AG1050 and AGLC bioinks with varying CELC concentrations. FESEM images reveal that the surface micro-morphologies of the AGLC bioinks change significantly with the addition of CELC to the AG1050 bioink. The pure AG1050 bioink sample, shown in Figure 10.6(a), displays a flat, homogeneous surface with no distinct texture. The high miscibility and blend homogeneity between alginate and gelatin are attributed to forming a

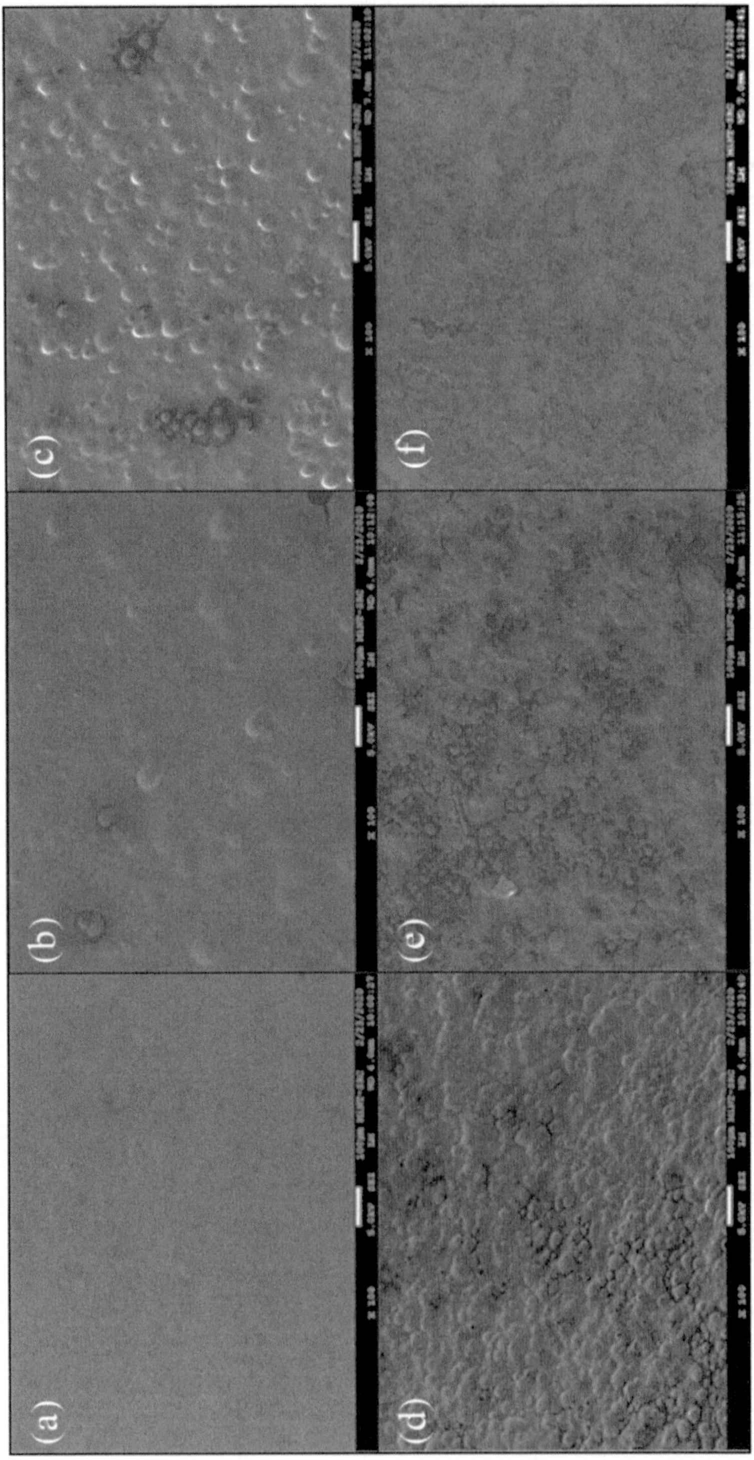

Figure 10.6 The surface morphologies of (a) AG1050, (b) AGLC105001, (c) AGLC105005, (d) AGLC105010 (e) AGLC105020 and (f) AGLC105040. Adapted with permission from Abdulmaged et al. 2022, by permission of the *Polymers* under the Creative Commons Attribution License.

polyelectrolyte complex, resulting from the interaction between the negatively charged carboxyl groups of alginates and the positively charged amino groups of gelatin.

The CELC samples exhibited a uniformly fine-grained surface, as illustrated in Figure 10.6 (b–f). The surface roughness of the AGLC bioinks increased with higher CELC concentrations. FESEM images in Figure 10.6 (b–d) reveal that at lower CELC concentrations (1%, 5%, and 10%), CELC formed discrete, tiny, island-like droplets scattered across the bioink surfaces. As CELC concentration increased, these droplets evolved into richer CELC domains that gradually coalesced and interconnected, as shown in Figure 10.6 (e–f). The pattern and distribution of CELC micro-domains on the AGLC bioinks surface varied with CELC concentration, which may influence the behavior of different cell types, as cells can detect and respond to changes in their microenvironment (Soon et al. 2014, Du et al. 2016, Wu et al. 2017).

10.3.6 MTT Assay

HeLa cell viability was assessed after 24 hours of incubation with varying concentrations of AGLC bioinks. The *in vitro* results (See Figure 10.7) indicated no significant difference in cellular activity or viability between cells treated with AGLC105001, AGLC105005, and AGLC105010 bioinks and the control samples, as illustrated in Figure 10.7. Specifically, AGLC105001 bioink samples achieved a cell viability of 101%, slightly higher than the control at 100%. AGLC105005 bioink samples presented a cell viability of 95% compared to the control. In contrast, AGLC105020 and AGLC105040 bioinks displayed lower cell viabilities of 83.5% and 79%, respectively.

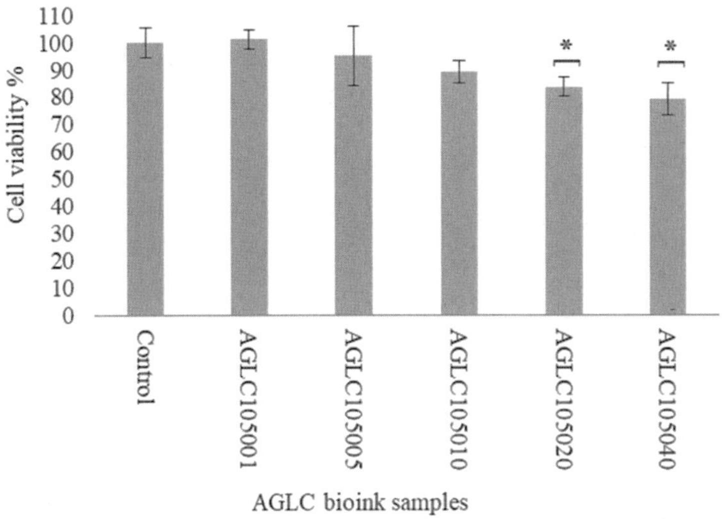

Figure 10.7 Cell viability % of HeLa cells after incubation with AGLC bioinks for 24 h. Adapted with permission from Abdulmaged et al. 2022, by permission of the *Polymers* under the Creative Commons Attribution License.

Figure 10.8 (a) illustrates the formation of a 3D cell structure on the surface of AGLC105001. Cell viability was assessed using the LIVE/DEAD® assay and visualized with a fluorescent microscope, as shown in Figure 10.8 (b). Figures 10.8 (c) and (d) demonstrate that cells on the AGLC scaffold established strong cell-cell adhesion in multi-stratified layers, likely due to the favorable mechanical properties of the composite hydrogel (Soon et al. 2011, Wong et al. 2016). Live cells were stained green with Calcein AM, while dead cells were stained red with Ethidium homodimer. The AGLC bioinks exhibited good biocompatibility, supporting cell viability and aggregation. This positive biocompatibility of AGLC105001 is primarily attributed to CELC, which may supply essential biological molecules to cells at the bioink interface (Hwang et al. 2002, Soon et al. 2014). However, despite CELC's known biocompatibility, higher concentrations of CELC (above 50%) were associated with a slight decrease in cell viability. Thus, excessively high CELC concentrations could be detrimental, leading to cytotoxicity and reduced cell viability (Wu et al. 2017).

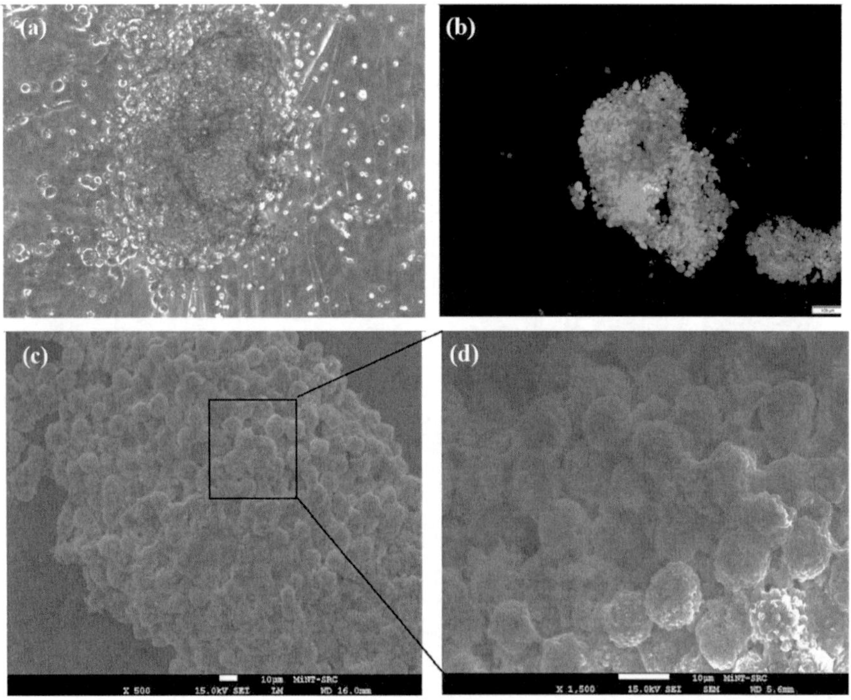

Figure 10.8 (a) Aggregates of cells adhered on the composite hydrogel, (b) the aggregates of cells were stained with LIVE/DEAD assay, (c) the electron microscopy of the cells' aggregates and (d) the enlarged excerpt of (c).

10.4 CONCLUSIONS

This study characterized and compared AGLC bioinks with varying CELC concentrations against the AG1050 bioink. Adjustments to the physicochemical properties of AGLC bioinks were achieved by adding CELC, and cross-linking with Ca^{2+} ions proved effective. FTIR analysis confirmed the successful incorporation of CELC functional groups into the AGLC bioinks, which is significant for tissue engineering due to CELC's cell adhesion micro-domains. FESEM imaging revealed that CELC addition altered the microstructure of AGLC bioinks, potentially affecting cell behaviors such as adhesion, migration, and proliferation. Increasing CELC concentration reduced hydrophilicity (wettability) of the AGLC bioinks, with a 26.13% increase in water contact angle compared to the CELC-free AG1050 bioink. Despite this, all AGLC bioinks maintained a hydrophilic nature, except for the bioink with 60% CELC (AGLC105060), which displayed hydrophobic properties. *In vitro* cell culture studies showed that bioinks with lower CELC concentrations were biocompatible and supported HeLa cell proliferation, as evidenced by the MTT assay. This work contributes to (i) the development of biomimetic Alginate-Gelatin-Cholesteryl ester liquid crystal bioinks suitable for microextrusion-based 3D printing, which can be printed and physically cross-linked with calcium ions for stable, well-defined structures and (ii) comprehensive characterization of these bioinks, covering wettability, microstructure, thermal stability, and chemical properties. Analysis indicates that AGLC105001, AGLC105005, and AGLC105010 bioinks are the most promising due to their thermal stability, wettability, non-cytotoxicity, and favorable surface morphology, suggesting they create optimal conditions for cell adhesion and proliferation. Overall, AGLC bioinks developed in this study show strong potential for tissue engineering applications. Future research should focus on assessing the biocompatibility of AGLC bioinks and their polymers in 3D structures by encapsulating various cell types, which is a crucial next step.

REFERENCES

Abdulmaged, A.I., C.F. Soon, B.A. Talip, S.A. Othman, G.P. Lim and K.S. Tee. 2020. Investigation on the printability of bioink based on alginate-gelatin hydrogel and liquid crystals. Bulletin of Electrical Engineering and Informatics. 9(4): 1718–1725.

Abdulmaged, A.I., C.F. Soon, B.A. Talip, S.A.A. Zamhuri, S.A. Mostafa and W. Zhou. 2022. Characterization of Alginate–Gelatin–Cholesteryl Ester Liquid Crystals Bioinks for Extrusion Bioprinting of Tissue Engineering Scaffolds. Polymers. 14: 1021.

Acharya, B.R. 2011. Applications of liquid crystals in chemical and biological detection. SPIE Newsroom.

Andersen, T., P. Auk-Emblem and M. Dornish. 2015. 3D cell culture in alginate hydrogels. Microarrays. 4(2): 133–161.

Ashammakhi, N., S. Ahadian, C. Xu, H. Montazerian, H. Ko, R. Nasiri, et al., 2019. Bioinks and bioprinting technologies to make heterogeneous and biomimetic tissue constructs. Materials Today Bio. 100008.

Axpe, E. and M.L. Oyen. 2016. Applications of alginate-based bioinks in 3D bioprinting. International Journal of Molecular Sciences. 17(12): 1976.

Berg, J., T. Hiller, M.S. Kissner, T.H. Qazi, G.N. Duda, A.C. Hocke, et al., 2018. Optimization of cell-laden bioinks for 3D bioprinting and efficient infection with influenza a virus. Scientific Reports. 8(1): 1–13.

Chimene, D., K.K. Lennox, R.R. Kaunas and A. K. Gaharwar. 2016. Advanced bioinks for 3D printing: a materials science perspective. Annals of Biomedical Engineering. 44(6): 2090–2102.

Chung, J.H., S. Naficy, Z. Yue, R. Kapsa, A. Quigley, S.E. Moulton, et al., 2013. Bioink properties and printability for extrusion printing living cells. Biomaterials Science. 1(7): 763–773.

Devi, N. and D.K. Kakati. 2013. Smart porous microparticles based on gelatin/sodium alginate polyelectrolyte complex. Journal of Food Engineering. 117(2): 193–204.

Di Giuseppe, M., N. Law, B. Webb, R.A. Macrae, L.J. Liew, T.B. Sercombe, et al., 2018. Mechanical behaviour of alginate-gelatin hydrogels for 3D bioprinting. Journal of the Mechanical Behavior of Biomedical Materials. 79: 150–157.

Du, L., X. Yang, W. Li, X. Luo, H. Wu, J. Zhang, et al., 2017. Construction of physical crosslink-based chitosan/liquid crystal composite hydrogel and evaluation on their cytocompatibility. Regenerative Biomaterials. 4(1): 39–45.

Duan, B., L.A. Hockaday, K.H. Kang and J.T. Butcher. 2013. 3D Bioprinting of heterogeneous aortic valve conduits with alginate/gelatin hydrogels. Journal of Biomedical Materials Research. 101A: 1255–1264.

Hwang, J.J., S.N. Iyer, L.S. Li, R. Claussen, D.A. Harrington and S.I. Stupp. 2002. Self-assembling biomaterials: liquid crystal phases of cholesteryl oligo (L-lactic acid) and their interactions with cells. Proceedings of the National Academy of Sciences. 99(15): 9662–9667.

Iwabata, K., U. Sugai, Y. Seki, H. Furue and K. Sakaguchi. 2013. Applications of biomaterials to liquid crystals. Molecules. 18(4): 4703–4717.

Jang, T.S., H.D. Jung, M.H. Pan, W.T. Han, S. Chen and J. Song. 2018. 3D printing of hydrogel composite systems: Recent advances in technology for tissue engineering. International Journal of Bioprinting. 4(1): 126.

Jia, J., D.J. Richards, S. Pollard, Y. Tan, J. Rodriguez, R.P. Visconti, et al., 2014. Engineering alginate as bioink for bioprinting. Acta Biomaterialia. 10(10): 4323–4331.

Jiang, T., J.G. Munguia-Lopez, S. Flores-Torres, J. Kort-Mascort and J.M. Kinsella. 2019. Extrusion bioprinting of soft materials: An emerging technique for biological model fabrication. Applied Physics Reviews. 6(1): 011310.

Kajave, N.S., T. Schmitt, T.U. Nguyen and V. Kishore. 2020. Dual crosslinking strategy to generate mechanically viable cell-laden printable constructs using methacrylated collagen bioinks. Materials Science and Engineering: C. 107: 110290.

Lai, J.Y., D.H.K. Ma, M.H. Lai, Y.T. Li, R.J. Chang and L.M. Chen. 2013. Characterization of cross-linked porous gelatin carriers and their interaction with corneal endothelium: biopolymer concentration effect. PLoS One. 8(1): e54058.

Lee, K.Y. and D.J. Mooney. 2012. Alginate: properties and biomedical applications. Progress in Polymer Science. 37(1): 106–126.

Lopes, S., L. Bueno, F.D. Aguiar Junior and C. Finkler. 2017. Preparation and characterization of alginate and Gelatin microcapsules containing *Lactobacillus rhamnosus*. Anais da Academia Brasileira de Ciências. 89(3): 1601–1613.

Mitov, M. 2017. Cholesteric liquid crystals in living matter. Soft Matter. 13(23): 4176–4209.

Murphy, S.V. and A. Atala. 2014. 3D bioprinting of tissues and organs. Nature Biotechnology. 32(8): 773–785.

Panwar, A. and L.P. Tan. 2016. Current status of bioinks for micro-extrusion-based 3D bioprinting. Molecules. 21(6): 685.

Park, J., S.J. Lee, S. Chung, J.H. Lee, W.D. Kim, J.Y. Lee, et al., 2017. Cell-laden 3D bioprinting hydrogel matrix depending on different compositions for soft tissue engineering: characterization and evaluation. Materials Science and Engineering: C. 71: 678–684.

Prévôt, M.E., S. Ustunel and E. Hegmann. 2018. Liquid crystal elastomers—a path to biocompatible and biodegradable 3D-LCE scaffolds for tissue regeneration. Materials. 11(3): 377.

Rutz, A.L., K.E. Hyland, A.E. Jakus, W.R. Burghardt and R.N. Shah. 2015. A multimaterial bioink method for 3D printing tunable, cell-compatible hydrogels. Advanced Materials. 27(9): 1607–1614.

Sarker, B., D.G. Papageorgiou, R. Silva, T. Zehnder, F. Gul-E-Noor, M. Bertmer, et. al., 2014a. Fabrication of alginate–gelatin crosslinked hydrogel microcapsules and evaluation of the microstructure and physico-chemical properties. Journal of Materials Chemistry B. 2(11): 1470–1482.

Sarker, B., R. Singh, R. Silva, J.A. Roether, J. Kaschta, R. Detsch, et al., 2014b. Evaluation of fibroblasts adhesion and proliferation on alginate-gelatin crosslinked hydrogel. PloS One. 9(9): e107952.

Sefat, F., M. Youseffi, S.A. Khaghani, C.F. Soon and F. Javid. 2016. Effect of transforming growth factor-β3 on mono and multilayer chondrocytes. Cytokine. 83: 118–126.

Shaw, G.S., K. Uvanesh, S.N. Gautham, V. Singh, K. Pramanik, I. Banerjee, et al., 2015. Development and characterization of gelatin-tamarind gum/carboxymethyl tamarind gum based phase-separated hydrogels: a comparative study. Designed Monomers and Polymers. 18(5): 434–450.

Sidiq, S. and S.K. Pal. 2016. Liquid crystal biosensors: new approaches. International Proceedings. Indian National Science Academy. 82: 75–98.

Skardal, A. and A. Atala. 2015. Biomaterials for integration with 3-D bioprinting. Annals of Biomedical Engineering. 43(3): 730–746.

Soon C.F., M. Youseffi, T. Gough, N. Blagden and M.C.T. Denyer. 2011. Rheological characterization of the time-dependent cholesteric based lyotropic liquid crystals, Materials Science and Engineering C. 31(7): 1389–1397.

Soon, C.F., S.A. Khaghani, M. Youseffi, N. Nayan, H. Saim, S. Britland, et al., 2013. Interfacial study of cell adhesion to liquid crystals using widefield surface plasmon resonance microscopy, Colloids and Surfaces B: Biointerfaces. 110: 156–162.

Soon, C.F., W.I.W. Omar, R.F. Berends, N. Nayan, H. Basri, K.S. Tee, et. al., 2014. Biophysical characteristics of cells cultured on cholesteryl ester liquid crystals. Micron. 56: 73–79.

Soule, E.R. and A.D. Rey. 2012. Modelling complex liquid crystal mixtures: from polymer dispersed mesophase to nematic nanocolloids. Molecular Simulation. 38(8–9): 735–750.

Sun, W., B. Starly, A.C. Daly, J.A. Burdick, J. Groll, G. Skeldon, et al., 2020. The bioprinting roadmap. Biofabrication. 12(2): 022002.

Sung, B. and M.H. Kim. 2018. Liquid-crystalline nanoarchitectures for tissue engineering. Beilstein Journal of Nanotechnology. 9(1): 205–215.

Tabriz, A.G., M.A. Hermida, N.R. Leslie and W. Shu. 2015. Three-dimensional bioprinting of complex cell-laden alginate hydrogel structures. Biofabrication. 7(4): 045012.

Traugutt, N.A., D. Mistry, C. Luo, K. Yu, Q. Ge and C.M. Yakacki. 2020. Liquid-crystal-elastomer-based dissipative structures by digital light processing 3D printing. Advanced Materials. 2000797.

Urbanski, M., C.G. Reyes, J. Noh, A. Sharma, Y. Geng, V.S.R. Jampani, et al., 2017. Liquid crystals in micron-scale droplets, shells and fibers. Journal of Physics: Condensed Matter. 29(13): 133003.

Valot, L., J. Martinez, A. Mehdi and G. Subra. 2019. Chemical insights into bioinks for 3D printing. Chemical Society Reviews. 48(15): 4049–4086.

Wong S.C., C.F. Soon, W.Y. Leong and K.S. Tee. 2016. Flicking technique for microencapsulation of cells in calcium alginate leading to the microtissue formation. Journal of Microencapsulation. 33(2): 162–171.

Wu, Z.L. and J.P. Gong. 2011. Hydrogels with self-assembling ordered structures and their functions. NPG Asia Materials. 3(6): 57–64.

Wu, H., Y. Shang, J. Zhang, L.H. Cheang, X. Zeng and M. Tu. 2017. The effects of liquid crystal-based composite substrates on cell functional responses of human umbilical cord-derived mesenchymal stem cells by mechano-regulatory process. Journal of Biomaterials Applications. 32(4): 492–503.

Wu, Z., Q. Li, S. Xie, X. Shan and Z. Cai. 2020. *In vitro* and *in vivo* biocompatibility evaluation of a 3D bioprinted gelatin-sodium alginate/rat Schwann-cell scaffold. Materials Science and Engineering: C. 109: 110530.

Zhang, Y.S., K. Yue, J. Aleman, K. Mollazadeh-Moghaddam, S.M. Bakht, J. Yang, et al., 2017. 3D bioprinting for tissue and organ fabrication. Annals of Biomedical Engineering. 45(1): 148–163.

Zhao, J., Y. Chen, S. Yang, S. Wu, R. Zeng, H. Wu, et al., 2016. Improving blood compatibility via surface heparin-immobilization based on a liquid crystalline matrix. Materials Science and Engineering: C. 58: 133–141.

Zhu, W., X. Ma, M. Gou, D. Mei, K. Zhang and S. Chen. 2016. 3D printing of functional biomaterials for tissue engineering. Current Opinion in Biotechnology. 40: 103–112.

Index

Made in the USA
Las Vegas, NV
17 September 2026

5400e319-04b0-4ff3-be09-7d7be0b2181fR05